Receptor–Ligand Interactions

The Practical Approach Series

SERIES EDITORS

D. RICKWOOD
Department of Biology, University of Essex
Wivenhoe Park, Colchester, Essex CO4 3SQ, UK

B. D. HAMES
Department of Biochemistry and Molecular Biology,
University of Leeds, Leeds LS2 9JT, UK

Affinity Chromatography

Anaerobic Microbiology

Animal Cell Culture (2nd Edition)

Animal Virus Pathogenesis

Antibodies I and II

Biochemical Toxicology

Biological Membranes

Biomechanics—Materials

Biomechanics—Structures and Systems

Biosensors

Carbohydrate Analysis

Cell Growth and Division

Cellular Calcium

Cellular Neurobiology

Centrifugation (2nd Edition)

Clinical Immunology

Computers in Microbiology

Crystallization of Proteins and Nucleic Acids

Cytokines

The Cytoskeleton

Diagnostic Molecular Pathology I and II

Directed Mutagenesis

DNA Cloning I, II, and III

Drosophila

Electron Microscopy in Biology

Electron Microscopy in Molecular Biology

Enzyme Assays

Essential Molecular Biology I and II

Fermentation

Flow Cytometry

Gel Electrophoresis of Nucleic Acids (2nd Edition)

Gel Electrophoresis of Proteins (2nd Edition)

Genome Analysis

HPLC of Macromolecules

HPLC of Small Molecules

Human Cytogenetics I and II (2nd Edition)

Human Genetic Diseases

Immobilised Cells and Enzymes

Iodinated Density Gradient Media

Receptor–Ligand Interactions

A Practical Approach

Edited by

E. C. HULME

Division of Physical Biochemistry
National Institute for Medical Research
Mill Hill, London NW7 1AA

at
OXFORD UNIVERSITY PRESS
Oxford New York Tokyo

Oxford University Press, Walton Street, Oxford OX2 6DP
Oxford is a trade mark of Oxford University Press

Published in the United States
by Oxford University Press, New York

A catalogue record for this title
is available from the British Library

Library of Congress Cataloging-in-Publication Data
Receptor-ligand interactions : a practical approach / edited by E. C.
Hulme.
(Practical approach series)
Includes bibliographical references and index.
1. Drug receptors—Research—Methodology.   2. Cell receptors—
—Research—Methodology.   3. Ligand binding (Biochemistry)
4. Receptor-ligand complexes.   5. Radioligand assay.   I. Hulme, E. C.
II. Series
RM301.41.R4   1992   615'.7—dc20   91—36849
ISBN 0–19–963090–9
ISBN 0–19–963091–7 (pbk.)

Typeset by Footnote Graphics, Warminster, Wilts
Printed in Great Britain by Information Press Ltd, Oxford, UK

Preface

The idea that cell-surface receptors are well-defined molecular entities to which drugs, hormones, or neurotransmitters bind in obedience to the Law of Mass Action was definitively formulated by A. J. Clark in a monograph entitled *The mode of action of drugs on cells* (1933, Edward Arnold, London). This concept was crucial to the development of quantitative pharmacology. Unfortunately, however, the direct measurement of drug–receptor interactions was several orders of magnitude beyond the sensitivity of the biochemical techniques available to the pioneers of pharmacology.

The concentrations of receptor–ligand complexes measured in binding studies on tissue preparations are characteristically less than one picomole per millilitre. Two factors are responsible for this; firstly, the low natural abundance of receptors in most tissues; secondly, the high affinities of many drugs for their receptors, given that the concentration of binding sites in a binding assay should be no more than the K_d of the ligand. The problem of availability of the receptor proteins is beginning to succumb to the use of high-level expression systems (for receptors for which cDNA clones are available). This will clearly facilitate the study of low affinity interactions. Nevertheless, the need to measure receptor binding in the picomolar range on natural-abundance sources is likely to be with us for the foreseeable future.

In a classic example of the unpredictability of scientific progress, the fundamental advance which enabled the direct detection of ligand binding to receptors was a spin-off from nuclear physics, namely the availability of an abundant supply of pure radioisotopes, particularly ^3H and ^{125}I. These could be incorporated into drugs, hormones, and neurotransmitters, extending the sensitivity of biochemical measurements into the femtomolar range.

Pioneering radiochemical development work was done at the Radiochemical Centre, Amersham (now Amersham International) in the UK and at New England Nuclear (now Du Pont–NEN) in the USA. Therefore, it seems appropriate that Chapter 1 of the present volume, which describes strategies for the radiolabelling of compounds of pharmacological interest, should have been contributed by an erstwhile representative of Amersham International. Appendices 1 and 2 contain a listing of radiolabelled agents available from Amersham and Du Pont–NEN which are used specifically to label receptor binding sites. It is a tribute to the pace of activity in this field that by the time that this book is published, there will have been many valuable additions to this list.

The use of naturally-occurring toxins has played a fundamental part in the development of receptor studies. The most famous example of this has been the use of α-bungarotoxin isolated from the venom of *Bungarus multicinctus* to characterize the nicotinic acetylcholine receptors of the neuromuscular junction, and the electroplax. This useful creature has been good enough to provide us with a second toxin, β-bungarotoxin, which, together with the dendrotoxins has provided a paradigm for the definition of another novel set of probes, in this case for potassium channels. This very topical subject is surveyed in Chapter 3. Of course, naturally-occurring polypeptide ligands for receptors are not always available, but the use of antibodies provides a more general approach. Chapter 2 outlines aspects of the progress which can be made by following the immunological route in the study of membrane-bound receptors.

The rest of the book is concerned with the performance and interpretation of binding studies. Such studies are now in the mainstream of pharmacology, both in academic laboratories and in the pharmaceutical industry.

Binding studies are easy to perform, but surprisingly difficult to perform well. Chapter 4 outlines the many considerations which need to be weighed up in the design and performance of the major variants of binding experiments if the conclusions are, as far as possible, to be uncontaminated by artifacts. Chapters 5–9 concentrate on the practical aspects, particularly tissue and membrane preparations (Chapter 5), and assay methods. Membrane filtration, which is now the mainstream method, is covered in Chapter 6, microcentrifugation in Chapter 7, ligand adsorption in Chapter 8, and gel filtration in Chapter 9.

Chapter 10 discusses the methodology of studies of the fast kinetics of ligand interactions with cell-surface receptors. It is fair to comment that such studies are still under-exploited, largely owing to the intrinsic difficulty of the experiments, and the complexities of their interpretation. Pharmacological analysis of receptor systems often concentrates on the equilibrium situation, ignoring the reality that, *in vivo*, receptor systems operate dynamically. Kinetic studies on receptors are likely to become a growth area as larger amounts of recombinant receptor proteins become available, allowing the exploitation of fluorescence and other spectroscopic techniques.

Finally, Chapter 11 reviews the extraction of quantitative information from binding studies. Many software packages for curve-fitting are now available, but few, if any of them, take into account the full complexity of binding data. Chapter 11 endeavours to chart a path through this particular minefield, and to provide guidance to the reader in choosing or setting up the appropriate analytical methods.

Receptor binding studies are now an established field of endeavour. Many of the concepts developed in this book owe their origins to earlier studies in enzymology, so that the methodological problems surveyed have an evergreen quality. The range of behaviour which can be displayed by even relatively simple receptor systems retains the ability to surprise even the experienced grinder-and-binder. However, forewarned is forearmed!

London   E. C. HULME
June 1991

Contents

3. Polypeptide neurotoxins as probes for certain voltage-dependent K$^+$ channels   37

J. Oliver Dolly

4. Strategy and tactics in receptor-binding studies   63

E. C. Hulme and N. J. M. Birdsall

Contents

8. Charcoal adsorption for separating bound and free radioligand in radioligand binding assays 247

Philip G. Strange

9. Gel-filtration assays for solubilized receptors 255

E. C. Hulme

10. Receptor-binding kinetics 265

Heino Prinz

Contents

Contents

Appendices

Contents lists of related volumes   449

Index   452

Contributors

N. J. M. BIRDSALL
National Institute for Medical Research, Mill Hill, London NW7 1AA, UK.

N. J. BUCKLEY
National Institute for Medical Research, Mill Hill, London NW7 1AA, UK.

J. O. DOLLY
Department of Biochemistry, Imperial College, London SW7 2AZ, UK.

E. C. HULME
Division of Physical Biochemistry, National Institute for Medical Research, Mill Hill, London NW7 1AA, UK.

K. G. McFARTHING
Serono Diagnostics Ltd, 21 Woking Business Park, Albert Drive, Woking, Surrey GU21 5JY, UK. (Formerly at Amersham International.)

H. PRINZ
Max-Planck Institut für Ernährungsphysiologie, Rheinlanddam 201, D-4600 Dortmund 1, Germany.

W. R. ROESKE
Departments of Pharmacology and Internal Medicine, The University of Arizona, Arizona Health Sciences Center, Tucson, Arizona 85724, USA.

P. G. STRANGE
The Biological Laboratory, The University, Canterbury, Kent CT2 7NJ, UK.

A. D. STROSBERG
Institut Cochin de Genetique Moleculaire, Laboratoire d'Immuno-Pharmacologie, 22 Rue Mechain, 75014 Paris, France.

J.-X. WANG
Departments of Pharmacology and Internal Medicine, The University of Arizona, Arizona Health Sciences Center, Tucson, Arizona 85724, USA.

WAN WANG
Departments of Pharmacology and Internal Medicine, The University of Arizona, Arizona Health Sciences Center, Tucson, Arizona 85724, USA.

Contributors

J. W. WELLS
Faculty of Pharmacy and Department of Pharmacology, University of Toronto, Toronto, Ontario, Canada M5S 1A1.

H. I. YAMAMURA
Departments of Pharmacology and Internal Medicine, The University of Arizona, Arizona Health Sciences Center, Tucson, Arizona 85724, USA.

R. L. YOUNG
Du Pont–NEN Products, Medical Products Department, 549 Albany St, Boston, Mass 02118, USA.

Abbreviations

A	absorbance
$A_{11}R$	angiotensin-2 receptor
ATP	adenosine 5'-triphosphate
B	total binding
βAR	β-adrenergic receptor
βBgTx	β-bungarotoxin
BHR	Bolton–Hunter reagent
Bq	Becquerel (1 radioactive decay per second)
CHAPS	3-(3-cholamidopropyl) dimethylammonio-1-propane sulphonate
CHO cells	Chinese hamster ovary cells
Ci	curies (1 Ci = 2.22×10^{12} d.p.m.)
CM	carboxymethyl
d.p.m.	radioactive decays per min
DTx	dendrotoxin
EDTA	ethylene diamine tetra-acetic acid
EGTA	ethylene glycol-bis (β-aminoethylether)N,N,N',N'-tetra-acetic acid
ELISA	enzyme-linked immunosorbent assay
HPLC	high-performance liquid chromatography
Ig	immunoglobulin
i.p.	intraperitoneally
K	association (affinity) constant
K_d	dissociation constant
kDa	kilodaltons
k_{12}	association rate constant
k_{21}	dissociation rate constant
L^*	labelled ligand
M_r	relative molecular mass
μm	micrometre
nAChR	nicotinic acetylcholine receptor
nH	Hill coefficient
NMR	nuclear magnetic resonance
NSP	N-succinimidyl-[2,3-^{3}H] propionate

PAGE	polyacrylamide gel electrophoresis
PBS	phosphate-buffered saline
PEG	polyethylene glycol
PEI	polyethylenimine
PMSF	phenylmethylsulphonyl fluoride
R	receptor
RIA	radioimmunoassay
SEM	standard error of mean
SA	specific activity
SSC	SSC($1\times$) 0.15 M NaCl, 0.015 M Na citrate(pH7.0)
SDS	sodium dodecyl sulphate
TBq	terabecquerel (10^{12} decays per second)
$t_{1/2}$	half-time
TE	Tris–EDTA buffer
TLC	thin-layer chromatography
Tm	melting temperature (DNA duplex)
WGA	wheat-germ agglutinin

1

Selection and synthesis of receptor-specific radioligands

KEVIN G. McFARTHING

1. Introduction

The vast majority of biological phenomena involve molecular communication between cells. The molecules which mediate this communication require the presence of a target molecule, or receptor, on the target cell which recognizes the signal and translates this, usually by a complex series of intracellular events, into a biological effect or response. The study of the distribution, concentration, structure, and function of these receptors is thus fundamental to the understanding of the biological response.

The most logical way to study the receptor is to use a labelled version of the specific molecule, and in the vast majority of cases this label is radioactive. The most commonly used radionuclides are tritium and iodine-125, followed by carbon-14, sulphur-35, and phosphorus-32. In this chapter I will attempt to summarize some of the criteria which should be addressed when planning the synthesis and use of a radiolabelled receptor probe, concentrating on the two most widely-used radionuclides, tritium and iodine-125. There is a wide variety of labelling methods available, and I will, therefore, discuss the features and drawbacks of each method to enable the reader to choose the most appropriate for the problem at hand. In addition, the properties of specific activity and stability will also be considered, as these are of fundamental importance to the use of a radioligand. Note that an excellent and comprehensive summary of receptor types and their selective ligands has been published in *Trends in Pharmacological Sciences* (1), and that a substantial amount of pertinent information is summarized in Appendices 1 and 2 to this volume.

2. Selection of radioligands

2.1 Selection of isotope

The physical characteristics of each isotope not only vary in the nature of their

Table 1. Physical characteristics of tritium and iodine-125

	Tritium	Iodine-125
Radiation	beta (100%)	electron capture, gamma
Half-life	12.43 years	60 days
Energy (max)	0.0186 MeV	0.035 MeV
Decay product	^{3}He (stable)	^{125}Te (stable)
Range of β particles in photographic emulsion	1 μm	—
Maximum specific activity	1.06 TBq/matom (28.76 Ci/matom)	80.9 TBq/matom (2186 Ci/matom)
Efficiency of measurement	40% (liquid scintillation)	75–90% (γ-counting)

emission, i.e. particulate and/or electromagnetic, but in a wide range of other parameters. These are summarized for tritium and iodine-125 in *Table 1*.

Tritium has a long half-life (12.4 yr) and thus does not require a correction for decay during the lifetime of the experiment and often that of the tracer. In practice the life of the tracer is dependent upon factors of chemical stability and the degree of self-radiolysis (see Section 4). Iodine-125 has a half-life of 60 days, and consequently an iodine-125 labelled tracer has a potentially shorter useful shelf-life than one labelled with tritium. In addition, calculations for decay have to be incorporated into the measurement of specific activity when either unreacted sodium [^{125}I]iodide, or the ^{125}I-tracer, have been stored for a significant length of time.

The different half-lives of tritium and iodine-125 also reflect an inverse difference in specific activity. The specific activity of iodine-125 (80.9 TBq/ mmol; 2186 Ci/mmol) coupled with γ-emission makes this isotope more useful in applications which require, or are improved by rapid, sensitive detection either by γ-counting or by autoradiography, whereas the properties of tritium (specific activity of 1.06 TBq/mmol; 28.76 Ci/mmol and low energy beta particles) make it more amenable to systems where the researcher requires high resolution autoradiography and is prepared to use liquid scintillation counting. The major advantage of tritium is that, in general, the incorporation of this isotope into a molecule gives little or no alteration of the biological structure and function, thereby giving as true a representation as possible of the interaction of the native molecule with its receptor. With few exceptions, the addition of iodine-125 to a receptor ligand strictly results in the formation of an analogue, although it must be stressed that such analogues may be equally useful for receptor analysis; very often the kinetics of the interaction are indistinguishable from those exhibited by the natural, unlabelled ligand.

The dual emission of iodine-125—γ rays and electron capture leading to X-ray emission—facilitates detection by both γ-counting and autoradiography. This, coupled with high specific activity, makes iodine-125 the isotope of choice in situations where sensitivity requirements are of the utmost.

The short path length and low energy of the beta particles emitted by tritium are an advantage to high resolution autoradiography, but they are the main contributory factors to the decomposition, by self-radiolysis, of tritiated compounds, as virtually all of the beta energy emitted is absorbed within the sample. Further consideration of this property and ways to reduce it are described below in the section on stability.

The isotope selected to radiolabel a receptor ligand will also be influenced by the molecular nature of the ligand. Molecular size is important; the smaller the ligand, the more likely it is that the incorporation of iodine-125 will disturb the interaction of the ligand with the receptor. Thus large proteins are usually labelled with iodine-125, whereas small neurochemicals, such as the benzodiazepines, would normally be labelled with tritium. The availability of the appropriate reactive groups will also influence the isotope due to the specific labelling methods used to incorporate the relevant radioisotope into the ligand. For example, compounds containing phenolic groups are readily labelled with iodine-125, and compounds with double bonds can be labelled with tritium using catalytic hydrogenation. These methods will be discussed in more detail below in Sections 2.2 and 2.3.

2.2 Selection of labelling reactions: iodine-125

2.2.1 General considerations

Synthetic methods for the incorporation of iodine-125 into a ligand will primarily be determined by whether iodine is an endogenous atom in the molecule to be labelled. With few exceptions, iodine will usually be a replacement or an additional atom; hence the considerations extend to whether the labelling is to be direct or indirect.

Direct labelling involves the addition of one or more atoms of iodine to the ligand; whereas indirect labelling, or derivatization, is effected by the chemical conjugation of an iodine-containing moiety to the ligand. The method to be used is often determined by the chemical nature of reactive groups in the ligand. Direct labelling methods substitute ^{125}I *ortho* to the hydroxyl group in phenolic rings, and also react with imidazole and carbon–carbon double bonds. Indirect methods usually involve reaction with amino or sulphydryl groups.

Direct methods usually involve the oxidation of I^- to I_2 (I^+I^-) followed by electrophilic substitution of I^+, usually into an aromatic ring. This oxidation is effected by several reagents, such as iodine monochloride, chloramine-T, lactoperoxidase/H_2O_2, and Iodogen. The chloramine-T reaction is shown in *Figure 1*. Examples and protocols for the use of these reactions are given below.

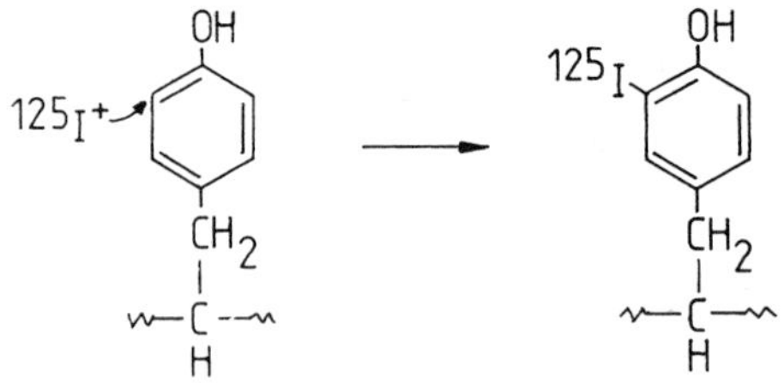

Figure 1. Chloramine-T iodination.

A common problem associated with direct oxidative iodination reactions is the unwanted oxidation of susceptible groups in the molecule other than the target site. This frequently occurs in peptides and proteins which contain methionine residues, as the thioether linkage is first converted to a sulphoxide, then, under prolonged and severe oxidation conditions, to a sulphone. These sulphones and sulphoxides are frequently biologically inactive. Methods which incorporate strongly oxidizing chemical reagents, such as those employing chloramine-T, are most likely to lead to oxidation, whereas more gentle procedures, such as the use of lactoperoxidase, are less likely to oxidize the ligand than they are to convert I^- to I^+. Conversely, the stronger oxidative power of these stronger oxidizing reagents gives a higher efficiency of reaction. In general, the higher the concentration of ligand in the labelling reaction, the higher will be the yield of labelled, non-oxidized tracer; this will be accompanied by a decrease in the incorporation of $Na[^{125}I]$, but it may be desirable if methods are available to purify labelled, non-oxidized, carrier-free tracer from all other products of the reaction. The term 'carrier-free' may, in this case, be defined as only containing radioactive ligand; this does not mean that, in every case, every molecule is radioactive, as the $Na[^{125}I]$ used in the reaction is unlikely to be of 100% isotopic abundance.

Direct methods of iodination are all characterized by being efficient, rapid, convenient, and simple to perform. Consequently, they minimize the radiological dose to the operator, although care must be taken to ensure that reactions are performed in well-ventilated fume hoods as the oxidation reaction may temporarily produce volatile $[^{125}I]$iodine.

Indirect methods of iodination employ the conjugation of an iodine-125 containing molecule to the ligand of choice. For example, the Bolton and Hunter reaction employs N-succinimidyl-3-(4-hydroxy 5-$[^{125}I]$iodophenyl) propionate to react with amino or sulphydryl groups in ligands, usually the ε-amino group of lysine in proteins or peptides. This reaction is shown in *Figure 2*. These methods may not always be the first method chosen in a synthetic strategy, as direct methods are usually more economical. However, this economy may turn out to be false as the advantages of indirect labelling become apparent. The Bolton and Hunter reaction involves the addition of a solution (aqueous or organic) of ligand to the conjugate followed by a short

4

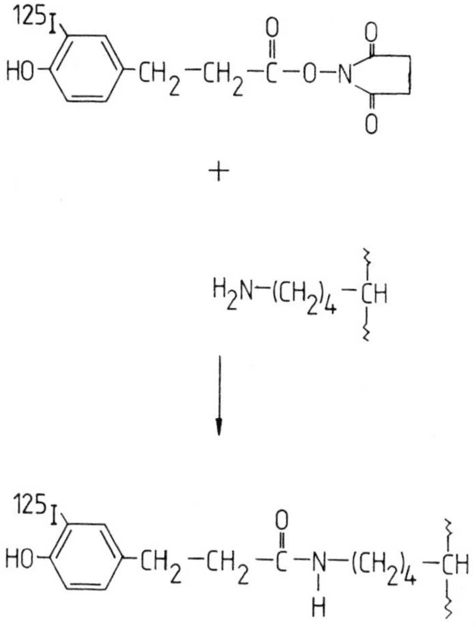

Figure 2. The Bolton and Hunter reaction.

incubation period; no oxidative conditions are employed, and different functional groups react to those modified in direct labelling. Conversely, the addition of a relatively large iodine-containing conjugate to a ligand will often interfere with receptor-binding properties to a much greater degree than will the addition of an iodine atom, thus indirect methods are more often used to label larger molecules such as proteins.

Occasionally, a receptor ligand will contain no groups that may be attacked by any of the common direct or indirect methods, as in the case of many steroids. These molecules must first be chemically conjugated to molecules such as tyramine, which can then be directly iodinated. These reactions must be carefully controlled to avoid the side-reaction of double bond addition.

Usually the structure of a receptor ligand will be known prior to labelling, and if the 'business end' of the molecule, i.e. that moiety which is crucial for binding to the receptor, is also known, iodination can be targetted to reactive groups in other areas of the molecule. For example the C-terminal portion of Substance P is essential for receptor binding, and the molecule can be labelled at the N-terminal portion without deleterious effects to these properties. Substance P contains no groups that are amenable to direct iodination, hence labelling is usually performed by the Bolton and Hunter reagent. Thus, the first choice in an iodination reaction is dependent upon the nature and location of the group to be modified.

A ligand may contain more than one reactive group amenable to iodination.

In this case either a different reaction is chosen which will only label one group, or purification methods are chosen which will separate the different isomers. The reasons why it is optimal only to incorporate one atom of iodine per molecule are discussed below in Section 4.

The presence of fragile oxidizable groups is the next criterion to be considered. If these groups can be oxidized without an influence on receptor-binding kinetics then direct methods offer more advantages than do indirect procedures. If oxidation is to be avoided, either mild direct or indirect methods are favoured.

Finally, empirical factors often play an important part in the selection of a method for ligand iodination. If the ligand is a large protein of undetermined tertiary structure, it will be impossible to specifically label one particular amino acid either by direct or indirect means. It then becomes necessary to use a selection of methods, probably using a variety of conditions for each method, and to subsequently use the receptor-binding assay itself as a measure of the functionality of the radiolabelled tracer.

2.2.2 Safety notes

This section is not intended to be a comprehensive guide to the handling of iodine-125, rather it is a collection of important considerations. For a comprehensive guide the reader is referred to *Amersham Review no. 18* by Bolton (2).

(a) It is very strongly recommended that any iodination reaction vial is not handled directly; the vial should either be clamped in an appropriate holder or firmly embedded in a rack.

(b) It is advisable to designate certain adjustable pipettes, such as those manufactured by Gilson, to be used solely for iodinations. This reduces the risk of radioactive contamination elsewhere in the laboratory.

(c) Disposable plastic or rubber gloves should be worn at all times, and disposed of within the iodination facility in a tin can, for example a paint tin.

(d) The iodination facility should be a well-ventilated fume cupboard, and not a laminar flow cabinet which pushes air outwards towards the user.

As a further illustration of point (a), *Table 2* shows the dose rates from 37 MBq, 1 mCi of iodine-125 at various distances from the source.

Table 2. Dose rates from iodine-125

	Dose rate from 37 MBq (1 mCi) in mrem/h, μGy/h		
	Principal photon emission		
(MeV)	At surface	At 10 cm	At 10 cm through 6 mm lead
0.027,0.031	7000,700	70,7	1.0,0.1

2.2.3 Iodination using chloramine-T

The following protocol is usually suitable for labelling proteins and peptides.

Protocol 1. Iodination using chloramine-T

1. Add 10 μl of 0.2 M sodium phosphate buffer, pH 7.4, to a glass or polystyrene container. Glass should be avoided if the protein or peptide is known to adhere non-specifically. Polypropylene Eppendorf or Sarstedt microcentrifuge tubes are suitable.
2. Add a maximum volume of 50 μl of the protein solution, preferably dissolved in 50 mM sodium phosphate buffer, pH 7.4, at a concentration of 0.5–2.0 mg/ml.
3. Add 0.5–1.0 mCi, 18.5–37 MBq of sodium [^{125}I]iodide (Amersham code IMS.30 or NEN; 100 mCi/ml, 3.7 GBq/ml) to the vial.
4. Add 10 μl of a 1.0 mg/ml solution of chloramine-T in 50 mM sodium phosphate buffer, pH 7.4.
5. Wait 15 seconds (or longer if appropriate) and add 20 μl of a 10 mM solution of cysteine in 50 mM sodium phosphate buffer, pH 7.4.

The reaction mixture is now ready to be purified by a method of your choice. Sephadex G-25 gel filtration is commonly used for purifying proteins, and this is described below.

Protocol 2. Purifying the reaction mixture using a Sephadex column

1. Equilibrate a prepacked column of Sephadex G-25 (Pharmacia, PD-10, Cat. No 17-0851-01) with 15 ml of 50 mM sodium phosphate, pH 7.4, containing 0.2% bovine serum albumin or gelatin, and 0.02% sodium azide as preservative.
2. Let the level of buffer in the top of the column fall to the glass sinter which defines the top of the gel bed.
3. Add the iodination mixture to the top of the column, and let the liquid level fall to the sinter.
4. Add two aliquots of 0.2 ml of column buffer, letting the level fall to the sinter each time.
5. Add buffer to the top of the column, and let the column flow by gravity.
6. Collect 12 fractions of 0.5 ml each.
7. Count samples of each fraction in a γ-counter.
8. Pool those fractions containing the first peak of radioactive counts, to give the final preparation of radiolabelled protein.

2.2.4 Iodination using the Bolton and Hunter reagent

The protocol shown below is usually suitable for labelling proteins and peptides using the Bolton–Hunter reagent (BHR). This reagent is supplied in dry benzene containing 0.2% dimethyl-formamide (DMF) (Amersham code IM.5861; see also NEN). The procedure involves three basic stages—evaporation of the solvent containing the Bolton and Hunter reagent, addition of the protein to be labelled, and purification of the labelled material.

Protocol 3. Iodination using BHR

1. Evaporate the solvent by blowing a gentle stream of dry nitrogen on to the surface of the solvent. This is best accomplished by inserting two hypodermic syringe needles through the seal in the top of the vial, and attaching a tube leading from the nitrogen source to one of the needles. The other needle then acts as an outlet for the gaseous solvent. This operation *must* be performed in a well-ventilated fume cupboard or similar facility.

   Note: An extra safety precaution may be taken by using a charcoal trap to absorb volatilized [^{125}I]iodine. This is usually available on request from the manufacturer of BHR.

2. Dissolve an appropriate weight of protein or peptide in an appropriate volume of 0.1 M sodium borate, pH 8.5. As a guide, 10 µg of protein or peptide in 10–20 µl will usually be sufficient to react with 37 MBq, 1 mCi of BHR. Lower amounts will result in markedly lower yields.

3. Add the protein or peptide solution to the dried BHR.

4. Incubate for 15 minutes at room temperature. Alternatively, incubate for 1 h at 0–4 °C.

5. Add 50 µl of 0.2 M glycine in 0.1 M sodium borate, pH 8.5, to quench the reaction.

6. Incubate for 5 minutes at room temperature. Alternatively, incubate for 20 minutes at 0–4 °C.

7. The labelled peptide or protein may now be purified by the method of choice. The Sephadex column method described in *Protocol 2* is also appropriate for proteins labelled using BHR.

2.2.5 Iodination using lactoperoxidase

The following protocol is again suitable for labelling proteins and peptides with iodine-125. The procedure involves the sequential addition of reagents, with hydrogen peroxide added last to initiate the reaction.

Protocol 4. Iodination with lactoperoxidase

1. Dispense 10 µg of the protein or peptide to be iodinated, in a volume of 20 µl or less, into a polypropylene vial (Sarstedt or Eppendorf tubes are suitable).
2. Add 10 µl of 0.2 M sodium phosphate buffer, pH 7.4.
3. Add 18.5–27 MBq, 0.5–1.0 mCi of Na ^{125}I (Amersham code IMS.30).
4. Add 10 µl of a 0.1 mg/ml solution of high activity lactoperoxidase (Sigma, code no. L 8257).
5. Add 10 µl of 0.003% hydrogen peroxide solution in deionized water.
6. Incubate the solution at room temperature for 30 minutes.
7. Add 10 µl of 1 mM cysteine in deionized water.
8. Purify the iodinated protein by a method of your choice. Sephadex G-25 is generally unsuitable for the purification of proteins as ^{125}I-labelled lactoperoxidase will co-elute in the void volume of the column.

An alternative method involves the use of lactoperoxidase immobilized to agarose beads, the oxidative power being supplied by glucose oxidase coupled to the same beads. The product is available from Bio-Rad (Enzymobeads; Cat. No. 170–6001 or 170–6003).

2.2.6 Iodination using Iodogen

Iodogen is the trade name (Pierce Chemical Co.) for 1,3,4,6-tetrachloro-3α, 6 α-diphenylglycoluril, which is supplied as a solid. The following method can be used for the iodination of proteins and peptides.

Protocol 5. Iodination using Iodogen

1. Dissolve the Iodogen in dichloromethane to a concentration of 1.0 mg/ml. If the reaction is to be performed in a polypropylene vial the Iodogen should be dissolved in methanol.
2. Dispense a 50 µl aliquot of Iodogen solution into a suitable vial. This will be sufficient to oxidize 1 mCi of Na ^{125}I, larger or smaller amounts should be adjusted proportionally.
3. Evaporate the solution by blowing a gentle stream of air or nitrogen on to the surface. Once the vessel has been coated with the Iodogen, it may be stored at −20°C, in the dark, for up to a month.
4. Add 10–20 µg of the protein to be iodinated in a volume of 10 µl, to 10 µl of 50 mM sodium phosphate, pH 7.4 in the Iodogen-coated vial.
5. Add 37 MBq, 1 mCi of Na ^{125}I (Amersham code IMS 30; 10 µl). Incubate 15 min at room temperature.

9

Protocol 5. *Continued*

6. Add 250 µl of 50 mM sodium phosphate buffer containing 50 µg/ml of tyrosine. This is to remove any excess iodine-125 which may remain, as it is possible to transfer a small amount of active Iodogen with the aqueous solution.

7. Transfer the reaction mixture to another vial prior to purification by the method of choice. Some purification methods, for example reverse phase HPLC, will allow tyrosine to be omitted from step 6, and permit the use of HPLC eluants to dilute the reaction mixture. *Protocol 2* is also appropriate for purifying proteins labelled using Iodogen.

2.3 Selection of labelling reactions: tritium

2.3.1 General considerations

Any discussion of the radiolabelling of molecules with tritium must be prefaced by a brief, but strong, note of caution on the special facilities and precautions needed before such experiments can be undertaken. Operations involving the use of tritium gas usually require a dedicated laboratory and special facilities; specialized, sealed manifolds; health physics safety monitoring is especially necessary as tritium contamination cannot be easily detected; and the high molar specific activities and small chemical scale usually required, necessitate the use of high (often GBq/Ci scale) amounts of tritium.

In addition, chemical methods which involve the use of tritiated sodium borohydride and tritiated methyl iodide also require similar facilities to those used for tritium gas. This is again due to the small chemical scale which necessitates the use of large amounts of tritiated material, often up to 1.85–3.7 TBq, 50–100 Ci, which may still result in very low (MBq, mCi) yields. Tritiated sodium borohydride and tritiated methyl iodide can also give off volatile radioactivity and consequently need to be handled in dedicated, well-ventilated facilities.

Whilst tritiated compounds are extremely important to the study of receptors, their synthesis, compared to ^{125}I-labelling, is thus difficult and complicated to set up. I will, therefore, limit this section to the discussion of the principles involved, and limit practical protocols to the use of *N*-succinimidyl [2,3-^{3}H]propionate (NSP) which is much easier and more straightforward to use.

There are four main types of tritium labelling:

(a) *Specific*—where the tritium atoms are in known positions within the molecule.

(b) *Uniform*—where the radioactive atoms are distributed throughout the molecule in a uniform manner. This is unusual with tritium.

(c) *General*—where the tritium atoms are distributed randomly throughout the molecule.

(d) *Nominal*—where the labelling position is indicated when uncertainty exists on the location of the tritium atoms.

There are four main methods used for tritium labelling of molecules, namely isotopic exchange reactions, chemical synthesis, biochemical methods, and conjugation. The characteristics of each method will be discussed briefly here, but for a comprehensive text on tritium, tritiated compounds, and tritium labelling methods, the reader is referred to Evans, 1974 (3).

Tritium exchange reactions are normally accomplished by the use of tritium gas or tritiated solvents, and usually with the involvement of metal catalysts. Exchange reactions most often lead to general labelling of the molecule, but may occasionally give rise to specific labelling. However, in both cases the labelling position and the quantitative distribution cannot be predicted with certainty. Exchange reactions are now most commonly effected using tritiated solvents, such as water or acetic acid, and a metal catalyst, such as palladium or platinum. The reactions may involve heating to over 100°C in sealed tubes, and may also take advantage of the weakly acidic character of some C–H bonds by the use of appropriate bases. The use of acids to protonate molecules can make such molecules more amenable to tritium exchange.

In all exchange reactions it is essential that the substrate is of very high chemical purity and is reasonably stable under the experimental conditions used.

Chemical synthesis of tritiated receptor ligands will be the method of choice in situations where a suitable intermediate exists and a specific labelling position is required. Wherever possible, the tritiation should be a simple one-step reaction and be performed as the final synthetic step. There are four major chemical methods for the tritiation of organic compounds:

- reduction of unsaturated compounds with tritium gas using palladium or platinum catalysts
- catalytic halogen–tritium replacement
- reduction of certain functional groups with tritiated metal hydrides
- reactions using tritiated methyl iodide

The chemical method selected will be determined by the functional groups available in the molecule under study, and which of those groups the researcher would prefer to modify with a view to obtaining as high a specific activity as possible. For example, catalytic tritiation will saturate double bonds; ketones, acids, and imines can be modified with tritiated metal hydrides; and tritiated methyl iodide is an ideal reagent for *N*-methylation.

The reduction of functional groups with metal hydrides generally results in highly specific positional labelling, as does methylation using methyl iodide because of the high stability of the tritium atoms in the methyl group. However, the catalytic methods used often result in random incorporation of tritium, due to double bond migration prior to saturation.

As with exchange reactions, it is essential that starting materials of the highest purity possible are used, especially with regard to isomeric and enantiomeric purity.

Indirect labelling is analogous to the use of conjugates such as the Bolton and Hunter reagent in iodine-125 labelling. The most common tritiated conjugate is *N*-succinimidyl-[2, 3-³H]propionate ([³H]NSP), which reacts with amino and sulphydryl functions in target molecules, usually proteins and peptides, to leave a tritiated propionyl side chain.

Biochemical methods of tritiation almost exclusively involve the use of purified enzyme preparations to convert an unlabelled substrate into the desired tracer compound. These methods are often very straightforward, providing that a purified enzyme is available—this is important as contaminating enzyme activities may degrade the product, introduce other functional groups, or compete for the tritiated substrate. Additionally, biochemical methods do not involve the use of tritium gas, thus obviating the need for special equipment. If stereospecific labelling is required, then this is facilitated by the use of the appropriate enzyme. Biochemical methods are being increasingly used, but are ultimately limited by the molecular structure of the substrates and the availability of purified enzymes, hence many pharmacologically active compounds which will be useful as receptor probes will not be amenable to tritiation using such techniques.

The selection of a tritiation method, therefore, is first decided by the pattern of labelling required, whether general or specific, and secondly by the functional groups present. In order to prepare functional radiolabel consistently from experiment to experiment, specific labelling will be required. In this way the pattern of decomposition through self-radiolysis will also be more consistent, and as such self-decomposition and degradation in the assay may produce artefacts in the receptor-binding experiment, those artefacts should at least be consistent. Patterns of labelling in tritium-labelled compounds can be determined by tritium NMR spectroscopy (4).

2.3.2 Tritiation using [³H]NSP

The tritiation of ligands using [³H]NSP involves evaporation of the solvent containing the NSP, addition of the ligand to be labelled, and purification of the labelled material.

Protocol 6. Tritiation with [³H]NSP

1. Evaporate the toluene containing the [³H]NSP by blowing a gentle stream of dry nitrogen on to the surface of the solvent. This is best accomplished

by inserting two hypodermic syringe needles through the seal in the top of the vial, and attaching a tube carrying the nitrogen to one of the needles. The other needle then acts as an outlet for the gaseous solvent. This operation *must* be performed in a well-ventilated fume cupboard or similar facility.

2. Add the ligand to be labelled to the vial, dissolved in 25 mM sodium phosphate, pH 8.0. As a guide it is recommended that a 1:1 molar ratio of ligand:[^{3}H]NSP is used. The ligand is normally dissolved in 10–20 μl of buffer for reaction with 37 MBq, 1 mCi of [^{3}H]NSP.

3. Incubate at room temperature for 2–4 hours.

4. The labelled ligand should now be immediately purified by the method of choice. The method described for the purification of proteins in *Protocol 2* is also appropriate for proteins labelled with [^{3}H]NSP.

---

3. Specific activity

The specific activity of a radiolabelled molecule may be defined as the unit radioactivity per mole, or fraction thereof. Thus the specific activity of iodine-125 and tritium-labelled compounds is usually expressed in TBq/mmol or Ci/mmol. The higher the specific activity of the labelled compound, the less mass needs to be used in a tracer experiment, thus not only reducing the radioactive dose to the operator but also optimizing the economic usage of the tracer. Also, and perhaps more importantly, high specific activities normally lead to an increase in sensitivity of detection of either low abundance or low affinity receptor sites.

In general, therefore, an isotope and a labelling method will be selected to give the highest possible specific activities. There are certain *caveats* which must be added to this sweeping statement. The addition of more than one iodine-125 atom to a molecule increases its instability—see Section 4—leading to the production of potentially-interfering radiolabelled impurities. This is also the case with tritium, but these effects may be limited by the use of radical scavengers. Also the addition of more than one atom of iodine-125 may lead to further structural changes which can interfere with the interaction of a ligand and its receptor.

The accurate measurement of the specific activity of a radiolabelled ligand is central to the study of the receptor, as it is used to directly quantify the number and affinity of those receptor sites. Specific activity may be measured in several ways. The measurement of the radioisotope content is straightforward and can be achieved with great accuracy using γ-counting or scintillation counting, taking care to consider the counting efficiency of the instrument in use. The measurement of the amount of ligand in the tracer preparation will usually be achieved in a manner appropriate to the nature of the ligand. For example, compounds with characteristic absorption spectra may be measured at the λ_{max}; tracers may also be quantified by physical methods, such as mass spectrometry; or by using specific antibodies in an immunoassay.

Some of these methods may involve placing large amounts of radioactivity in laboratory equipment which may result in potential contamination and high operator dose, and therefore both of these factors should be considered when attempting to quantify the amount of labelled ligand.

It should be stressed that measurements of specific activity may refer to a preparation of radiolabelled ligand, which may often include unreacted, un-labelled starting material, and perhaps a heterogeneous mixture of labelled molecules. This is undesirable and to be avoided if at all possible. It is preferable to use a purification method which is able to separate ligand labelled at the desired position(s) from all other components of the mixture, as has been demonstrated in our own laboratories with regulatory peptides. Specific activity measurements will then refer directly to a single molecular species.

Another approach to specific activity is slightly more empirical, and that is to consider the concept of effective specific activity. Most preparations of radioligand will produce fragments, some radiolabelled and some not, with continued storage, making the mixture increasingly heterogeneous. Those fragments may still be able to bind to the receptor, and the non-radioactive fragments may effectively decrease the specific activity. As each fragment will be chemically different from the parent ligand, the chemical or physical methods initially used to determine specific activity may ironically be the receptor-binding assay itself, and will have to be achieved using a standard preparation of receptor in which the number of binding sites and their affinity are known. Alternatively—and preferably!—the radiolabelled ligand may be repurified on the original system.

4. Stability of radioligands

A radioactive atom is inherently unstable. When it decays, other molecular events also occur and may have deleterious effects on the molecule, produc-ing fragments which are often labelled and which may interfere with the receptor-binding assay.

The analysis of stability may be accomplished by either biochemical or chemical methods. With prolonged storage of the radioligand an increase in non-specific binding in a receptor assay will indicate accumulation of labelled impurities, and the preparation may subsequently be analysed by chemical methods such as TLC or HPLC. TLC is a convenient and rapid technique for analysis and the radioactive spots can be detected by autoradiography or by linear analyser scanning.

Certain steps may be taken to minimize chemical decay and radiolysis, and these are discussed below.

4.1 Iodine-125

The decay of iodine-125 is accompanied by a complex series of events termed 'coulombic explosion'. The effect of this on a ligand is usually irreversible chemical damage, for example peptide chain cleavage. This can give rise to labelled impurities which may have reduced affinity or even antagonistic properties. In order to reduce this effect, the labelling reaction should aim to introduce no more than one atom of radioactive iodine into the molecule. When this atom decays, the molecule is disrupted but is no longer radioactive, and will, therefore, not be detected in the assay. Thus, the specific activity does not decrease in parallel with the radioactive half-life of iodine-125. The nature and properties of the decay fragments are still important as they may still potentially interfere, and if such properties are observed, the radioligand must be re-purified before use in the binding assay.

Radiolysis is the phenomenon of molecular damage caused by ionizing radiations. This appears to be less of a problem with iodine-125 than with other, more weakly-emitting nuclides, due to the energy and nature of the emission of iodine-125. The steps which a researcher should take, therefore, to reduce the instability of radioligand preparations refer primarily to the chemical nature of the ligand itself and the fact that it is being stored at a very low chemical concentration. For example, we have shown that ^{125}I-labelled peptides are best stored lyophilized or frozen in the presence of a carrier protein, such as albumin or gelatin, and stabilizing sugars such as lactose.

The small chemical scale will also necessitate further careful treatment; for example, exposure to UV light and rapidly changing temperature conditions should be avoided.

In addition, it should be remembered that bacteria and fungi are not deterred by the presence of radioactivity and will just as easily contaminate a preparation of radioactive ligand!

4.2 Tritium

The efforts which researchers can make to minimize the degradation of small chemical weights of ligand, as described briefly in Section 4.1, can also be made for tritiated ligands. However, the weak β-emission of tritium means that virtually all of the emitted energy is absorbed within the sample, hence the problem of radiolysis is much greater with tritium than with iodine-125. The β-particles penetrate only a few micrometres. Thus, the more dilute the ligand, the more likely it is that the resulting damaged species will be unlabelled, solvent molecules. It is also obvious that the higher the specific activity, the greater will be the rate of radiolysis. Secondary effects of tritium decomposition are also important, as these include the formation of energized ions and free radicals which are very reactive indeed.

In principle the same precautions can be taken to minimize loss due to primary radiolysis and secondary decomposition. The ligand should be dissolved in a suitable solvent which is capable of reacting with secondary species such as free radicals. The use of compounds such as ethanol which act as radical scavengers can often increase the stability of a wide range of compounds. Secondary decomposition is also temperature-dependent and can be minimized by storage at lower temperatures. If compounds are to be stored frozen, they should be frozen rapidly to minimize concentration (molecular clustering) effects.

Finally, there is no universal panacea for the storage of tritiated compounds, as low temperatures and the presence of radical scavengers will not guarantee stability, and other conditions may warrant investigation. Again, a full review of this topic is given by Evans (4). Commercially available preparations are subjected to stability trials which investigate the optimum storage conditions and ensure the longest stability practicable.

5. Purity and purification of radioligands

Most labelling reactions will give a wide range of products including, for example, oxidation products, unreacted intermediates, degraded or damaged ligand, and a spectrum of ligand labelled to various degrees. Molecules structurally unrelated to the ligand may still interfere with the interaction of that ligand and its receptor, and thus must be removed before a receptor-binding assay is performed. Ligand molecules which have been chemically damaged by the labelling reaction may also display different binding kinetics to those of the natural ligand. It is, therefore, advisable to purify the single molecular species of radioligand from all other components of the reaction mixture so that the results of the receptor-binding assay reflect the true interaction of ligand and receptor with no contribution from artefacts of the labelling reaction.

In some cases, for example large proteins, it is almost impossible to separate a single molecular species of radioligand, and it is inevitable that a spectrum of products will be used in the assay. The results of the receptor assay need to be interpreted with caution, especially where the presence of more than one affinity constant is indicated.

The purification method selected for the radioligand will be determined primarily by the chemical characteristics of the molecule. As a general guide the purification method which is used for the unlabelled starting material is the best place to start the search for a method applicable to the radioligand.

There are, however, further points to be considered when purifying radioactive compounds. Separation methods which require several manipulative steps are not advisable unless precautions are taken to reduce the risk of radiation dose and contamination. The same is true for methods which involve free and open access of the radioactivity to air such as TLC; this method also involves scraping the plate to remove the sample with subsequent solvent extraction, and during this step any aerosols formed may disperse the silica-borne activity posing a further hazard. Where possible, alternatives to TLC, such as HPLC, should be sought. Indeed, HPLC is the method of choice in our own laboratories due to its excellent resolving power, speed, convenience, and reproducibility. In addition, it is virtually a closed system, the column and injection units can be placed inside the appropriate level of shielding, and pumps, solvents, and control units can be accessible outside the shielding.

The speed of separation is thus important for reducing handling time. It is also advisable to remove the radioligand from other reactive components, such as oxidizing agents, despite stopping the reaction, as soon as possible to maximize yield by the prevention of further degradation. Methods such as small-scale gel filtration and ion exchange are thus attractive, although these methods are rarely able to resolve specifically-labelled molecules.

The major advantage of tritium as a tracer isotope in receptor-binding assays is that the tracer is usually chemically indistinguishable from the native molecule; this is a disadvantage when one attempts to purify the radioligand as it is virtually impossible to separate it from unlabelled material. The inevitable result is a decrease in specific activity. Hence, when a ligand is labelled with tritium it is important to maximize incorporation of the isotope through optimization of reaction conditions to give as high a specific activity as possible. If an exact structure is not required, then labelling with [^{3}H]NSP may be considered as this should change the hydrophobicity of the molecule sufficiently to enable separation by, for example, reverse phase HPLC.

The addition of an iodine atom to a molecule will lead to a change in the properties of that molecule, making it possible to separate radioiodinated ligand from unlabelled and damaged molecules. Indeed it is possible using HPLC to separate peptides labelled with iodine-125 at different positions; for example ^{125}I-(tyr^1)β-endorphin and ^{125}I-(tyr^{27})β-endorphin. Using similar methods, work in our own labortories has used reverse phase HPLC to separate the different monoiodinated isomers of a number of peptides, including vasoactive intestinal polypeptide, β-endorphin, glucagon, neurotensin, insulin, neuropeptide Y, insulin-like growth factor-1, growth hormone-releasing factor, adrenocorticotropic hormone, corticotropin releasing factor, endothelin-1, big endothelin, and α-bungarotoxin. In addition, reverse phase HPLC can be used to prepare a variety of molecules carrier-free at a specific activity of approximately 74 TBq/mmol, approximately 2000 Ci/mmol, i.e. one atom of iodine-125 per molecule. These include steroids, drugs, and nucleotides.

Labelling of ligands with the Bolton and Hunter reagent increases the hydrophobicity of the molecule and again facilitates separation on the basis of slightly differing chemical properties.

The chemical scale of labelling reactions are usually very small for two reasons; firstly to optimize the incorporation of radioisotope; and secondly because the exquisite sensitivity of radioactivity measurements is such that one labelling reaction on a very small chemical scale can yield sufficient ligand for a large number of assays. This small chemical scale poses further problems in purification of the ligand as traditional chemical purification methods, such as crystallization, are not possible, and indeed would be highly dangerous without the proper facilities! The small scale involved may lead to large losses of radioligand through adsorption to glass or plastic surfaces, and it is, therefore, wise to use conditions which either reduce or avoid this. For example carrier protein such as 0.2% (w/v) bovine serum albumin should be included in buffers when purifying iodinated peptides and proteins by gel filtration.

References

1. Watson, S. and Abbott, A. (ed.) (1991). *Trends in Pharmacological Sciences: Receptor Nomenclature Supplement*. Elsevier, Amsterdam.
2. Bolton, A. E. (1985). *Radioiodination Techniques*, Review 18 (2nd edn). Amersham International plc.
3. Evans, E. A. (1974). *Tritium and its Compounds*. Butterworths, London.
4. Evans, E. A. (1984). *Review 16*, Amersham International plc.
5. Evans, E. A., Warrell, D. C., Elridge, J. A., and Jones, J. R. (1985). *Handbook of Tritium NMR Spectroscopy*. John Wiley, Chichester.

2

Anti-receptor antibodies as ligands

A. D. STROSBERG

1. Introduction

In this chapter we will discuss three approaches to obtaining anti-receptor antibodies: through direct immunization, from patients' sera, and via anti-idiotypes. In each case we provide some of the specific procedures that have been used by various laboratories and discuss typical results.

2. Preparation of anti-receptor antibodies through direct immunization

2.1 Relationship between the epitope and the ligand-binding site

Antibodies raised against receptors may act as specific ligands by interacting with the binding site for agonists or antagonists. Such antibodies may actually compete with hormone and neurotransmitters, and this property provides a method for evaluating their affinity and specificity. However, it is often observed that the large immunoglobulin molecules, while binding specifically to the receptor, are unable to competitively inhibit the effect of the natural ligands.

On binding to the receptor, antibodies may trigger mechanisms normally activated by hormones and neurotransmitters: receptor redistribution, stimulation or inhibition of adenylate cyclase, receptor internalization, inhibition of receptor-mediated cell–cell interactions, etc. Some of these phenomena may be due to the multivalent immunoglobulin molecules causing aggregation of the receptor, rather than any direct effect on the ligand-binding site. Hence, formal proof for direct interaction at the ligand binding site requires the use of monovalent Fab antibody fragments.

The vast majority of anti-receptor antibodies are not directed towards the binding site, which usually only constitutes a small part of the molecule. When the initial immunogen is membrane-bound or cell-bound receptor, epitopes recognized by the resultant antibodies may be restricted to regions exposed at the surface of the cell. When purified receptor is injected, every part of the immunogen may induce a response and antibodies may be directed to epitopes located in the extracellular, transmembrane, or intracellular portions. Antibodies not directed against the ligand-binding site may still constitute very useful reagents: they may serve in visualization experiments (immunofluorescence, electron microscopy), in investigation of the function of various portions of the molecule, in analysis of specific sites for phosphorylation, for interaction with effector proteins, etc.

2.2 Procedures for preparation of anti-receptor antibodies

Several methods have been used to obtain anti-receptor antibodies (1,2). These range from simple immunization with cells bearing receptors at their surface to injection of affinity-purified receptor. Most recently, anti-receptor antibodies have been obtained by immunization with synthetic peptides derived from the nucleotide sequence of receptor genes. Anti-receptor peptide antibodies may display properties similar to those of antibodies raised against the whole protein.

2.2.1 Immunization with cell-bound or membrane-bound receptor

When the initial immunogen is extremely heterogeneous, as is the case with whole cells or membranes, most of the effort will be directed towards developing very specific methods for demonstrating the appearance of anti-receptor antibodies.

Immunization of a mouse may be carried out using two injections, at least one month apart, of 10^7 cells, or their equivalent membranes, which should bear at least 6×10^4 receptor molecules per cell. For a molecular weight of 50 kDa these minimal figures correspond to one picomole or 0.5 μg of receptor protein. To ensure that the receptor molecules remain exposed at the cell surface the cells may be fixed, for example with glutaraldehyde.

2.2.2 Immunization with ligand–receptor complexes

Without troubling to set up sophisticated specific affinity chromatography procedures, semi-purified receptor may be obtained by applying the following method:

(a) Prepare a ligand-conjugate, by coupling the small hormone, neurotransmitter, or growth factor to a carrier protein to which antibodies are available.

(b) On this conjugate, absorb receptor from a cell lysate or from solubilized membranes.

(c) Immunoprecipitate the complex and inject into mice, or rabbits if larger amounts are available.

Some specific methods are described by Kaveri *et al.* (3) who used an albumin–alprenolol conjugate to isolate β-adrenergic receptors from solubilized A431 human epidermal cells. After cell fusion and selection, three monoclonal antibodies were obtained which recognized photoaffinity-labelled receptors, immunoprecipitated ligand-binding activity, and identified the receptor polypeptide chains in immunoblots. None of these antibodies had any agonist-like activity.

An even simpler procedure is applicable when anti-ligand antibodies which do not inhibit binding of the ligand to the receptor are available—for example with receptors to which protein toxins bind, as is the case for the nicotinic acetylcholine receptor. In this instance anti-α bungarotoxin antibodies may be used to immunoprecipitate the toxin–receptor complex. The precipitate may then be injected to obtain anti-receptor antibodies.

2.2.3 Immunization with purified receptor

Specific procedures may be developed to obtain purified receptor, by classical protein chemistry or specific affinity chromatography, using as final steps high performance liquid chromatography or isoelectric focusing (for general references see Strosberg (4). A full account is given in *Receptor Biochemistry, a Practical Approach* (5). The quantities of protein obtained after several successive steps of purification are often quite limited, and usually difficult to manipulate without considerable loss of material. One way to circumvent this complication is to transfer the protein from the polyacrylamide gel to a nitrocellulose membrane (see ref. 5, Chapter 10) which is then introduced under the skin of a mouse. More drastic is the direct injection of a portion of polyacrylamide gel containing the receptor. Intrasplenic immunization or boosting prior to sacrifice of the animal is used with much success in a number of laboratories (6).

2.2.4 Immunization with synthetic receptor peptides

Peptide sequences may be derived from the sequence of the gene coding for the receptor. Those selected for immunization should preferably be chosen from parts of the receptor protein likely to be exposed at the surface of the molecule, as may be deduced from hydropathicity plots (see ref. 5, Chapter 12). Immunization is performed with free peptide, of at least 15 residues, or with peptide–carrier protein conjugates, a less attractive choice; since it induces anti-carrier protein and anti-linkage antibodies.

2.2.5 Screening methods

i. ELISA

One may usually expect the appearance of antibodies two to three weeks after the first round of immunizations. A straightforward procedure, but one often prone to artefacts, for detecting the presence of antibodies is an enzyme-linked immunosorbant assay (ELISA). In this, serum, or the immunoglobulin fraction to be tested, is added to plastic tubes or plates coated with whole cells or membranes; binding of any antibody is detected by adding a second enzyme-linked antibody directed against the immunoglobulin of the first species followed by a colourimetric assay for the enzyme. While a positive ELISA is often encouraging, it only indicates that antibodies to a cell or membrane component are present. Results obtained with IgM antibodies should be viewed with particular caution because these are known to produce non-specific results. This class of immuoglobulin is also quite fragile and does not withstand frequent freezing and thawing. More specific procedures have to be applied to demonstrate the presence of anti-receptor antibodies.

ii. Immunoprecipitation
Immunoprecipitation of solubilized receptor may often help to define the specificity of antibodies. Thus, if antibodies are able to bring down receptor bound to radiolabelled ligand, one may conclude that the epitopes recognized by the antibodies are not close to the ligand-binding site. Removal of ligand-binding activity from a soluble cell lysate may serve as an indication of the presence of specific antibody; this is often an easier procedure than dissociation of the presumed receptor–antibody complex and measurement of ligand-binding activity of the redissolved receptor.

iii. Immunoblotting
Immunoblotting (Western blot) is a useful procedure for rapidly screening relatively large numbers of sera or immunoglobulin fractions for the presence of antibody which recognizes the known receptor polypeptide chains (see ref. 5, Chapter 10). However, if the proteins are denatured by the detergent sodium dodecyl sulphate (SDS) before and during electrophoresis then the technique may not work. To increase confidence in the specificity of the recognition one may include affinity-labelled receptor in the same experiment.

An interesting example is provided by the study of André *et al.* (7) in which monoclonal antibodies against the muscarinic acetylcholine receptor were prepared. Two very different antibodies were obtained: the M23 antibody reacted with the receptor in Western blot, but did not immunoprecipitate any carbamylcholine-binding activity nor did it stimulate any agonist-like effects; the M35 antibody, an IgM, reacted very weakly with the muscarinic receptor in immunoblot, but it was also able to immunoprecipitate ligand-binding activity from calf brain. M35 mimicked agonist stimulation of intact guinea-pig myometrium: just like carbamylcholine, it caused a rise in intracellular cyclic GMP content, an inhibition of cyclic AMP accumulation due to pros-tacyclin, and induced uterine contractions. The first two effects were blocked by the antagonist atropine (8).

The same immunization procedure may thus generate both antibodies directed against the native (M35) and denatured (M23) receptor. Both types are useful for studying the receptor in different circumstances: for example

M35 could be used to visualize redistribution of the receptor at the cell surface after treatment with agonists; M23 could identify receptor polypeptide chains after electrophoresis in denaturing conditions. M35 is commercially available (Chemunex, France).

3. Anti-receptor antibodies in autoimmune disease

A number of autoimmune diseases are accompanied by the presence, in the serum, of antibodies directed against membrane receptors (reviewed in Strosberg and Schreiber, ref. 9). It is not always easy to demonstrate the presence of such antibodies, nor to explain their role in the pathogenesis of the disease. The following possibilities may be explored:

(a) Some autoimmune diseases are characterized by auto-antibodies against a variety of the patient's own components. This is the case in Lupus erythomatosus, where antibodies may be found against nucleic acids, proteins of the cytoskeleton, and other molecules.

   In such situations one may first look for antibodies that bind proteins at the surface of receptor-bearing cells, by using procedures such as immunofluorescence or ELISA on immobilized cells.

(b) Diseases in which a particular organ is the target of auto-antibodies also may constitute appropriate choices for the search for anti-receptor antibodies. This is the case in various forms of thyroiditis in which the Long Acting Thyroid Stimulating (LATS) antibodies have been found (10). Another example is idiopathic cardiomyopathy or the cardiac form of Chagas' disease in which one may find auto-antibodies against the catecholamine binding β-adrenergic receptors (11, 12).

(c) Diseases in which a particular function mediated by receptors is deficient provide the best opportunity for finding anti-receptor antibodies, although such symptoms are rare. In myasthenia gravis the autoantibodies are directed against the nicotinic acetylcholine receptor and these cause the receptor destruction; new-born infants can suffer from the disease through transfer of the maternal anti-receptor antibodies (13). In type B insulin-resistant diabetes, anti-insulin receptor antibodies may cause the blocking or decrease of insulin receptors, thus explaining the inefficiency of insulin treatment (14).

4. The anti-idiotype route to produce anti-receptor antibodies

4.1 Introduction

In a number of situations, anti-receptor antibodies cannot be obtained by either immunization or screening patients. This is the case when receptors are present at the cell surface in amounts too low to induce antibodies, or when they cannot be purified for lack of affinity chromatography or other specific methods, or finally when no autoimmune disease has been identified in which anti-receptor antibodies appear to be produced. An alternative route has been developed: first antibodies are made against a ligand, then anti-idiotypic antibodies are raised against the initial anti-ligand immunoglobulins. Amongst the anti-idiotypes, those directed against the ligand-binding site of the first antibody are identified by their ability to inhibit recognition of the ligand by the anti-ligand antibodies. A small fraction of such inhibitory anti-idiotypic antibodies also recognize the physiologic receptor for the ligand, sometimes even to the extent of being able to inhibit binding. Moreover, in some cases, non-inhibitory anti-idiotypic antibodies may also interact with the receptor.

The theoretical basis for interactions between anti-idiotypic antibodies and idiotypes was elaborated as the 'Network Theory' by Niels Jerne (15). Its extension to include idiotope-bearing receptors was first proposed for the insulin receptor by Sege and Peterson (16). In *Figure 1* we have represented two types of interactions between anti-idiotypes and receptors (17).

Anti-idiotypic anti-receptor antibodies have been prepared in a number of systems (for recent reviews, see Strosberg, refs 18, 19; also listed in *Table 1*). In several systems, these antibodies were shown to immunoprecipitate solubilized receptors or modulate effector functions mediated by the receptors (for example adenylate cyclase). The nicotinic acetylcholine receptor was purified by immunoaffinity chromatography on insolubilized anti-idiotype. Efforts are underway to use anti-idiotypic anti-receptor antibodies for the selection of bacterial clones expressing mammalian receptor protein generated by recombinant DNA technology.

4.2 Procedures for the preparation of anti-idiotypic anti-receptor antibodies

The steps involved in obtaining anti-idiotypic anti-receptor antibodies are described below. These include:

(a) The preparation and characterization of specific anti-ligand antibodies, the antibodies which display binding properties similar to those of the relevant receptor are selected.

(b) The preparation and selection of anti-idiotypic antibodies, especially those directed against the binding site of the anti-ligand antibodies.

(c) The selection of anti-idiotypic antibodies directed against the receptor.

We will now discuss these procedures with reference to anti-idiotypic antibodies directed against the β-adrenergic catecholamine receptor (βAR), the angiotensin II receptor ($A_{II}R$), and the nicotinic acetylcholine receptor (nAChR).

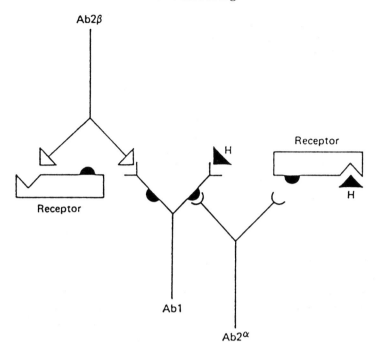

Figure 1. Hormone receptor binding of anti-idiotypic antibodies (Ab2) directed against antihormone antibodies (Ab1) can be explained by two structural models: in the first model, the anti-idiotype, called Ab2β, bears on its Fab region the complementary structural counterpart of the hormone-binding site, shared by the Ab1 and by the receptor; in the alternative model, the anti-idiotype, called Ab2α, recognizes a conformational homology on both Ab1 and receptor. (From Marullo, S. and Strosberg, A. D. (1986). In *Concepts in Immunopathology*, Vol. 3, ed. J. M. Cruze, pp. 176–192. Karger, Basel.)

4.2.1 Preparation of anti-ligand antibodies

i. Coupling procedures

A large number of procedures have been described for the preparation of anti-ligand antibodies. In order to obtain antibodies which best mimic the physiological receptors for the ligand in question, one should work with immunogens which are as small and rigid as possible—the larger the antigen, the higher the probability that antibodies will be produced against epitopes which are irrelevant to the binding of the ligand to the receptor. However, when small ligands are studied it is often necessary to couple these to carrier proteins.

The coupling may result in a large decrease in affinity of the antibody or the receptor for the modified ligand. The use of different coupling procedures, preferably at various sites on the ligand, may help define the epitope essential for binding. For this purpose, various bifunctional reagents are available

Table 1. Anti-idiotypic antibodies which recognize receptors (References are in Strosberg (1989): see reference 18)

Immunogen	Receptor (R)	Antibody-induced signal	Cells
Hormones			
Insulin	Insulin R	oxidization and uptake of glucose; inhibition of lipolysis	adipocytes
Retinol	RBP	retinol uptake	intestinal epithelial cells
Alprenolol	β-adrenergic R	adenylate cyclase activation or inhibition	nucleated erythrocytes A431 carcinoma cells
Thyrotropin	thyrotropin R	adenylate cyclase activation; increased iodine transport; follicle formation	thyroid cells
Angiotensin	angiotensin R		hepatocytes
Chemotactic peptide	chemotactic peptide R	chemotaxis, increased Ca^{2+}	neutrophils
Leukotriene B4	leukotriene B4 R		PMN leukocytes
Prolactin	Prolactin R	stimulation synthesis and secretion of lactalbumin	
Neurotransmitters			
Nicotinic agonist bis-Q	acetylcholine R	inhibition of Cs^+ flux in presence of carbachol	neuroglioma cell hybrids mouse brain cells
Opioid peptide	opiate R	inhibition of cAMP synthesis	ileum cells
Morphine	opiate R	inhibition of electrically-induced contractions	parotid gland cells
Substance P	(Substance P)-R	enhancement of phospholipid turnover; inhibition of spasmogenic action	ileum cells
Lymphotropic viruses, factors, and lymphokines			
Reovirus	mammalian reovirus R	blocking of viral binding to cells	lymphoid cells
Retrovirus	leukemogenic retrovirus R		leukaemia cells (T and B)
Interleukin 1	interleukin 1 R	proliferation of thymocytes	thymocytes
Factor H (β1H globulin)	(factor H)-R	stimulation of release of factor 1	B cells
Interferon αA	interferon αA-R	anti-viral activity	kidney, amnion cells

which allow selective coupling of ligands at predetermined positions; for example, peptide ligands can be coupled via amino-, carboxyl-, sulphydryl-side chains. The characteristics (affinity, titre) of the immune response varies with the density of ligands present on the carrier protein; it is, therefore, of interest to determine the efficiency of the coupling reaction, i.e. the coupling yield. The simplest way to do this is to use radioactive ligand, when available, in tracer concentration. One should, however, verify that the labelling does not affect the group involved in the coupling reaction; for example, one should avoid radioiodinating tyrosine residues on peptides to be coupled via a tyrosine residue. Other techniques can be used to quantitate the coupled ligand; these are based on UV absorbance or intrinsic fluorescence of the ligand, or, for peptides, on change in amino acid composition of the carrier. In *Table 2* we present a few combinations of carriers and coupling procedures (20–22).

ii. Immunization
In a typical immunization schedule, the animals are primed with 10–50 nmol of ligand (hapten or peptide), in complete Freund's adjuvant, by multiple intradermal or footpad injections. Then animals are boosted monthly with the same dose of ligand in incomplete Freund's adjuvant (intradermal or subcutaneous injections) and bled 7–10 days after each boost. When monoclonal antibodies are to be prepared, mice receive an intravenous or intrasplenic boost of antigen in saline 3–4 days before fusion.

iii. Screening of the antibodies
A first screening of sera or hybridoma supernatants is most easily performed by an enzyme immunoassay using the ligand coupled to a carrier protein different from that used for immunization. When highly radioactive ligands for the receptor are available, the affinity of the antibodies present in the hybridoma supernatants can be determined by a radioimmunoassay (23).

Table 2. Hapten–carrier conjugates

Coupling reaction

Bifunctional coupling reagents	Chemical group	References
Glutaraldehyde	amino	20
Carbodiimide	carboxyl	21
m-maleimidobenzoyl-*N*-hydroxysuccinimide	sulphydryl	22

Carrier proteins	Advantages	Disadvantages
Keyhole limpet haemocyanin (KLH)	very immunogenic	expensive; low solubility in water
Thyroglobulin	available	
Bovine serum albumin (BSA)	many sulphydryls	sticky, ubiquitous

Structural analogues of the ligand (agonists and antagonists, as well as analogues without any affinity to the receptor) can be used to determine the fine specificity of the anti-ligand antibodies, in comparison with the membrane receptor. In the course of our studies of the β-adrenergic and angiotensin II systems, we observed two kinds of results. In one case (24, 25) polyclonal as well as some monoclonal anti-alprenolol antibodies displayed binding properties similar to those of the receptor; in the second case (26) only one out of six monoclonal anti-angiotensin II antibodies bound agonists (but not antagonists) with the same relative affinities as the angiotensin II receptor.

4.2.2 Preparation and characterization of anti-idiotypic antibodies

i. Immunization

Numerous reports have described the preparation and characterization of anti-idiotypic antibodies induced by injection of anti-ligand antibodies or antigen–antibody complexes (for recent books on idiotypes, see Reichlin and Capra, ref. 27). To avoid raising antibodies against the constant regions, one prefers to genetically match the donor and recipient animals. The easiest species in which to do this is the mouse, since syngeneic strains are available. When using rabbits, one should select individuals with the same genetic markers. To enhance the immune response, the immunoglobulins may be injected after coupling to keyhole limpet haemocyanin, a very immunogenic carrier protein. Xenogenic anti-idiotypes may also be used, but they require preliminary extensive absorption on normal immunoglobulins to eliminate anti-isotype and anti-allotype antibodies before testing for the presence of anti-idiotype antibodies.

Usually 100 μg of affinity-purified anti-ligand antibodies is injected twice, three weeks apart. Instead of injecting monoclonal antibodies, one may also immunize with glutaraldehyde-fixed hybridoma cells (28) which theoretically present the antibodies with their Fab portions turned outwards from the cells.

ii. Polyclonal or monoclonal antibodies

As we stated before, when launching a research programme with the ultimate goal of obtaining anti-receptor anti-idiotypic antibodies, it is advisable to start working with polyclonal antibodies in rabbits, to test the feasibility of the approach, and to design the various assays. However, since one requires well-defined anti-ligand antibodies, monoclonals should also be made, at the same time, in mice or in rats (29). A second stage of the programme will involve the preparation of monoclonal anti-idiotypic antibodies, which provide the important advantage of consistency since the anti-idiotypic response may vary considerably in time with respect to its anti-receptor activity (30). Such variation has been explained by the appearance of anti-anti-idiotypic antibodies which neutralize the anti-receptor response. In line with this explanation, based on the Network Theory of Jerne, anti-idiotypic antibodies may occasionally be obtained during the preparation of the initial anti-ligand antibodies as was reported for antibodies against the acetylcholine receptor (31) and against the insulin receptor (32).

In *Figure 2* we present examples of such variations: the first (top) shows how immunization by insulin can be followed by the appearance not only of anti-insulin antibodies but also of antibodies directed aainst these anti-insulin antibodies. These anti-idiotypic antibodies contain a subpopulation which recognizes the insulin receptor. The synthesis of each antibody varies considerably from week to week. In the bottom figure we show the cyclical appearance of anti-idiotypic and anti-anti-idiotypic antibodies recognizing anti-catecholamine antibodies or the β-adrenergic receptor, or recognizing catecholamines (30).

iii. Detection of anti-idiotypic antibodies

Detection of anti-idiotypic antibodies is usually performed by three types of assays (*Table 3*), (a) binding to anti-ligand antibodies, an absolute prerequisite, (b) inhibition of binding of ligand to these antibodies, a desirable feature not always observed, and (c) finally isolation on affinity gel containing idiotype (see an application in *Figure 2*). In a syngeneic system, screening is more difficult since most assays are based on recognition by a second antibody which cannot distinguish between the idiotype and the anti-idiotype. Several methods exist to circumvent this problem:

(a) The anti-ligand antibody can be labelled biosynthetically, by radioiodination or by coupling (for example with glutaraldehyde) to an enzyme such as horseradish peroxidase, alkaline phosphatase, or β-galactosidase. The anti-idiotypic antibodies may then be detected in a sandwich ELISA or radioimmunoassay (RIA) by their ability to bind both to the coated idiotype and to the labelled idiotype in solution.

(b) Fab fragments prepared from the idiotype-bearing antibodies can be used to coat the plastic wells in an ELISA; anti-idiotypic antibodies are then detected after addition of enzyme-labelled Fc-specific antibodies.

Since we cannot *a priori* predict the type of interactions between the receptor and the putative anti-receptor anti-idiotypic antibodies, all the anti-idiotypic antibodies detected in the assays described above will have to be tested for receptor recognition before studying their ability to interfere with the ligand binding to the anti-ligand antibodies.

4.2.3 Selection of anti-idiotypic antibodies which recognize the receptor

Several tests must be used to ascertain the anti-receptor specificity of the anti-idiotypic antibodies (see *Table 4*).

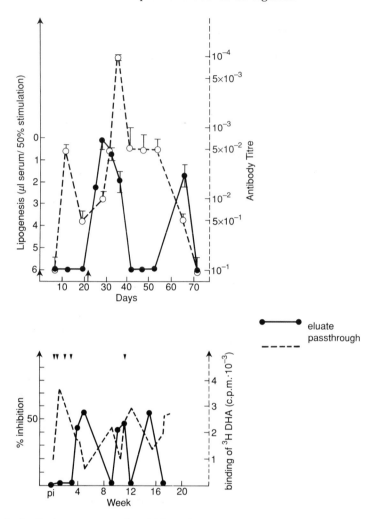

Figure 2. Cyclical appearance of anti-idiotypic responses.
Top: development of anti-insulin (– – – –) and anti-idiotypic anti-receptor antibodies
(———) following immunization to insulin.
Bottom: development of anti-idiotypic anti-β-adrenergic receptor (———) and anti-catecholamine antibodies (– – –) following immunization to anti-alprenolol antibodies.
(From Strosberg A. D. (1987). In *Excerpta Medical International Congress Series*, ed.
W. Pruzanski and M. Seligmann, pp. 21–34. Elsevier, Amsterdam.)

Antibody fractions (eluates) were purified by affinity chromatography on either
alprenolol–Sepharose (alprenolol is a catecholamine antagonist) or on anti-catecholamine antibody–Sepharose. Pass-through from this last gel may contain anti-anti-idiotypic antibodies which may be purified on the alprenolol gel. We measured
either inhibition of or binding of radiolabelled antagonist to the catecholamine β-adrenergic receptor.

Table 3. Detection of anti-idiotypic antibodies

Binding to anti-ligand antibody
Inhibition of binding of ligand to idiotype
Isolation on affinity-gel containing idiotype

Table 4. Detection of anti-receptor anti-idiotype antibody

Binding to receptor-bearing cells of various origins
Immunoprecipitation and/or immunoblotting of receptor
Inhibition (or stimulation) of binding of ligand to receptor
Stimulation or inhibition of biological activity modulated by receptor

---

i. Binding to different cells (or membranes) bearing the receptor
This can be done either in suspension or on solid-phase by ELISA or RIA. The binding specificity must be controlled using preimmune or unrelated IgG and also by inhibition of the signal after preincubation of the anti-idiotypic antibodies with the anti-ligand antibodies used for immunization. Polyclonal anti-idiotypic antibodies raised against anti-alprenolol antibodies were shown to bind to β-adrenergic receptor-bearing turkey erythrocytes and not to receptor-deficient human erythrocytes; this specific binding was inhibited by anti-alprenolol antibodies (24). Anti-idiotypic IgG were incubated for 1 h at 37 °C with turkey or human erythrocytes (2×10^6 cells) in phosphate buffer containing 0.5% bovine immunoglobulins. After three washings, cells were further incubated for 1 h with ^{125}I-(Fab')$_2$ fragments of goat anti-rabbit IgG antibodies. Radioactivity bound to the cells was then counted in a γ-counter.

ii. Immunoblotting experiments
These allow a comparison of the protein specifically recognized by the anti-idiotypic antibodies with that identified as the receptor either through affinity labelling or other procedures.

Anti-idiotypic antibodies against anti-angiotensin II (anti-AII) antibodies recognized a protein of the same molecular weight (63 kDa) as the AIIR, from two different tissues (33). Rat liver membranes (1 mg) or rat anterior pituitary homogenates (three glands) were loaded across the entire width of a SDS/polyacrylamide gel; after migration, electrophoretic transfer of the proteins was performed overnight in methanol/glycine/Tris buffer, to a nitrocellulose filter. After incubation in a blocking buffer (phosphate 0.01 M, NaCl/0.1% Tween 20/0.25% gelatin) for 7 h at room temperature, strips were incubated overnight at 4 °C with 1/40 dilutions of anti-idiotypic serum. Fixation was followed by staining using peroxidase-conjugated goat anti-rabbit IgG antibodies and H_2O_2/chloronaphthol as peroxidase substrate (see the chapter by Wheatley in ref. 5 for further details of electroblotting).

iii. The anti-idiotypic antibodies

These should be tested both for their ability to inhibit the ligand binding to the receptor and to immunoprecipitate the solubilized receptor–ligand complex (these tests are generally mutually exclusive).

Anti-alprenolol anti-idiotypic antibodies were able to inhibit [^{3}H]-dihydroalprenolol binding to turkey erythrocyte membranes (24); after incubation of membranes with antibodies for 1 h at 30°C, hormone binding was noncompetitively inhibited by more than 60%. On the other hand, anti-AII anti-idiotypic antibodies failed to block hormone binding to rat liver membranes, but did immunoprecipitate more than 50% of the anti-AII binding capacity from the same membranes after treatment with 50 mM Tris–HCl, pH 7.4/5 mM MgCl$_2$/200 mM sucrose/20% glycerol/10 mM CHAPS (3-[(3-cholamidopropyl)-dimethylammonio]-1-propanesulphonate) (33).

iv. Affinity purification using anti-idiotypic antibodies

While it has been previously shown that anti-idiotypic antibodies may be used to detect β-adrenergic receptors by immunoblotting (34), a recent report describes the actual isolation and purification of a nicotinic receptor from rat brain using a monoclonal anti-idiotype raised against a monoclonal antibody specific for L-nicotine (35). The following procedures were used to perform the immunopurification.

First, monoclonal antibodies were raised against L-nicotine, and one of these was used as an immunogen for preparing anti-idiotypes. Nine such monoclonal anti-idiotypes were selected for their ability to inhibit the binding of the anti-nicotine antibody to immobilized nicotine–polylysine. One anti-idiotype was further characterized as to ligand and receptor specificity, and 2 mg of antibody were then coupled to 2 g of cyanogen bromide-activated Sepharose 4B (Pharmacia). Approximately 10 mg of detergent-solubilized membrane protein was applied to the column in a buffer, 1 mM in detergent (CHAPS was used). After washing, the receptor protein was eluted simply by lowering the pH to 4. The eluate was concentrated by flow dialysis, assayed immediately for nicotine binding, or dialysed, lyophilized and stored at −70°C. The extent of immunopurification of the receptor was shown to be 10 000 fold, comparing favourably with the 8000-fold purification by affinity chromatography on L-6-hydroxymethyl nicotine. The two types of preparation yielded the same major receptor component of 62 kDa and a minor component of 57 kDa. While this preliminary study obviously requires additional characterization of the material purified by the anti-idiotype antibody, it illustrates the potential of this novel method for preparing anti-receptor antibodies.

v. The ability of antibodies to modulate the biological activity of the receptor should be determined

Thus, we demonstrated that anti-alprenolol anti-idiotypic antibodies were

32

able to stimulate the β-adrenergic receptor-coupled adenylate cyclase activity of turkey erythrocyte membranes (24 and see ref. 36 for specific protocols). After a 90 min preincubation time, at 0°C, of membranes with 5 mg/ml anti-idiotypic IgG, adenylate cyclase activity was assayed for 20 min at 30°C in the absence or presence of the hormone adrenalin: the antibodies stimulated the enzyme activity even in the presence of saturating concentrations of adrenalin.

As mentioned above, in most of the systems studied so far, the anti-receptor activity transiently appears during the course of the anti-idiotypic response. In the β-adrenergic system, this activity was indeed cyclical, regulated by spontaneous anti-anti-idiotypic antibodies with ligand-binding activity (see *Figure 2*). As a practical consequence, several bleedings from animals immunized with anti-ligand antibody must be tested for anti-receptor activity; investigators who intend to produce monoclonal anti-receptor anti-idiotypic antibodies, must time the fusion appropriately.

5. Immunocytochemical approaches to the study of anti-receptor antibodies

Anti-receptor antibodies may constitute exquisite tools for performing immunocytochemical analyses of membrane receptors *in situ*. Quite a variety of procedures and techniques have been used.

5.1 Labelling procedures

Binding of the antibody to the receptor on cells may be visualized under the light microscope by using a second anti-antibody which is labelled with an enzyme capable of generating a coloured product. The presence of the colour will serve as indication of binding. When the label is fluorescent, the binding of the antibody is visualized under the fluorescence microscope. The electron microscope is used when the label is an electron-dense protein, such as ferritin, or when the second antibody is 'loaded' with colloidal gold particles.

5.2 Types of analyses

Antibodies may be used to investigate the following properties:

(a) Distribution of the receptor under various conditions: where is the receptor located in the absence of ligand; does it redistribute in the presence of agonist and/or antagonist; does it get internalized and if so, in clathrin-coated or clathrin-negative invaginations and vesicles? To avoid redistribution of the receptor, induced by the antibody itself, one studies binding of the antibodies only after the cells have been fixed following the treatment with ligand.

(b) Different antibodies may be used to evaluate changes of conformation induced in the receptor by ligand binding.

(c) Agonistic or antagonistic antibodies, when available, may serve to mimic modifications induced by the specific ligands.

6. Conclusion

Even when the above criteria of specificity are fulfilled by some anti-idiotypic antibodies, the structural basis of their reactivity towards the receptor remains to be established: are they real internal images of the ligand or do they recognize structural homologies shared by the anti-ligand antibodies and the receptor? Whatever the answer to these questions, anti-receptor anti-idiotypic antibodies are fruitful probes for the biochemical identification and characterization of receptors.

Acknowledgements

We thank Drs P.-O. Couraud, J.-G. Guillet, and A. Johnstone for helpful discussions, and Mrs C. Peyrat for secretarial assistance. The work performed in the author's laboratory was supported by grants from the Centre National de la Recherche Scientifique (CNRS), the Institut National de la Santé et de la Recherche Médicale (INSERM), the Association pour la Développement de la Recherche sur le Cancer (ADRC), the Fondation pour la Recherche Médicale Française (FMRF), the Ligue Nationale Française contre le Cancer, the University Paris VII, the Tobacco Research Council, and the Ministère de la Recherche et de l'Enseignement Supérieur (MRES).

References

1. Strosberg, A. D., Chamat, S., Guillet, J. G., Schmutz, A., Durieu, O., Delavier, S., and Hoebeke, J. (1984). In *Monoclonal Antiidiotypic Antibodies: Probes for Receptor Structure and Function* (ed. J. C. Venter and J. S. Lindstrom), pp. 151–162. Allan R. Liss, NY.
2. Strosberg, A. D. (1987). In *Molecular Biology of Receptors* (ed. A. D. Strosberg), pp. 139–163. Ellis Horwood, Chichester.
3. Kaveri, S. V., Cervantes-Olivier, P., Delavier-Klutchko, C., and Strosberg, A. D. (1987). *Eur. J. Biochem.,* **167,** 449.
4. Strosberg, A. D. (1984). In *Membrane Receptor Purification and Characterization Techniques* (ed. C. Venter) pp. 1–13. Alan R. Liss, NY.
5. E. C. Hulme (ed.) (1990). *Receptor Biochemistry, a Practical Approach.* IRL Press, Oxford.
6. Nilsson, B. O., Svalander, P. C., and Larsson, A. (1987). *J. Immunol. Meth.,* **99,** 67.

7. André, C., Guillet, J. G., De Backer, J. P., Vanderheyden, P., Hoebeke, J., and Strosberg, A. D. (1984). *EMBO J.*, **3**, 17.
8. Leiber, D., Harbon, S., Guillet, J. G., André, C., and Strosberg, A. D. (1984). *Proc. Natl. Acad. Sci. USA*, **81**, 4331–4334.
9. Strosberg, A. D. and Schreiber, A. B. (1984). In *Antibodies to Receptors* (ed. M. F. Greaves), pp. 15–42.
10. Doniach, D., Bottazzo, G. F., and Khoury, E. L. (1980). In *Autoimmune Aspects of Endocrine Disorders* (ed. A. Pinchera, D. Doniach, G. F. Fenzi, and L. Baschieri), pp. 25–55. Academic Press, London
11. Magnusson, Y., Marullo, S., Höyer, S., Waagstein, F., Andersson, B., Vahina, Guillet, J. G., Strosberg, A. D., Hjalmarson, A., and Hoebeke, J. (1990). *J. Clin. Invest.*, **86**, 1658–63.
12. Sterin-Borda, L., Cantore, M., Pascual, J., Borda, E., Cossio, P., Arana, R., and Passeron, S. (1986). *Int. J. Immunopharmacol.*, **8**, 581–588.
13. Zakarija, M., McKenzie, J. M., and Banovac, K. (1980). *Ann. Intern. Med.*, **93**, 28–32.
14. Kahn, C. R., Baird, K. L., Fliers, J. S. Grunfeld, C., Harmon, J. T., Harrison, L. C., Karlsson, F. H., Kajuga, M., King, G. L., Lang, U. C., Poskalny, J. M., and Van Obberghen, E. (1981). *Recent Prog. Horm. Res.*, **37**, 477–533.
15. Jerne, N. (1974). *Ann. Immunol.* (Inst. Pasteur), **125C**, 373.
16. Sege, K. and Peterson, P. A. (1978). *Proc. Natl. Acad. Sci. USA*, **75**, 2443.
17. Marullo, S. and Strosberg, A. D. (1986). In *Concepts in Immunopathology* (ed. Cruze), Vol. 3, pp. 176–192. Karger, Basel.
18. Strosberg, A. D. (1989). In *Methods in Enzymology* (ed. J. J. Langone), Vol. 178, pp. 179–91. Academic Press, London.
19. Strosberg, A. D. (1989). In *Methods in Enzymology* (ed. J. J. Langone), Vol. 178, pp. 265–275. Academic Press, London.
20. Reichlin, M. (1980). In *Methods in Enzymology* (ed. H. Van Vunakis and J. J. Langone), Vol. 70, p. 159. Academic Press, London.
21. Bauminger, S. and Wilchek, M. (1980). In *Methods in Enzymology* (ed. H. Van Vunakis and J. J. Langone), Vol. 70, p. 151. Academic Press, London.
22. Yoshitake, S., Yamada, Y., Ishikawa, E., and Masseyeff, R. (1979). *Eur. J. Biochem.*, **101**, 395.
23. Farr, R. S. (1958). *J. Infect. Dis.*, **103**, 239.
24. Schreiber, A. B., Couraud, P. O., André, C., Vray, B., and Strosberg, A. D. (1980). *Proc. Natl. Acad. Sci. USA*, **77**, 7385.
25. Chamat, S., Hoebeke, J., and Strosberg, A. D. (1984). *J. Immunol.*, **133**, 1547.
26. Couraud, P. O. (1986). *J. Immunol.*, **136**, 3365.
27. Bona, C. A. (1987). In *Modern Concepts in Immunology* (ed. M. Reichlin and J. D. Capra), Vol. 2. J. Wiley, Chichester.
28. Guillet, J. G., Chamat, S., Hoebeke, J., and Strosberg, A. D. (1984). *J. Immunol. Meth.*, **74**, 163.
29. Priestly, J. V. (1987). In *Neurochemistry, a Practical Approach* (ed. A. J. Turner and H. S. Bachelard), p. 65. IRL Press, Oxford.
30. Couraud, P. O., Lü, B. Z., and Strosberg, A. D. (1983). *J. Exp. Med.*, **157**, 1369.
31. Cleveland, W. L., Wasserman, N. H., Sarangarajan, R., Penn, A. S., and Erlanger, B. F. (1983). *Nature*, **305**, 56.
32. Shechter, Y., Maron, R., Elias, D., and Cohen, I. R. (1982). *Science*, **216**, 542.

33. Couraud, P. O. (1987). *J. Immunol.*, **138**, 1164.
34. Guillet, J. G., Kaveri, S., Durieu-Trautmann, O., Delavier-Klutchko, C., Hoebeke, J. and Strosberg, A. D. (1985). *Proc. Natl. Acad. Sci. USA*, **82**, 1781.
35. Abood, L. G., Langone, J. J., Bjercke, R., Lux, X., and Banerjee, S. (1987). *Proc. Natl. Acad. Sci. USA*, **84**, 6587.
36. Benovic, J. (1989). In *Receptor Biochemistry, a Practical Approach* (ed. E. C. Hulme), p. 125. IRL Press, Oxford.

3

Polypeptide neurotoxins as probes for certain voltage-dependent K$^+$ channels

J. OLIVER DOLLY

1. Introduction

A paucity of specific probes for membrane-bound, neuronal proteins has restricted their molecular characterization, particularly where functional assays applicable to broken cell preparations are not available. Thus, the discovery of toxins acting selectively on nicotinic acetylcholine receptor or Na$^+$ channel proteins, for example, has resulted in the elucidation of their detailed structural and functional properties.

The present chapter describes the use of relatively novel snake-venom toxins, the dendrotoxins (DTx), and β-bungarotoxin (β-BgTx) to define receptor-binding sites in brain. Detailed characterization has established that these sites are located on a particular class of K$^+$ channels.

The work described here illustrates many of the problems which are faced by the investigator in trying to characterize a useful, receptor-specific ligand starting from a complex mixture of polypeptides, some of which possess biological activity. Because of this illustrative slant, the chapter is somewhat more descriptive than other chapters in this book. It should be noted that a number of the methods introduced here, in specific contexts, are described with greater generality in companion volumes in the *Practical Approach* series (references 1 and 2), or in other chapters of the present book.

Until recently, information on K$^+$ channels has been gained almost exclusively from electrophysiological recordings of membrane K$^+$ currents in a variety of tissues (3). However, over the last decade, several drugs and naturally-occurring toxins have been identified that can discriminate between different types of K$^+$ currents (4); consequently, these conductances can now be classified into several groups (for example voltage-dependent, Ca^{2+}-activated, ATP-sensitive, neurotransmitter, and second messenger-operated), with several variants of each frequently present. Heterogeneity of voltage-activated K$^+$ channels in neurons has proved difficult to define by neurophysiological means because their voltage profiles or activation/ inactivation time courses often overlap. Nevertheless, two main classes of outward, voltage-dependent K^+ currents, namely delayed rectifier and rapid, transient 'A' type, have been distinguished by inhibition with tetraethyl ammonium and aminopyridines, respectively. Recently, a family of neuronal 'A' currents has become apparent from both electrophysiological and biochemical investigations using α-dendrotoxin (α-DTx), and related polypeptides as additional inhibitors (5). Its members vary in their susceptibility to blockade by α-DTx, with the slowly-inactivating variety being most sensitive; on the other hand, homologues of α-DTx (β and γ) appear to inhibit non-activating variants preferentially (6). Likewise, subtypes of proteins exhibiting high affinity for α-DTx and/or β-bungarotoxin (β-BgTx) were detected in brain where they have characteristic locations. Identity of these proteins with K^+ channels has been established from purification (7, 8) and reconstitution (9) studies, whilst determination of their subunit structure (7, 8) together with immunological characterization (10, 11) unveiled homology with a K^+ channel protein cloned from *Drosophila* and rodent brain, and expressed in *Xenopus* oocytes (12, 13). In this chapter, details are given of the experimental protocols underlying these major advances with this single group of K^+ channels that play an important role in controlling nerve cell excitability and synaptic transmission.

2. Purification and radiolabelling of toxins acting on voltage-sensitive K⁺ channels

2.1 Dendrotoxins

These are basic, single-chain polypeptides that are found in the venom of certain mamba snakes, primarily *Dendroaspis polylepis* and *angusticeps* (14). Isolation of the toxins (detailed in *Protocols 1* and *2*) is accomplished by gel filtration of the venom (J. Leakey Ltd, P.O. Box 1141, Nakuru, Kenya, East Africa) followed by cation-exchange column chromatography (*Figure 1*); for convenience, a volatile buffer such as ammonium acetate is normally used. Average yields (batches of venom vary somewhat) of the various toxins obtained by these procedures are shown in *Table 1*. With the venom of *D. angusticeps* (Eastern green mamba), fractionation on Sephadex G-50 yields four main protein peaks; the third and largest contains dendrotoxins and is readily identifiable (14). When the latter is applied to a column of CM-Sepharose, numerous proteins are eluted with a linear gradient of ammonium acetate (*Figure 1A*). When pooled conservatively, peak 9 contains homogeneous α-DTx; a single protein seen on native cathodic and SDS–PAGE, sizing (15) or cation exchange HPLC (see below) is used as evidence of purity, whilst determination of the amino acid composition establishes its identity (14). Other homologues of α-DTx active on K^+ channels (6, and

Table 1. Yields of K$^+$ channel toxins from chromatographic fractionation of mamba venoms: *D. angusticeps* (A) and *polylepis* (B)

Toxin (peaks)		Volume (ml)	Total protein from 2 g venom (mg)	% Recovery (per g venom)
A.				
α-DTx	(9)	127	73.6	3.7
β-DTx	(10)	58	10.9	0.5
isoforms	(11)	76	10.9	0.5
γ-DTx	(12)	68	18.5	0.9
δ-DTx	(13)	79	15.9	0.8
B.				
Toxin B	(4)	56	86	4.3
Toxin E	(7)	24	16	0.8
Toxin I	(8)	168	210	10.5
Toxin K	(10)	76	71	3.5

A, B. Toxins obtained by separation on a CM–Sepharose column of the toxin-containing fractions (pooled) from gel filtration of *D. angusticeps* and *polylepis* venoms, as detailed in *Figures 1A* and *B*, respectively.

below) elute in peaks 10 and 11 (β-isoforms), 12 (γ), and 13 (δ). For details of procedure see *Protocol 1*.

Protocol 1. Purification of α-, β-, γ-, and δ-DTx (*Figure 1A*)

1. Dissolve 2 g *D. angusticeps* venom in 20 ml 0.1 M ammonium acetate, pH 4.5.
2. Centrifuge to remove particulate matter.
3. Apply the supernatant to a Sephadex G-50 (superfine) column (2.5 × 150 cm) equilibrated at 4°C with the above solution.
4. Elute at a flow rate of 18 ml/h.
5. Monitor U.V. absorbance at 280 nm.
6. Collect 4 ml fractions and pool those containing dendrotoxins (third and largest peak).
7. Pump (15 ml/h) these fractions on to a CM-Sepharose CL-6B column (2.5 × 37 cm) equilibrated with 0.1 M ammonium acetate, pH 4.5.
8. Collect 4 ml fractions.
9. Wash with 150 ml of the equilibration buffer.
10. Apply a linear gradient of ammonium acetate (0.1 M, pH 4.5, to 1.1 M, pH 7, total volume 2 litres).

Protocol 1. *Continued*

11. Measure A_{280nm} and conductivity of fractions (*Figure 1A*).

12. Pool fractions from A containing α- (peak 9), β-isoforms (peaks 10 and 11), γ- (peak 12), and δ-DTx (peak 13).

13. Gel filter on Sephadex G-10 column (2.5 × 60 cm) in 0.01 M ammonium acetate.

14. Lyophilize eluates until salt-free.

A final HPLC step on a TSK SP-5PW cation-exchange column is necessary to obtain samples of β₁-, β₂-, and γ-DTx that are deemed pure by the aforementioned criteria. Amino acid compositions, preferably with partial sequence (6), are essential for correct identification of these DTx homologues.

With the venom of *D. polylepis* (Black mamba), the second major peak obtained from the Sephadex G-50 gel filtration contains several polypeptides structurally related to α-DTx. These can, again, be separated by cation-exchange chromatography (*Figure 1B* and *Protocol 2*). Toxin I, the polypeptide with highest affinity for K⁺ channels (15), and the most abundant, elutes in pure form in peak 8; another biologically-active homologue, toxin K (14), being more basic is eluted later (peak 10) in the salt gradient. Whereas both of these toxins are electrophoretically homogeneous and give a single protein peak when subjected to HPLC (as detailed below for β-BgTx), other protease inhibitor homologues, toxins B and E, which lack the ability to facilitate transmitter release (14) are present together with contaminants in peaks 4 and 7, respectively. A further cation-exchange HPLC step (15, 16) is needed to obtain toxins B and E in pure form.

Protocol 2. Purification of toxins I, K, B, and E (*Figure 1B*)

These are isolated, chromatographically, from the venom of *D. polylepis* as detailed in *Protocol 1* with some modifications.

1. Pool the second major protein peak from the Sephadex G-50 column.

2. Fractionate on a CM-Sepharose CL-6B column as specified in *Protocol 1*, but first equilibrate with 0.05 M ammonium acetate, pH 6.

3. Wash the column with 150 ml of this buffer (not included in the 4 ml fractions plotted in *Figure 1B*).

4. Apply a linear gradient of ammonium acetate (0.05 M, pH 6, to 1.2 M, pH 7; 2 litres).

5. Pool fractions containing toxin B (peak 4), E (peak 7), I (peak 8), and K (peak 10).

6. Desalt by gel-filtration and lyophilize (see *Protocol 1*).

7. Further purify toxins B and E by cation-exchange HPLC, as detailed in *Protocol 3* below.

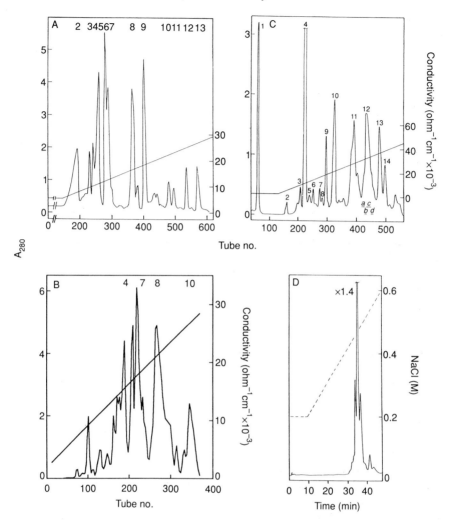

Figure 1. Purification from snake venoms of toxins that act on K⁺ channels. A. α-, β-, γ-, and δ-DTx; for details see *Protocol 1*. B. Toxins I, K, B, and E; for details see *Protocol 2*. C. β-BgTx (see *Protocol 3*). D. β₁-isoform of β-BgTx (see *Protocol 3*). Figures 1A, 1C, and 1D were kindly provided by Dr A. L. Breeze and *Figure 1B* by Ms V. Scott.

2.2 β-Bungarotoxin (β-BgTx)

This is a di-chain polypeptide (M_r ~21 000) from the venom of the Formosan banded krait, *Bungarus multicinctus*; it comprises a 13 500 M_r protein (homologous to phospholipases A_2) disulphide-linked to a 7500 M_r polypeptide that is structurally related to α-DTx. It blocks transmitter release in brain slices and at the neuromuscular junction (14), by a mechanism involving saturable binding to ecto-receptors on the neuronal plasma membrane that directly affect K^+ permeability (5); additionally phospholipolysis, due to its Ca^{2+}-dependent enzymic activity, ensues. Although only limited evidence is available for an inhibitory action on a fraction of the voltage-activated K^+ conductance at motor nerve terminals (reviewed in ref. 4) and dorsal root ganglion neurons (17), β-BgTx has proved extremely useful in differentiating subtypes of K^+ channel protein (5, 16, and below). β-BgTx is isolated from the venom of *B. multicinctus* (Sigma, Poole, Dorset, UK) by direct cation-exchange chromatography (*Figure 1C* and *Protocol 3*). The resultant peak 12 contains several isoforms of β-BgTx; although all of these are biologically-active, successful radioiodination of only $β_1$-isotoxin has been achieved reproducibly in our hands (see Section 2.3). Thus for this purpose, pooled fractions from the centre of this peak (i.e. 12b) are chromatographed on TSK SP-5PW HPLC (*Figure 1D*) or Pharmacia Mono-S HR 5/5 FPLC columns. Conservative pooling of the major central peak yields material that meet all the standard criteria of purity (16); amino acid analysis establishes its identity as the $β_1$-isoform.

Protocol 3. Purification of β-BgTx (*Figure 1C*)

1. Dissolve 1.1 g *B. multicinctus* venom in 15 ml 0.05 M ammonium acetate, pH 6.

2. Centrifuge to remove insoluble material.

3. Apply directly to a CM-Sepharose CL-6B column as in *Protocol 1*, using the elution gradient specified in *Protocol 2*.

4. Carefully pool peak 12 as indicated in *Figure 1C*.

5. Desalt and lyophilize.

6. To isolate $β_1$-isoform of β-BgTx from the others present in peak 12 by cation-exchange HPLC, dissolve the sample in 0.02 M sodium phosphate buffer, pH 6.5; load aliquots containing 1–2 mg protein at 0.5 ml/min on to a TSK SP-5W column (7.5 mm × 7.5 cm); elute (see *Figure 4D*) with a linear gradient of NaCl (0.2–0.6 M, 20 ml total volume).

7. Pool the major peak containing $β_1$-BgTx.

8. Desalt and freeze-dry.

2.3 Radioiodination procedures

2.3.1 α-DTx

This is radiolabelled using a modification of the chloramine-T method (*Protocol 4*) that allows retention of the toxin's biological activity (15).

Protocol 4. Radioiodination of α-DTx

1. Dispense Na^{125} I (20 μl; approx. 2 mCi) into a silanized microfuge tube containing 5 μl of 100 mM phosphate buffer, pH 7.4.
2. Add α-DTx (20 μg in 10 μl of water).
3. Initiate the reaction by adding 5 μl chloramine-T (0.22 mM final concentration; freshly prepared in 100 mM phosphate buffer, pH 7.4), and vortex rapidly.
4. Quench the iodination within 20 sec by diluting to 1 ml of 20 mM sodium phosphate buffer (pH 7.4), containing 0.02% (w/v) Triton X-100 and 0.005% (w/v) NaN$_3$.
5. Withdraw duplicate 10 μl aliquots and add each to 990 μl of the above buffer containing 9 mM K$_2$S$_2$O$_5$ (for determination of specific radioactivity, see below).
6. Apply the remainder to a CM-Sepharose CL-6B column (2 ml) equilibrated in the aforementioned 20 mM phosphate buffer.
7. Remove unreacted ^{125}I by washing with about 8 ml of this buffer.
8. Elute the iodinated toxin (^{125}I–α-DTx) by the inclusion of 0.6 M NaCl (a linear gradient does not give a satisfactory separation of ^{125}I–α-DTx from the native toxin).
9. Store the radioactive fractions at 4°C, with the appropriate shielding; after 6 weeks the level of non-saturable binding increases dramatically.
10. To measure the specific activity of the labelled toxin add 1 ml 1% (w/v) BSA plus 10% TCA to 10 μl aliquots of the reaction mixture.
11. Sediment the precipitated protein by centrifugation.
12. Wash the pellets with 10% TCA and count in a γ-counter.
13. Use the radioactive content of the precipitate (protein bound) to calculate specific activity using the original amount of protein added. Values of 400–600 Ci/mmol are obtained, representing approximately 0.2–0.3 mol ^{125}I per mol of α-DTx.

Electrophoresis of the radioiodinated preparation on native cathodic gels revealed a major labelled band, together with trace amounts of a less basic species (15). When the iodination was repeated under conditions that yielded stoichiometric labelling, using NaI together with a trace of Na^{125}I, the central toxicity of the sample (when injected intraventricularly into rat brain) was identical to that of native α-DTx, confirming that its biological activity is not altered by radioiodination.

2.3.2 β-BgTx

This is radiolabelled by the same procedure as that used for α-DTx (16) except that the buffer (100 mM Hepes, pH 7.4, 5 mM $CaCl_2$, 50% (v/v) glycerol), quantity of toxin (60 μg), and final concentration of chloramine-T (0.15 mM) are different. Chromatographic separation of ^{125}I-labelled β-BgTx (^{125}I-β-BgTx) and determination of its specific radioactivity (200–400 Ci/mmol) are exactly as described for ^{125}I-α-DTx. This toxin is rather sensitive to the conditions employed for radioiodination; however, use of pure $β_1$-isoform together with the conditions specified above are successful. The resultant ^{125}I-β-BgTx preparation gives acceptably low levels of non-saturable binding to synaptic membranes for 6 weeks after preparation, when stored at 4°C; in contrast, labelling performed using lactoperoxidase/H_2O_2 as the oxidizing system yielded material that showed a high initial level of non-saturable binding (50–70% at 1 nM), approaching 100% within 2–4 days.

2.3.3 δ-DTx

Use of the chloramine-T procedure proved inferior to lactoperoxidase-catalysed iodination of δ-DTx (*Protocol 5*), in terms of stability of the product (18).

Protocol 5. Lactoperoxidase-catalysed iodination of δ-DTx

1. Add 5 μl lactoperoxidase (0.61 pmol) in 100 mM phosphate buffer, pH 7.4, 5 μl δ-DTx (10 μg, 1.4 nmol), and 10 μl Na^{125}I (approx. 1 mCi, 0.45 nmol) to a silanized plastic microtube.

2. Start the reaction by adding 5 μl H_2O_2 (4.4 nmol) in phosphate buffer.

3. Add a second 5 μl aliquot of H_2O_2 30 min later.

4. Quench the reaction after 15 min further by diluting to 1 ml with the buffer.

5. Withdraw aliquots for determination of specific radioactivity (300–600 Ci/mmol), as described for α-DTx in *Protocol 4*.

6. Separate ^{125}I-labelled δ-DTx (^{125}I-δ-DTx) from the reaction mixture by chromatography on a CM-Sepharose CL-6B column, as detailed in *Protocol 4*.

---

3. Biochemical assays for the action of toxins on neuronal ion channels

Isolated brain synaptosomes retain the functional characteristics of the nerve terminal *in vivo*, notably, the ability to generate a negative plasma membrane potential, maintain a low cytosolic concentration of Ca^{2+} ($[Ca^{2+}]_c$) when polarized and respond to depolarizing inputs by increasing the cycling rate of Na^+, Ca^{2+}, and K^+ with an elevation of $[Ca^{2+}]_c$ and Ca^{2+}-dependent efflux of neurotransmitter. As a consequence, the actions of different neurotoxins on the various ionic permeabilities of synaptosomes can be readily investigated by quantifying changes in these parameters. Thus, fluorometric determination of transmitter release or $[Ca^{2+}]_c$, in the absence and presence of selective inhibitors of different K^+ channels, gives an indirect measurement of the involvement of the latter in nerve terminal excitability. On the other hand, quantitation of ^{86}Rb efflux allows direct monitoring of membrane K^+ permeability and its perturbation by toxins.

3.1 Isotopic determination of toxin-induced reduction in synaptosomal K^+ permeability

This is accomplished (*Protocol 6*) by loading synaptosomes with $^{86}Rb^+$ (which substitutes for K^+ and is accumulated via the Na^+/K^+ ATPase system), layering aliquots on to filter beds, and connecting to a low volume, high flow-rate multi-channel pump. Such an arrangement allows superfusion with rapid, simultaneous switching between three solutions, simultaneously for each of twelve samples and collection of effluent at 10 sec intervals for scintillation counting. The breakthrough and rate of exchange of the second and third solutions are indicated by the inclusion of $[^3H]$water and $[^{14}C]$sucrose, respectively. As this system allows the accurate measurement of $^{86}Rb^+$ efflux, small changes in the resting K^+ permeability of synaptosomes induced by drugs can be reliably quantified (19). In addition, subsequent depolarization in the presence and absence of K^+ channel blockers (for example 4-aminopyridine, tetraethylammonium) allows different components of the evoked K^+ efflux to be dissected. Indeed, α-DTx is found to reduce a fraction of the ($100\,\mu M$ dosage) veratridine-stimulated, synaptosomal K^+ conductance (*Figure 2A*). This finding is consistent with the toxin's ability to inhibit fast-activating, aminopyridine-sensitive voltage-dependent K^+ currents in somatic membranes (5), as established from intracellular recordings. It is noteworthy that this $^{86}Rb^+$ method allows the demonstration of an α-DTx-sensitive K^+ channel in central nerve terminals, a locus not readily amenable to analysis by electrophysiological techniques.

Protocol 6. Determination of K^+ permeability (*Figure 2A*)

1. Isolate synaptosomes from guinea-pig cerebral cortex by centrifugation of the homogenized tissue in iso-osmolar sucrose and on a discontinuous Ficoll gradient.

2. Perform all assays in the following physiological medium:
   - NaCl   122 mM
   - KCl   3.1 mM

Protocol 6. *Continued*

- KH$_2$PO$_4$ 0.4 mM
- NaHCO$_3$ 5 mM
- Tes–Na   20 mM
- MgSO$_4$ 1.2 mM
- glucose   5 mM
- BSA   1 mg/ml
- adjust to pH 7.4

3. Resuspend synaptosomes (4 mg protein) in 0.5 ml physiological medium as above, but not including KCl and CaCl$_2$.

4. Load with ^{86}RbCl (200–500 μCi/ml of 3.1 mM concentration) by incubating for 1 h at 37°C.

5. Dilute suspension to 4 ml with Ca^{2+} and K$^+$-deficient medium.

6. Load aliquots (100 μg synaptosome protein) on to 12 Rainin 13 mm PVDF 0.45 μm sealed syringe filters by first back-pumping into the respective superfusion lines. Continue subsequent superfusion (3 ml/ min) for 5 min prior to sample collection every 10 sec; include 1.3 mM CaCl$_2$ at this time.

7. Simultaneous exchange between any one of three buffers is achieved by use of teflon discs mounted on a stainless steel bar which compress either the first and third or the second buffer inlet for each superfusion line; selection between the first and third buffers is achieved by a second switch.

8. Add [^{3}H]water (0.01 μCi/ml) and U-[^{14}C]sucrose (0.05 μCi/ml) to the second and third buffers, respectively, to determine the breakthrough and rise time of drugs.

9. Expose labelled synaptosomes to 100 nM α-DTx for 2 min.

10. Measure efflux of ^{86}Rb$^+$ evoked by depolarizing agents.

3.2 Fluorometric assay of synaptosomal [Ca^{2+}]$_c$ and release of endogenous transmitter: effects of K$^+$ channel probes

Using the Ca^{2+}-selective dye, Fura-2, [Ca^{2+}]$_c$ in cerebrocortical synaptosomes can be readily quantified (20, see the chapter by Gurney in reference 2 for a detailed account, and *Protocol 7*). Isolated nerve terminals accumulate this fluorescent probe upon incubation with the acetoxymethyl ester (*Protocol 7*).

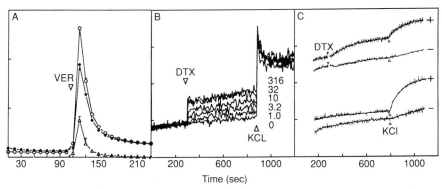

Figure 2. Effects of K⁺ channel toxins on isolated central nerve terminals. A. K⁺ permeability; following 2 min exposure to 100 nM α-DTx, the efflux of $^{86}Rb^+$ evoked by depolarization with veratridine is clearly diminished (y axis scale: 1% of remaining $^{86}Rb^+$/sec/division): ○, Control; ●, toxin present; △ difference plot. B. $[Ca^{2+}]_c$. In the experiments shown, 1.3 mM $CaCl_2$ was added 55 min after the start of the incubation. The data presentation begins 5 min after the addition of $CaCl_2$ and shows the effect of varying concentrations of α-DTx (nM) and KCl (to 33.5 mM) (y axis scale 100 nM Ca^{2+}/division). C. Glutamate efflux: Measurement of glutamate efflux is shown in the absence (−) or presence (+) of Ca^{2+}, with and without 316 nM α-DTx; where indicated KCl to 33.5 mM was added. y axis scale 8 nmol/mg division; traces are Y-shifted from a common origin for clarity. A and B kindly provided by Dr G. R. Tibbs; C, adapted from reference 23.

Protocol 7. Quantification of $[Ca^{2+}]_c$ in synaptosomes with Fura-2 (*Figure 2B*)

1. Load with Fura-2 by resuspending synaptosomes (1.3 mg/ml) in medium devoid of $CaCl_2$ and incubate in the presence of 5 μM Fura-2 acetoxymethyl ester (Fura-2/AM, Calbiochem, USA) for 35 min at 37 °C.

2. Remove external dye by pelleting the synaptosomes (Eppendorf microfuge 12000 g, 60 sec), resuspend in fresh medium (0.67 mg/ml), and transfer to a continuously stirred cuvette in a thermostatted Perkin-Elmer LS 5B fluorimeter interfaced with an IBM-PC compatible computer.

3. Determine $[Ca^{2+}]_c$ as follows:

● Record emission at 505 nm with cycling of the excitation wavelength between 240–380 nm.

● Measure maximum and minimum 340/380 nm ratios by the sequential addition of 0.3% (w/v) sodium dodecylsulphate and 7.5 mM EGTA, pH adjusted to 8.0 with Tris base.

● Calculate $[Ca^{2+}]_c$ in Lotus 123 using a K_d of 224 nM (cf. reference 2).

Addition of α-DTx (21) produces a dose-dependent increase in $[Ca^{2+}]_c$ (*Figure 2B*) at a lower concentration range than observed with the homologues β-, γ-, and δ-DTx (22). As other identified blockers of K^+ channels, 4-aminopyridine and charybdotoxin, gave a similar result, it can be deduced that the effect of these toxins on $[Ca^{2+}]_c$ arises from an inhibition of outward K^+ conductance(s) and a consequential depolarization that activates voltage-sensitive Ca^{2+} channels.

The role of such K^+ channels in the overall regulation of excitability of central nerve terminals can also be assessed from observing the effects of these toxins on transmitter release. A convenient, continuous assay for synaptosomal efflux of L-glutamate involves its oxidation by glutamate dehydrogenase and measurement (detailed in *Protocol 8*) of the concomitant formation of NADPH in a fluorimeter (23).

Protocol 8. Measurement of glutamate efflux (*Figure 2C*)

1. Resuspend (0.67 mg/ml) and preincubate synaptosomes at 37°C for 1 h.
2. Transfer to a continuously stirred cuvette in a fluorimeter interfaced as detailed in *Protocol 7*, step 2.
3. Add $NADP^+$ (1 mM), $CaCl_2$ *or* EGTA (1.3 mM), and 50 units glutamate dehydrogenase (Sigma, cat. no. G 2626).
4. Determine efflux of glutamate by change in fluorescence at 460 nm following excitation at 340 nm. Measurement of resting and K^+-stimulated efflux is shown in the absence or presence of Ca^{2+}, with and without inclusion of α-DTx (see *Figure 2C*).

From the typical traces shown in *Figure 2C* for guinea-pig cerebrocortical synaptosomes, it is apparent that α-DTx raises the amount of glutamate released in a Ca^{2+}-dependent manner. Collectively, use of the outlined assays has revealed that the toxins tested cause substantial alterations of these diverse parameters, highlighting a prominent role for α-DTx-sensitive K^+ channel(s) in the functioning of central nerve terminals.

4. Biochemical identification and localization of DTx receptor sub-types/putative K+ channel proteins in rodent brain

Purified synaptosomes from rat (15) or guinea-pig cerebral cortex (21), and synaptic plasma membranes are convenient preparations for the characterization of the toxins' binding sites. For quantitation of their distribution in the central nervous system, cryostat sections of rat brain are employed; to prepare the latter (24), a light prefixation of the brain tissue is necessary (*Protocol 9*).

Protocol 9. Preparation of cryostat sections of rat brain

1. Anaesthetize rats with 60 mg sodium pentobarbitone in 1 ml, i.p.
2. Perform intracardiac perfusion with 0.1% formaldehyde (approximately 250 ml in 10 mM sodium phosphate buffer, pH 7.3, containing 0.9% NaCl).
3. Dissect slabs of brain as required and freeze in isopentane cooled with solid CO_2.
4. Cut cryostat sections (10 μm thick) at −20°C.
5. Thaw-mount sections on to subbed slides (dipped in 0.5% gelatine and 0.05% chromic potassium sulphate).
6. Dry at room temperature for at least 1 h.
7. Store over silica gel at −20°C. Sections can be kept for up to 3 months before use. (See Chapter 5 of this volume for further details.)

4.1 Detection of saturable binding of ^{125}I-labelled α-DTx and β-BgTx to synaptosomal membranes and brain sections

Toxin binding is measured by incubation of the membrane preparations (50–200 μg of protein in 250 μl of the buffers specified in *Figure 3* legend) at 22°C for 45 min with ^{125}I–α-DTx or ^{125}I–β-BgTx at the concentration specified and, for competition studies, with varying concentrations of test ligand. Bound- and free-toxin are then separated by centrifugation through silicone 'oil' (see also Chapter 7 of this volume), followed by γ-counting of the pellet. Non-saturable binding is determined, likewise, by inclusion of approximately 100-fold excess of unlabelled toxin. All plasticware used should be silanized to minimize absorption of the toxin which can be appreciable, particularly when low concentrations are exposed to untreated tubes/pipettes.

Brain sections are labelled (24, 26) with each toxin as outlined above and as shown in *Protocol 10*.

Protocol 10. Detection of toxin binding in brain sections (*Figure 3*)

1. Label brain sections (24, 26) with each toxin, as outlined above in text, using Krebs phosphate medium containing 1 mg/ml BSA.
2. Wash samples by immersing in three changes (5 min each) of the ice-cold buffer.
3. Rinse in deionized water and dry quickly in a stream of cold air.
4. Expose toxin-labelled sections to ^{3}H-sensitive Ultrofilm, at 4°C for 1–10 days (depending on the radioactive content).

Protocol 10. *Continued*

5. Develop resultant film for 5 min at 18°C in undiluted D19 developer, wash thoroughly, and fix in 20% Hypam.

6. Produce enlarged photographic darkfield images by printing directly from the autoradiograms.

7. Identify cell layers/regions using an atlas for rat brain, if necessary by staining with pyronin Y/methyl green.

8. Quantitate the binding sites in various areas by analysing the autradiograms with a suitable scanner. We use a Biomed Instruments 1D, 2D soft laser densitometer controlled by an Apple IIe microcomputer.

9. Visualize the digitized data with a program such as the 2D/1D PCIV that allows pictorial representations of the image, or direct read-out of optical densities at specific points.

10. Determine the radiation dose producing each images in d.p.m. $\times$ days $\times$ mm^{-2} (24).

11. Calibrate by preparing and processing ^{125}I-labelled brain-paste standards (350–160000 d.p.m./mg protein).

12. Construct a standard curve (log$_e$ of optical density vs log$_e$ of radiation dose).

13. Use this to convert optical densities of brain section autoradiograms to fmoles of ^{125}I–α-DTx- or ^{125}I–β-BgTx-bound/mm^2, using the specific radioactivity of each labelled toxin.

14. Quantify the extent of toxin binding in a whole brain section by removing the sample from the slide (after exposure) with a dampened Whatman GF/C glass microfibre filter disc (2.5 cm) followed by γ-counting.

15. Measure the protein content (24) by dissolving the sample on each disc in 0.45 M NaOH for 24 h and removing the glass fibre by centrifugation; assay the supernatant.

Assays performed on synaptosomal membranes (*Figure 3A*) or whole brain slices (*Figure 3B*) with a series of concentrations of ^{125}I-labelled α-DTx or β-BgTx showed that saturable binding of each occurred. Although the K_d values (~0.5 nM) calculated from these Scatchard plots were similar for both toxins, the content of sites obtained for β-BgTx represents a fraction of that

---

Figure 3. Heterogeneity of receptors for α-DTx and β-BgTx in rat synaptosomal membranes and brain slices revealed by direct toxin binding and competition assays. A,B. Scatchard plots of the saturable binding of ^{125}I–α-DTx (●) and ^{125}I–β-BgTx (■) to cerebrocortical synaptosomes (A), and cryostat sections of rat brain (B), measured in Krebs phosphate medium as detailed in *Protocol 10*. Adapted from references 5, 24, and 26. C. Inhibition of ^{125}I–β-BgTx (0.7 nM) binding to synaptic membranes in Krebs phosphate buffer containing 0.74 mM SrCl$_2$ by varying concentrations of unlabelled β-BgTx (●) or α-DTx (■). D. Antagonism of ^{125}I–α-DTx (0.7 nM) binding to synaptosomal

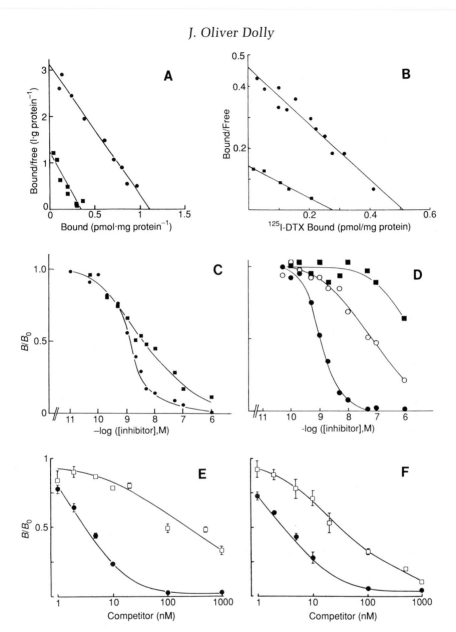

membranes by unlabelled α-DTx (●) and β-BgTx in Krebs (■) or 50 mM imidazole–HCl buffer, pH 7.4, containing 90 mM NaCl, 5 mM KCl and 1.5 mM $SrCl_2$ (○). C and D are adapted from reference 16; non-displaceable binding measured in the presence of excess β-BgTx has been subtracted from the values of fractional, saturable binding plotted. E,F. Inhibition of $^{125}I–α$-DTx binding by unlabelled α-DTx (●) and β-BgTx (□) in cryostat brain sections. Values shown (± standard errors) were calculated using several sets of densitometric readings of autoradiographic images (see text) in corpus callosum (E) and amygdala (F); E and F are from reference 26. Amounts of ^{125}I-labelled toxin bound in the presence of competing toxin (B) are expressed relative to the total (B_o).

for α-DTx in each preparation. Competition studies revealed that ^{125}I-β-BgTx binding is blocked efficiently (*Figure 3C*) and in a non-competitive manner (16) by α-DTx; the binding of ^{125}I-α-DTx is inhibited only partially by β-BgTx, even at high concentrations (*Figure 3D*). Note that this antagonism by unlabelled β-BgTx is more pronounced (compared to that observed with Krebs buffer) in imidazole solution (and involves a complex, non-competitive mechanism), a finding substantiated by additional, low-affinity binding of ^{125}I-β-BgTx detected in the imidazole buffer (5, 16). Apparent heterogeneity in these receptors is also detectable autoradiographically (26). Interestingly, the effectiveness of β-BgTx in reducing ^{125}I–α-DTx binding varies significantly in sections from different rat brain regions, for example the corpus callosum and amygdala (*Figure 3E,F*). Based on these and other findings (5, 16), it can be surmised that two or more sub-types of high-affinity binding sites for α-DTx exist in brain or synaptic membranes (in a stoichiometry of ~3:1), and the less abundant variety only binds β-BgTx with high affinity (K_d ~0.5 nM). Another site exhibiting lower affinity for β-BgTx seems to exist on all the receptors but its properties remain to be established, particularly the influence thereon of different buffers. The interpretation of binding data is considered in detail in Chapters 4 and 11 of this volume.

4.2 Localization of receptor subtypes for α-DTx and β-BgTx in the central nervous system

Examination of autoradiograms prepared from sequential sections through rat brain demonstrated a widespread distribution of α-DTx binding sites in both grey and white matter regions (*Figure 4A*). In contrast, saturable binding of β-BgTx showed a more discrete distribution (*Figure 4C,D*), with high site densities being present in synaptic-rich regions (for example the molecular layer of the hippocampus). It is noteworthy that an excess of α-DTx abolished all ^{125}I–β-BgTx binding (24), whereas β-BgTx produces partial blockade of ^{125}I–α-DTx labelling; the extent of this varies in different regions (*Figure 3E,F*) and, noticeably, some β-BgTx-resistant α-DTx binding sites reside in white matter areas (*Figure 4B*). On the basis of extensive quantitative analysis of these light-microscope pictures, and additional electron-microscope autoradiographic studies, it can be deduced that the common receptor subtype binding both toxins avidly resides predominantly on synaptic membranes, whilst other α-DTx sites also occur there and on axonal plus somatic membranes. Notably, these locations are equivalent to sites where α-DTx has been found to block K$^+$ currents (5). An obvious, major advantage of these autoradiographic techniques is that they allow the distribution of K$^+$ channels/toxin receptors to be readily mapped and in a quantitative manner (24, 26), an achievement not possible by electrophysiological means; moreover, they allow recognition of areas/neurons containing K$^+$ channels that are worthy of detailed electrophysiological analysis.

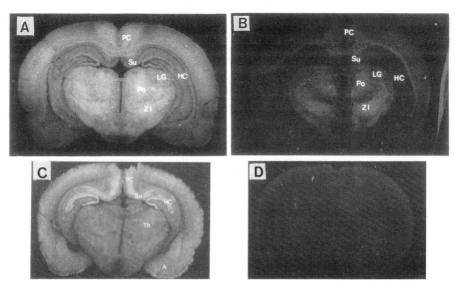

Figure 4. Autoradiographic localization of receptors for ^{125}I-labelled α-DTx and β-BgTx in cryostat sections of rat brain. A,B. Autoradiograms shown are from sections at the level of the hippocampal region that were labelled with 2 nM ^{125}I–α-DTx alone (A) or in the presence of 1 μM β-BgTx (B). C,D. as above except labelling was performed with 2 nM ^{125}I–β-BgTx alone (C) or with the inclusion of 1 μM β-BgTx (D). Experimental details are given in the text and in *Protocol 10*. Identified brain regions shown: HC, hippocampal formation; Su, subiculum; LG, lateral geniculate body; Po, posterior thalamic nuclear group; PC, posterior cingulate cortex; Zi, zona incerta; Th, thalamus; and A, amygdala. Taken from references 24 and 26.

5. Isolation of α-DTx receptor/K⁺ channel proteins from mammalian brain: determination of oligomeric properties and subunit structure

Rat synaptosomal membranes were used in the initial studies (27) on the solubilization and characterization of receptors for α-DTx and β-BgTx; however, synaptic membranes purified from bovine cortex (8) is the preferred preparation for large-scale receptor purification. During the separation of these membranes and subsequent receptor isolation, it is imperative to minimize proteolysis by the inclusion of protease inhibitors (listed below) and maintain a low temperature throughout.

5.1 Quantitation of toxin binding to solubilized receptors

[125]I–α-DTx binding to solubilized membrane extracts (detailed later), is assayed by rapid (centrifugation) gel filtration (*Protocol 11*).

Protocol 11. Quantitation of toxin binding to solubilized receptor

1. Pour Sephadex G-100 (fine) columns (total volume of 2 ml) in syringes containing a disc of Whatman No. 1 filter paper.

2. Wash with one bed volume of 25 mM imidazole–HCl, pH 7.4, 30 mM KCl and 0.2% (w/v) Triton X-100.

3. Centrifuge for 2 min at 100 g.

4. Form receptor–toxin complex by incubating an appropriate aliquot (150 μl) of detergent extract with 100 μl of a solution of 7.5 nM ^{125}I–α-DTx plus 250 μg/ml cytochrome C for 30 min at 23 °C, in a total volume of 250 μl.

5. Set up incubations supplemented with 500 nM (final) unlabelled toxin for measurement of non-displaceable binding.

6. Carefully apply an aliquot (200 μl) of each reaction mixture to the columns before centrifugation, as above.

7. Collect the receptor–toxin complex in the void volume (approx. 200 μl) and quantify by γ-counting.

Binding of ^{125}I–β-BgTx is similarly determined using a modified (to yield maximum activity) incubation mixture (25 mM imidazole–HCl, pH 7.4, 100 mM KCl, 1.5 mM SrCl$_2$ and MnCl$_2$, 0.1% (w/v) phosphatidylcholine, 0.2% (w/v) Lubrol PX) and Sephacryl S-200 columns washed in the latter buffer containing 25 mM KCl.

5.2 Solubilization of α-DTx and β-BgTx receptors from synaptic plasma membranes in a stable form

The optimal procedure for solubilization of the toxin-binding sites entails agitation of membranes (~10 mg/ml) at 4 °C for 1 h with extraction buffer:

- 4% (w/v) Lubrol PX in 62.5 mM imidazole–HCl, pH 7.4
- 250 mM KCl
- 2.5 mM EDTA
- 25 μg/ml soybean trypsin inhibitor
- 50 μg/ml bacitracin
- 0.25 mM benzamidine
- 0.5 mM phenylmethanesulphonyl fluoride (PMSF)

This is followed by 2.5-fold dilution with cold deionized water and centrifugation at 30 000 g for 20 min and then at 100 000 g for 45 min at 4 °C; the resultant supernatant (crude extract) is used. With the exception of Tween, all of the detergents tested solubilized the receptors with retention of α-DTx and β-BgTx binding activities (*Table 2A*). However, a stable preparation

($t_{1/2} > 10$ days) of α-DTx binding activity is achieved only with Lubrol PX or CHAPS; Lubrol is also most effective for β-BgTx binding though its $t_{1/2}$ is appreciably shorter (~42 h). The stability of binding activities for both toxins in Triton X-100 was unacceptably low, though this can be doubled by the inclusion of 10% (w/v) glycerol or sucrose. The presence of K^+ in the extraction buffer is a prerequisite for maintaining the receptors in active form (*Table 2B*); although Rb^+ can substitute for K^+ in the case of α-DTx binding, it is less effective with β-BgTx acceptor activity. If K^+ is omitted during solubilization, but subsequently added back, activity cannot be recovered; likewise, extraction in K^+-containing medium followed by removal of K^+ leads to inactivation of the binding sites. It is not unexpected that K^+ channel proteins contain a K^+ binding site; its occupancy is clearly essential for toxin-binding activity. Importantly, binding characteristics of the receptors in the solubilized state are shown to remain unchanged from those in the mem-

Table 2A. Yields and stabilities of α-DTx and β-BgTx binding activities extracted with different detergents from synaptic plasma membranes

Detergent	Specific toxin binding (% of that in membranes)			Stability ($t_{1/2}$ at 4°C)	
	^{125}I–DTx[a]	^{125}I–DTx[b]	^{125}I–β-BgTx[b]	^{125}I–DTx[a]	^{125}I–β-BgTx[b]
Triton X–100	43	51	50	30 h	12 h
Lubrol PX	45	69	72	>270 h	42 h
CHAPS	48	45	46	>270 h	—
Nonidet P40	40	—	—	20 h	—
Brij 58	—	43	50	—	—

[a,b] Bovine and rat synaptic membranes were used, respectively.

Table 2B. Cation dependence of detergent solubilization of toxin-binding activities

Ion	Concentration (mM)	Relative specific binding (%)		
		^{125}I–DTx[a]	^{125}I–DTx[b]	^{125}I–β-BgTx[b]
K^{+*}	250	100	100	100
K^+	100	89	52	10
Rb^+	250	105	68	18
Cs^+	250	35	6	0
Na^+	250	0	0	0
Li^+	250	0	0	0

* Data expressed relative to these values.
[a,b] Bovine and rat synaptic membrane extracted in Lubrol and Triton, respectively.
Results kindly provided by D. Parcej and C. Donegan.

branes; K_d values determined for ^{125}I-α-DTx and ^{125}I-β-BgTx together with K_Is for their inhibition by toxin homologues (27) are very similar in both preparations. The proteinaceous natures of these receptors are readily established from inactivation by heating or trypsinization. Sedimentation analysis on sucrose gradients (performed in D_2O and H_2O to allow correction for bound detergent, see references 1, 2), together with determination of Stokes radius by gel filtration, show (27) that binding sites for both toxins reside on large proteins of similar size ($S_{20,w} = 13.2$; $M_r \sim 400\,000$).

5.3 Purification of α-DTx receptors

Conventional separation methods such as gel filtration, anion-exchange, or hydrophobic chromatography are of limited use in the purification of α-DTx receptors (~5-fold enrichment being achieved). Affinity chromatographic methods, using toxin I or wheat-germ lectin immobilized on CNBr-activated Sepharose 4B, have proved successful. The most effective resin is prepared by coupling toxin I to Sepharose gel containing a low content of active groups (8).

Protocol 12. Preparation of toxin I–Sepharose columns

1. Activate Sepharose 4B (28) with 8 mg CNBr/ml (of settled gel) to yield approx. 0.5 µmol cyanate esters/ml.

2. Wash with 50% (v/v) acetone/0.05 M HCl, then 1 mM HCl.

3. Resuspend resin in 2 volumes of coupling buffer (50 mM triethanolamine-HCl, pH 7.8, 0.5 M NaCl).

4. React with 0.3 mg of toxin I/ml of gel for 2 h at room temperature.

5. Gently centrifuge to sediment the gel.

6. Measure absorbance of supernatant at 280 nm to determine the quantity of toxin I coupled (usually 50–80%).

7. Block any remaining free reactive groups by exposure to 0.5 M ethanolamine–HCl, pH 8.5, for 4 h at room temperature.

8. Wash resin sequentially in a sintered glass funnel with 1 litre each of coupling buffer, 0.1 M sodium acetate buffer (pH 5) in 1 M NaCl, water, and 0.1 M Tris–HCl, pH 8.5, in 1 M NaCl.

9. Store in 0.1 M sodium acetate buffer, pH 6, containing 0.5 M NaCl and 0.05% (w/v) NaN₃.

Wheat-germ agglutinin (WGA) is covalently linked to CNBr-activated Sepharose 4B (Pharmacia) at 5 mg/ml of resin, according to the manufacturer's instructions.

Table 3. Affinity chromatography purification of α-DTx receptors from bovine cortex

	Protein (mg)	^{125}I-α-DTx binding (pmoles)	Specific activity (pmoles/mg of protein)	Recovery (%)	Purification -fold
Lubrol extract of synaptic membranes	728	209	0.29	100	1
Toxin I–Sepharose 4B column: breakthrough	703	5	0.007	2	0
DTT* eluate	0.02	23	1165	11	3966

* Assayed after gel filtration on Sephadex G-75.

A Lubrol PX extract (116 ml) of plasma membranes was loaded (25 ml/h) on to a 10 ml column of toxin I–Sepharose-4B, previously equilibrated with:

- imidazole–HCl, pH 7.4   25 mM
- KCl   100 mM
- EDTA   1 mM
- benzamidine   0.2 mM
- PMSF   0.5 mM
- Lubrol PX   0.2% (w/v)

After washing the resin with 100 ml of the same buffer, α-DTx receptors were eluted with the equilibration buffer containing 10 mM dithiothreitol (DTT).

When crude detergent extract of bovine synaptic membranes is loaded on to a column of toxin I–Sepharose, approximately 100 pmoles of receptor becomes bound per milligram of immobilized toxin; in contrast, the majority of extraneous protein is not adsorbed (*Table 3*). Following washing to remove further unwanted protein, α-DTx binding proteins can be dissociated from the gel with 0.6 M KCl (7, 8); however, this yields insufficient enrichment. Receptor activity can be eluted both in a high state of purity (*Figure 5A*) giving an approximate 3966-fold purification and a reasonable recovery (~11%; *Table 3*), by a novel procedure based on the reduction of essential disulphide bridges in the immobilized toxin (and possibly in the receptor, also). The reducing agent (10 mM dithiothreitol) used in the elution buffer is removed by rapid gel filtration on Sephadex G-75 prior to assay of α-DTx binding. A final purification step exploits the glycosylated nature of the receptor protein; although certain lectin resins (Sepharose-4B attached to *Ricinus communis* agglutinin 1 or concanavalin A) tested failed to bind the receptor; chromatography on immobilized WGA with elution by *N*-acetylglucosamine (~45% recovery) is effective in separating contaminating proteins (*Figure 5A*). The resultant material binds both α-DTx and β-BgTx, indicating that these procedures isolate the receptor subtypes. Some success has been reported (29) on their subfractionation by chromatography on immobilized β-BgTx, though it is difficult to obtain a preparation of this affinity resin that gives adequate recovery of receptor.

5.4 Determination of subunit composition by analysis of purified receptor and cross-linked complex with α-DTx

Analysis by SDS polyacrylamide gel electrophoresis of the protein samples, obtained from the toxin I and lectin gels, shows two major proteins (*Figure 5A*); a sharp band with M$_r$ of 37000 together with a diffuse band with M$_r$ ≈78000, indicative of microheterogeneity. Evidence that component(s) of

Figure 5. Electrophoretic analysis of purified α-DTx receptor and its cross-linked toxin complex. SDS polyacrylamide gel (8%) electrophoresis was performed under reducing conditions and proteins were revealed by silver staining (8) or autoradiography (25, 30; see reference 1, Chapter 10 for details). A. Protein staining patterns are shown for samples of α-DTx receptor from bovine synaptic membranes at different stages of purification. Crude extract (track 6), breakthrough (5) and dithiothreitol eluate of toxin I–Sepharose column (2); eluate of WGA-affinity column (1), buffer blank (3), and (4) molecular weight markers (kDa)—β-galactosidase (116), serum albumin (67), ovalbumin (43), glyceraldehyde 3-phosphate dehydrogenase (36), carbonic anhydrase (29), and trypsinogen (24). Provided by D. Parcej. B. Autoradiogram prepared from samples of guinea-pig synaptosomal membranes that were labelled with 2 nM ^{125}I–α-DTx in the absence (track 1) or presence (track 2) of 1 μM α-DTx and cross-linked with dimethylsuberimidate (1 mg/ml) for 1 h at 22°C. Arrows indicate the mobilities of marker proteins, in descending order (M$_r$ in kDa): β-galactosidase (116); phosphorylase B (97.4); serum albumin (67); ovalbumin (43); carbonic anhydrase (29).

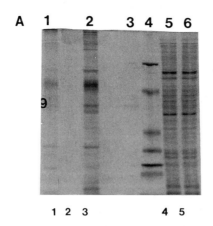

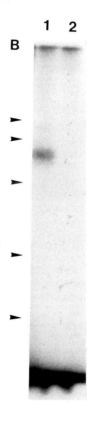

the latter comprise the receptor is provided by cross-linking of the ^{125}I–α-DTx complex in rodent (18, 25) or chick (30) synaptosomal membranes with a bifunctional agent, dimethylsuberimidate followed by SDS gel electrophoresis/autoradiography (*Figure 5B*). This yields a radiolabelled component with M_r of 75 000 (±2000; after allowance for one toxin assumed to be attached to each receptor) and, thus, within the error of such gels, correspond to, at least, part of the broad band seen in the pure receptor. Structural similarity of the larger subunit to A-type K^+ channel cloned from *Drosophila* shaker has been gained from observed antibody cross-reactivity (10, 11). Involvement or otherwise of the 37 000 M_r protein with the receptor/K^+ channel complex remains to be established; such information will help to ascertain if this family of K^+ channels have hetero-oligomeric structures containing two very different sizes of subunits and/or are composed of multiple copies of one or more of the subunits present in the diffuse 78 000 M_r electrophoretic band. Clearly, information acquired by the molecular neurobiological research strategies outlined here, used in conjunction with molecular genetics and electrophysiological approaches, will eventually give insight into the molecular structures underlying the characteristic properties of variants of fast-activating, voltage-sensitive K^+ channels.

Acknowledgements

I thank research colleagues who have helped develop and successfully use the techniques outlined herein, particularly those who provided illustrations (A. Breeze, C. Donegan, Z. Muniz, D. Nicholls, D. Parcej, A. Pelchen-Matthews, V. Scott, and G. R. Tibbs). Also, my thanks are due to Rosemary Davis for typing the manuscript. This research is supported by The Wellcome Trust, MRC, and SERC.

References

1. E. C. Hulme (ed.) (1990). *Receptor Biochemistry, a Practical Approach*. Oxford University Press, Oxford.
2. E. C. Hulme (ed.) (1990). *Receptor–Effector Coupling, a Practical Approach*. Oxford University Press, Oxford.
3. Moczydlowski, E., Lucchesi, K., and Ravindran, A. (1988). *J. Membrane Biol.*, **105**, 95.
4. Dolly, J. O. (1990). In *Neuromuscular Transmission: Basis and Clinical Aspects* (ed. A. Vincent and D. Wray), p. 107. Manchester University Press.
5. Dolly, J. O., Stansfeld, C. E., Breeze, A., Pelchen-Matthews, A., Marsh, S. J., and Brown, D. A. (1987). In *Neurotoxins and Their Pharmacological Implications* (ed. P. Jenner), p. 81. Raven Press, New York.
6. Benishin, C. G., Sorensen, R. G., Brown, W. E., Krueger, B. K., and Blaustein, M. P. (1988). *Molec. Pharmacol.*, **34**, 152.
7. Rehm, H. and Lazdunski, M. (1988). *Proc. Natl. Acad. Sci. USA*, **85**, 4919.

8. Parcej, D. N. and Dolly, J. O. (1989). *Biochem. J.*, **257**, 899.

9. Rehm, H., Pelzer, S., Cochet, C., Chambaz, E., Tempel, B. L., Trautwein, W., Pelzer, D., and Lazdunski, M. (1989). *Biochemistry*, **28**, 6455.

10. Barbas, J. A., Rubio, N., Pedroso, E., Pongs, O., and Ferrus, A. (1989). *Molec. Brain Res.*, **5**, 171.

11. Rehm, H., Newitt, R. A., and Tempel, B. L. (1989). *FEBS Lett.*, **249**, 224.

12. Stühmer, W., Stocker, M., Sakmann, B., Seeburg, P., Baumann, A., Grupe, A., and Pongs, O. (1988). *FEBS Lett.*, **242**, 199.

13. Christie, M. J., Adelman, J. P., Douglass, J., and North, R. A. (1989). *Science*, **244**, 221.

14. Harvey, A. L., Anderson, A. J., Mbugua, P. M., and Karlsson, E. (1984). *J. Toxicol.*, **3**, 91.

15. Black, A. R., Breeze, A. L., Othman, I. B., and Dolly, J. O. (1986). *Biochem. J.*, **237**, 397.

16. Breeze, A. L. and Dolly, J. O. (1989). *Eur. J. Biochem.*, **178**, 771.

17. Petersen, M., Penner, R., Pierau, Fr.K., and Dreyer, F. (1986). *Neurosci. Lett.*, **68**, 141.

18. Muniz, Z. M., Diniz, C. R., and Dolly, J. O. (1990). *J. Neurochem.*, **54**, 343.

19. Tibbs, G. R., Nicholls, D. G., and Dolly, J. O. (1989). In *Ion Transport* (ed. D. J. Keeling and C. D. Benham). Academic Press, New York.

20. Kauppinen, R. A., McMahon, H. T., and Nicholls, D. G. (1988). *Neuroscience*, **27**, 175.

21. Tibbs, G. R., Nicholls, D. G., and Dolly, J. O. (1989). *FEBS Lett.*, **255**, 159.

22. Muniz, Z. M., Tibbs, G. R., Maschot, P., Bougis, P., Nicholls, D. G., and Dolly, J. O. (1990). *Neurochem. Internat.*, **60**, 105.

23. Tibbs, G. R., Dolly, J. O., and Nicholls, D. G. (1989). *J. Neurochem.*, **52**, 201.

24. Pelchen-Matthews, A. and Dolly, J. O. (1988). *Brain Res.*, **441**, 127.

25. Mehraban, F., Breeze, A. L., and Dolly, J. O. (1984). *FEBS Lett.*, **174**, 116.

26. Pelchen-Matthews, A. and Dolly, J. O. (1989). *Neuroscience*, **29**, 347.

27. Black, A. R., Donegan, C. M., Denny, B. J., and Dolly, J. O. (1988). *Biochemistry*, **27**, 6814.

28. Kohn, J. and Wilchek, M. (1982). *Biochem. Biophys. Res. Commun.*, **107**, 878.

29. Rehm, H. and Lazdunski, M. (1988). *Biochem. Biophys. Res. Commun.*, **153**, 231.

30. Black, A. R. and Dolly, J. O. (1986). *Eur. J. Biochem.*, **156**, 609.

4

Strategy and tactics in receptor-binding studies

E. C. HULME and N. J. M. BIRDSALL

1. Introduction

1.1 Information obtainable from binding studies

The questions answerable by receptor-binding studies fall into two major categories:

Firstly, there are fundamental questions which one may ask about:

- the molecular pharmacology of receptors,
- the mechanism of ligand binding, and
- the nature and mechanism of receptor–effector interactions.

Secondly, there are applied questions which concern the cell biology and anatomical distribution of receptors, particularly:

- their concentrations in different tissues,
- their distribution on different cell types,
- their ontogenetic development, and
- how their synthesis, membrane insertion, coupling, degradation, and recycling are regulated in a given cell type.

Ligand binding provides a direct approach to the *in vitro* investigation of receptors. Such studies first became a practical proposition with the availability of receptor-specific ligands radiolabelled to high specific activity with tritium or iodine. Over the past 15 years, this has led to an explosive growth in both the practice and the literature of binding studies. Excellent overviews have been provided by Cuatrecasas and Hollenberg (1), and by Yamamura, Enna, and Kuhar (2). Binding assays have been widely applied to receptor sub-classification as well as in functional and localization studies (3–5).

1.2 Basic principles

In outline, the principle of receptor-binding assays is deceptively straightforward. The basic binding assay protocol is as follows:

Protocol 1. Binding assay procedure

1. Choose and make a tissue preparation containing the receptor.
2. Select a suitable labelled ligand.
3. Incubate the receptor preparation with an appropriate concentration of a labelled ligand for a defined time at a defined temperature.
4. Separate the bound from the free ligand, using an appropriate separation technique.
5. Measure the bound and free ligand concentrations.
6. Repeat steps (3)–(5) with the addition of unlabelled ligands or modulating agents as dictated by the aims of the experiment.
7. Analyse the data to extract quantitative estimates of rate constants and/or affinity constants.
8. Relate the estimates of the binding parameters to pharmacologically-determined values.

The receptor preparation may be:

- a whole animal
- a whole tissue or tissue slice
- a whole cell suspension
- a membrane fraction
- a solubilized preparation
- a purified preparation

For work on non-enriched preparations the ligand is usually radiolabelled. Fluorescent labels are valuable when the receptor concentration is high enough to permit their use, despite their lower sensitivity of detection. They are suitable for the extraction of highly time-resolved data, for example in stopped-flow studies (Chapter 10). *In vivo*, whole animal studies involve injection of a tracer ligand usually followed by tissue isolation and assay, for example by scintillation counting, or autoradiography. This rather specialized area will not be considered further here.

Separation of bound from free ligand usually entails:

(a) centrifugation or filtration assays for particulate preparations;
(b) gel-filtration, precipitation, or adsorption assays for solubilized preparations.

These techniques are discussed in detail in later chapters of this book. Equilibrium dialysis has also been applied to receptor-binding assays, but is usually not practical for multiple assays and is not widely used in receptor studies. The same information can be obtained by equilibrium gel filtration (Chapter 9). A comprehensive account of the application of equilibrium dialysis to protein-binding studies is given by Klotz in *Protein Function, a Practical Approach* (6).

A very recent development, which may eventually find application in receptor-binding studies, is the scintillation proximity assay. This method entails the covalent coupling of the binding protein to the surface of microspheres containing a scintillator. Radiolabelled ligand molecules which become bound to the immobilized binding macromolecule are sufficiently close to the fluor-containing bead to cause light emission, whilst free ligand in solution is not. In principle the technique thus allows binding measurements to be made on a continuous basis without the need to sample the reaction mixture, or physically to separate bound from free ligand. Its disadvantage is that it demands covalent immobilization of the receptor, without loss or modification of its activity, and probably also requires a relatively highly purified preparation of the protein. So far the technique has been largely limited to antibody-based assays. However, as larger quantities of purified recombinant receptors become available (cf. reference 7), the potential advantages of this approach may make it an attractive option for some applications. Further information about the technique is available from Amersham International plc.

Unlabelled ligands, if present in a receptor-binding assay, may either compete directly with the tracer or may modulate the tracer binding by non-competitive interaction, for example by interaction with a site distinct from but linked to the tracer binding site. A non-competitive binding site may be a separate, allosteric site located on the receptor itself, or may be on an effector molecule, or on another interacting macromolecule.

While the basic protocol may seem straightforward, many complexities arise:

(a) From the open-ended nature of binding experiments, which have large numbers of variable experimental parameters.

(b) From the complex nature of the trade-offs which need to be made between ligand affinities, ligand concentrations, concentrations of binding sites, receptor stability, ligand stability, and incubation times if artefacts are to be avoided.

(c) From the fact that the extraction of numerical values for rate constants, binding constants, and numbers or concentrations of sites usually requires the application of curve-fitting, and other statistical techniques.

2. Basic strategy

2.1 Introduction

The development of a valid binding assay falls into four main phases:

- initial choices
- exploration

- validation
- exploitation and analysis

This programme proceeds recursively, so the choices to be made are inter-active at all stages. To illustrate the basis for decision making, the basic theory and experiments necessary to establish a binding assay in the simplest case will be outlined in this section. More detailed consideration of the issues raised, and experimental details, will be reserved for later parts of the chapter (Sections 3 and 4) and Chapter 11. In this way, both the character of the information to be gained from binding studies and the nature of the experi-mental approach should become clearer.

It is assumed that:

(a) a suitable tissue preparation is available (Chapter 5);
(b) a labelled tracer ligand has been chosen which is known, by pharmaco-logical criteria, to interact with the receptor of choice (Chapter 1 and Appendices 1 and 2 to this volume);
(c) a number of unlabelled antagonists of the receptor are available;
(d) the pharmacologically-determined affinity constants of both the labelled and unlabelled ligands are known.

2.2 Mathematical basis

A certain familiarity with the theoretical background is necessary for the interpretation of binding studies. However, the algebra of binding equations rapidly becomes complex, and off-putting. For this reason, the more complex equations have been banished to tables, where they can be referred to if necessary. In the text, we have provided rough-and-ready computer programs which illustrate receptor-binding properties in a much more graphic fashion than the equations themselves can do. Much more detail is given in Chapter 11.

2.2.1 Tracer binding under non-depletion conditions

In the simplest case, the binding of a labelled ligand L^* to a receptor R is a simple bimolecular association reaction:

$$R + L^* \underset{k_{21}}{\overset{k_{12}}{\rightleftharpoons}} RL^*, \tag{1}$$

k_{12} is the ASSOCIATION rate constant (*on-rate*) of the ligand–receptor interaction. The probability that a receptor and a ligand molecule will associ-ate to give rise to a molecule of the receptor–ligand complex within a given time interval is proportional to

$$k_{12} \cdot R \cdot L^* \tag{1a}$$

k_{21} is the DISSOCIATION rate constant (*off-rate*) of the receptor–ligand complex. The probability that a molecule of the complex will break down within a given time interval to give free ligand and free receptor is proportional to:

$$k_{21} \cdot RL^*. \tag{1b}$$

Under any given condition, the rate of change of the concentration of the receptor–ligand complex with time is given by the difference between the rates of formation and breakdown of receptor–ligand complexes,

$$\frac{d(RL^*)}{dt} = k_{12} \cdot R \cdot L^* - k_{21} \cdot RL^*, \tag{2}$$

where R and L^* are the free concentrations of receptor and ligand. Since no catalytic processes are involved, both the receptor and the ligand are subject to conservation conditions

$$R + RL^* = R_t \tag{2a}$$

$$L^* + RL^* = L_t^*, \tag{2b}$$

where R_t and L_t^* are the total receptor and total labelled ligand concentrations. For the present, we assume that $R_t \ll L_t^*$ so that the free and total ligand concentrations are effectively the same, i.e. that there is no ligand depletion.

Equation 2 is readily solved, subject to the conservation conditions (see reference 4). The most useful results obtained are the following:

(a) When the free receptor is mixed with the labelled ligand to initiate the binding process, the time-course of formation of the receptor ligand complex is given by:

$$RL^* = RL_{eq}^*(1 - e^{-(k_{12} \cdot L^* + k_{21})t}). \tag{3}$$

This expression shows that RL^*, initially zero, rises exponentially to its equilibrium value, RL_{eq}^*, with a time-course governed by the net rate constant $k_{12} \cdot L^* + k_{21}$. What this means in practice is shown in Section 2.3.1 and *Figure 1a*.

(b) If, after a finite level of binding has been attained, the free ligand is suddenly removed, for example by dilution, or if, alternatively, a second unlabelled ligand is added at a concentration sufficient to fully occupy the binding site even in the presence of L^*, so that L^* can dissociate but not rebind, the receptor–ligand complex will begin to dissociate with a time-course governed by the following equation

$$RL^* = RL_0^* \cdot e^{-k_{21} \cdot t}. \tag{4}$$

This expression shows that the complex, present at an initial concentration RL_0^*, disappears monoexponentially at a rate governed by the dissociation rate constant k_{21} alone. An example of such a dissociation time-course is given in Section 2.3.5 and *Figure 1b*. The importance of equations 3 and 4 is that they can be used experimentally to determine the rate constants of the receptor–ligand interaction.

If multiple independent classes of binding sites are present, the resultant net association and dissociation time-courses are the sums of exponential terms of the above kinds.

At binding equilibrium, the rates of formation and breakdown of the receptor–ligand complex must be equal, i.e. the rate of change of the concentration of the complex becomes zero.

$$\frac{d(RL^*)}{dt} = 0, \text{ so } k_{12} \cdot R \cdot L^* = k_{21} \cdot RL^*. \tag{5}$$

Thus
$$RL^* = (k_{12}/k_{21})R \cdot L^* = K \cdot R \cdot L^* \tag{6}$$

K is known as the ASSOCIATION or AFFINITY CONSTANT of the binding reaction. It is equal to the ratio of the association to the dissociation rate constant in this simplest case.

The DISSOCIATION CONSTANT of the binding reaction is defined as the inverse of the association constant, i.e.

$$K_d = 1/K = k_{21}/k_{12}. \tag{7}$$

At equilibrium, the concentration of the receptor–ligand complex is given by the following expression:

$$RL^* = \frac{R_t \cdot K \cdot L^*}{(1 + K \cdot L^*)} \tag{8}$$

Re-written in terms of the dissociation instead of the association constant, this becomes:

$$RL^* = \frac{R_t \cdot L^*}{(K_d + L^*)}. \tag{9}$$

Equations 8 and 9 are forms of the *simple Langmuir isotherm*. They predict a hyperbolic, saturable dependence of the concentration of the receptor–ligand complex on free ligand concentration. Examples of such profiles are shown in Section 2.4.

Note that when $L^* = K_d = 1/K$, $RL^* = R_t/2$, i.e. WHEN THE FREE LIGAND CONCENTRATION IS EQUAL TO THE K_d, THE RECEPTOR BINDING SITES ARE HALF-SATURATED WITH LIGAND. Conversely, the free ligand concentration at 50% receptor saturation, the EC_{50} concentration, is a MEASURE of K_d (or $1/K$).

In the case of multiple independent classes of binding sites, the resultant binding curve is simply a sum of Langmuir isotherms. The equations for association, dissociation, and equilibrium binding are summarized in *Table 1*.

Table 1. Simple bimolecular receptor–ligand interaction with no ligand depletion

Reaction scheme	$R + L^* \underset{k_{21}}{\overset{k_{12}}{\rightleftharpoons}} RL^*$
Minimal depletion conditions:	$R_t < 0.1 \, K_d$
Ligand association (Initial value of $RL^* = 0$)	$RL^* = RL^*_{eq}(1 - e^{-(k_{12} \cdot L^* + k_{21})t})$
Linearizing plot:	$-\log_e(1 - RL^*/RL^*_{eq})$ vs t
Slope:	$+(k_{12} \cdot L^* + k_{21})$
y-Intercept:	0

Replot slope vs L^* from several experiments to obtain straight line with slope k_{12} and y-intercept k_{21}

Ligand dissociation	$RL^* = RL^*_0 e^{-k_{21} \cdot t}$
Linearizing plot:	$\log_e(RL^*)$ vs t
Slope:	$-k_{21}$
y-Intercept:	RL^*_0
Equilibrium binding (Langmuir Isotherm) NB $RL^* = RL^*_{eq}$	$RL^* = \dfrac{R_t \cdot K \cdot L^*}{(1 + K \cdot L^*)} = \dfrac{R_t \cdot L^*}{(K_d + L^*)}$
Linearizing plot:	RL^*/L^* vs RL^* (bound/free vs bound: Scatchard plot)
Slope:	$-K$ (or $-1/K_d$)
x-Intercept:	R_t

A plot of RL^* vs RL^*/L^* is called the Eadie–Hofstee plot.

Note
L^* = free ligand concentration; RL^*_0 = bound ligand at $t = 0$
R_t = total receptor concentration; RL^*_{eq} = bound ligand at equilibrium.

Methods for linearizing the expressions are also given. These will be used later.

2.2.2 Simple bimolecular association with ligand depletion

Ligand depletion occurs when the concentration of the receptor–ligand complex formed during the binding reaction becomes a significant fraction of the total ligand concentration. This leads to somewhat more complex equations describing association kinetics (the integrated rate equation) and equilibrium binding. The resulting expressions are summarized in *Table 2*. Under non-depletion conditions, they reduce to the simpler forms given in *Table 1*.

Under actual experimental conditions, an element of ligand depletion is frequently unavoidable. For this reason, the use of the more general expressions is the preferred option in data analysis (Chapter 11). As will be shown later (Section 4.10), ignoring depletion leads to erroneous interpretation of ligand association data.

Table 2. Simple bimolecular receptor–ligand interaction with ligand depletion

Reaction scheme
$$R + L^* \underset{k_{21}}{\overset{k_{12}}{\rightleftharpoons}} RL^*$$

Ligand association
(Integrated Rate Equation)
Initial value of $RL^* = 0$
$$RL^* = \frac{a \cdot b(e^{(a-b)k_{12} \cdot t} - 1)}{(a \cdot e^{(a-b)k_{12} \cdot t} - b)}$$

    where $a = RL^*_{eq}$; $b = R_t \cdot L^*_t / RL^*_{eq}$

Linearizing plot:
$$\log_e \frac{(b \cdot (RL^* - a))}{(a \cdot (RL^* - b))} \text{ vs } t$$

Slope: $(a - b)k_{12}$

y-Intercept: 0

Use of this plot requires accurate knowledge of R_t

Ligand dissociation   as in *Table 1*

Equilibrium binding $RL^* = \dfrac{(L^*_t + R_t + K_d) - ((L^*_t + R_t + K_d)^2 - 4R_t \cdot L^*_t)^{1/2}}{2}$

NB $RL^* = RL^*_{eq}$

Note
L^* = free ligand concentration.
R_t = total receptor concentration.
L^*_t = total ligand concentration.
RL^*_{eq} = bound ligand at equilibrium.
Linearizing plot: no convenient linearization in terms of L^*_t. Use Scatchard plot of RL^*/L^* vs RL^*. Best to fit binding function directly to data, after rewriting in terms of x and p (see Section 2.2.7).
NB $EC_{50} = K_d + R_t/2$ where EC_{50} = total concentration of ligand at 50% receptor saturation (see ref. 1).

2.2.3 Competitive interaction

If an unlabelled competing ligand (designated A) is present during the association process, the rate of approach to equilibrium will be reduced. In the simplest case, two ligands compete directly for the same set of binding sites:

$$RA \underset{k_{31}}{\overset{k_{13}}{\rightleftharpoons}} A + R + L^* \underset{k_{21}}{\overset{k_{12}}{\rightleftharpoons}} RL^*. \tag{10}$$

The form taken by the tracer ligand association time-course in the presence of the competing ligand is given in *Table 3*. Note that this equation does not take into account depletion of the tracer or competitor. Numerical integration of the rate equation under depletion conditions is straightforward, but is not considered here.

Under non-depletion conditions, the level of equilibrium binding of the tracer ligand is also reduced in a hyperbolic fashion as the concentration of

Table 3. Competitive interaction

Reaction scheme	$RA \underset{k_{31}}{\overset{k_{13}}{\rightleftharpoons}} A + R + L^* \underset{k_{21}}{\overset{k_{12}}{\rightleftharpoons}} RL^*$

*Tracer ligand association/
dissociation: no depletion*

$RL^* = PA{\cdot}e^{at} + PB{\cdot}e^{bt} + RL^*_{eq}$

$$a,b = \frac{-c1 \pm (c1^2 - 4{\cdot}c2)^{1/2}}{2}$$

$c1 = k_{13}{\cdot}A + k_{31} + k_{12}{\cdot}L^* + k_{21}$

$c2 = k_{21}k_{13}{\cdot}A + k_{31}k_{21} + k_{31}k_{12}{\cdot}L^*$

$PA = (d(RL^*_0)/dt - b(RL^*_0 - RL^*_{eq}))/(a - b)$

$PB = (a(RL^*_0 - RL^*_{eq}) - d(RL^*_0)/dt)/(a - b)$

RL^*_0 and RL^*_{eq} are values of RL^* at time zero, and at equilibrium. R_0 is concentration of free receptor at time zero.

$d(RL^*_0)/dt = k_{12}{\cdot}R_0{\cdot}L^* - k_{21}{\cdot}RL^*_0$

$$RL^*_{eq} = \frac{R_t{\cdot}K{\cdot}L^*}{(1 + K{\cdot}L^* + KA{\cdot}A)} \text{ (see below)}$$

*Linearizing plot for
time-course*

No simple linearization possible. Fit directly to association/dissociation time-course using a multi-exponential function.

Equilibrium binding

NB $RL^* = RL^*_{eq}$

Multiple sites (no depletion, so $L^* = L^*_t, A = A_t$)

$$RL^* = \sum_i \frac{Ri_t{\cdot}Ki{\cdot}L^*}{(1 + Ki{\cdot}L^* + KiA{\cdot}A)}$$

Linearizing plot:
(Single site)

$(1 - RL^*/RL^*_0)/A$ vs $(1 - RL^*/RL^*_0)$

NB $RL^*_0 = RL^*$ when $A = 0$

Slope:

$-KA/(1 + K{\cdot}L^*)$

x-Intercept:

1

Multiple sites (with tracer ligand depletion so $L^* = (L^*_t - RL^*)$)

$$RL^* = \sum_i \frac{Ri_t{\cdot}Ki(L^*_t - RL^*)}{(1 + Ki(L^*_t - RL^*) + KiA{\cdot}A)}$$

Solve by iterative procedure—see program COMPDEP *Protocol 4*. Fit directly to measurements of RL^* vs A.

Note
Ri_t = total concentration of ith receptor species.
Ki = affinity constant for binding of L^* to ith receptor species.
KiA = affinity constant for binding of competitor to ith receptor species.

the competing ligand increases. The equilibrium value of RL^* is given by:

$$RL^* = \frac{R_t \cdot K \cdot L^*}{(1 + K \cdot L^* + KA \cdot A)}. \tag{11}$$

Here K and KA are the association constants of the tracer and competitor, and L^* and A their free concentrations at equilibrium. The form which the inhibition curve takes is shown in Section 2.5. From it, the value of KA may be derived, i.e. the affinity of an unlabelled ligand may be measured by its effect on the binding of a labelled ligand. This is very important for the practise of binding studies.

Tracer ligand depletion can seriously distort the form of the equilibrium competition curve. Equilibrium binding, under conditions of tracer depletion, is given as a functional equation in *Table 3*. This equation is most efficiently solved by an iterative procedure (used in the computer program COMPDEP, see *Protocol 5*). Artefacts caused by tracer depletion, and tactics for detecting, minimizing, and circumventing them are discussed in Sections 4.10.2 and 4.10.3 of this chapter. As usual, it is desirable to analyse experimental data using the more generalized mathematical expression. This is comprehensively discussed in Chapter 11. Note that it is again possible to sum expressions for binding to a single set of sites to obtain equations appropriate for the description of multiple independent populations of binding sites.

2.2.4 Receptor isomerization

It is frequently discovered that the equations for simple bimolecular association and dissociation do not adequately describe tracer ligand kinetics, even when depletion has been taken into account. There are several possible reasons for this. Relatively trivial explanations involve the presence of multiple independent populations of binding sites with non-homogeneous kinetic properties, or the existence of enantiomers of the ligand (8). However, it may also imply receptor isomerization, or ternary complex formation, for instance, by recruitment of an effector molecule after the initial ligand-binding step. The subject of receptor binding kinetics is described in detail in Chapter 10, but for the sake of completeness, the equation governing the simplest form of the isomerization process is given here.

The requisite process is:

$$R + L^* \underset{k_{21}}{\overset{k_{12}}{\rightleftharpoons}} R1L^* \underset{k_{32}}{\overset{k_{23}}{\rightleftharpoons}} R2L^*, \tag{12}$$

and the equation governing it, which is mathematically a variant of that for tracer binding in the presence of a competitor is given in *Table 4*.

Table 4. Receptor isomerization

Reaction scheme	$R + L^* \underset{k_{21}}{\overset{k_{12}}{\rightleftharpoons}} R1L^* \underset{k_{32}}{\overset{k_{23}}{\rightleftharpoons}} R2L^*$

Ligand association/
dissociation
(no depletion, so $L^* = L_t^*$)

$$RL^* = R1L^* + R2L^* = PA \cdot e^{at} + PB \cdot e^{bt} + RL_{eq}^*$$

$$a,b = \frac{-c1 \pm (c1^2 - 4 \cdot c2)^{1/2}}{2}$$

$$c1 = k_{12} \cdot L^* + k_{21} + k_{23} + k_{32}$$

$$c2 = k_{12} \cdot L^* \cdot k_{23} + k_{12} \cdot L^* \cdot k_{32} + k_{32} \cdot k_{21}$$

$$PA = (b(RL_0^* - RL_{eq}^*) - d(RL_0^*)/dt)/(b - a)$$

$$PB = (d(RL_0^*)/dt - a(RL_0^* - RL_{eq}^*))/(b - a)$$

$$d(RL_0^*)/dt = k_{12} \cdot R_0 \cdot L^* - k_{21} \cdot R1L_0^*$$

Equilibrium binding

$$RL_{eq}^* = \frac{R_t \cdot K \cdot L^*}{(1 + K \cdot L^*)}$$

$$K = \frac{k_{12}}{k_{21}} (1 + \frac{k_{23}}{k_{32}})$$

Note
RL^* = total concentration of receptor–ligand complex.
R_0 = concentration of free receptor at time zero.
$R1L_0^*$ = concentrations of initial, non-isomerized receptor–ligand complex at time zero.

2.2.5 Non-competitive mechanisms: allosteric interactions; the ternary complex model

In more complex cases, an unlabelled ligand may bind non-competitively to a second (allosteric) site on the receptor binding subunit, to another subunit in the case of an oligomeric receptor, or to an effector subunit which may be permanently or transiently associated with the binding subunit. In the simplest case, the allosteric interaction is described by the following model:

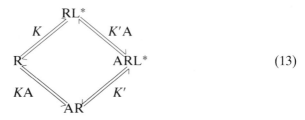

(13)

The interaction is no longer competitive, but instead allows the formation of a ternary complex, ARL*. An important point to note is that the affinity constant of L* for the free receptor is not in general equal to that of L* for the receptor–modulator complex AR. If $K' > K$, the interaction between A and L* is positively co-operative. If $K' < K$, the interaction is negatively co-operative. The effect of increasing the concentration of A is thus to change the apparent affinity of the receptor for L*, but not the total concentration of binding sites. Experimentally, instances of both negative and positive co-operativity are found.

The equation governing tracer ligand binding in the presence of a non-competitive ligand is given in *Table 5*. The effect is to replace the singular tracer ligand affinity constant K by $(K + K' \cdot KA \cdot A)/(1 + KA \cdot A)$. The apparent tracer affinity constant thus displays a continuous hyperbolic dependence on the modulator concentration, changing from K in the absence of modulator to K' in the presence of receptor-saturating concentrations of modulator, so that it attains a limiting value. An illustration of this behaviour is given in Section 2.5.4 of this chapter.

The modulation of ligand binding as a result of a change in the ionization of a group on the receptor is a special case of this and can be described, mathematically, by the equation describing non-competitive interaction. In this case, KA is the ionization constant of a group on the free receptor. $\text{Log}_{10}(KA)$ is the pK of this group. As an experimental approach, this is proving to be very useful in defining subtle differences in the way ligands interact with receptors.

It should be noted that the presence of a non-competitive interaction may exert a profound effect on tracer kinetics. This is because the rate of association of the tracer ligand with the AR complex, or its rate of dissociation from the ternary complex ARL* may be very different from the corresponding rates of association with or dissociation from the free receptor. Commonly,

---

Table 5. Non-competitive interaction

Reaction scheme	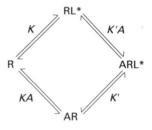

Equilibrium binding (no tracer ligand depletion, so, $L^* = L^*_t$; $A = A_t$)	$RL^* + ARL^* = \dfrac{R_t \cdot Kapp \cdot L^*}{(1 + Kapp \cdot L^*)}$
	$Kapp = \dfrac{K + K' \cdot KA \cdot A}{1 + KA \cdot A} = \dfrac{K(1 + KA' \cdot A)}{(1 + KA \cdot A)}$
Linearizing plot:	Scatchard plot
Slope:	$-Kapp$
x-Intercept:	R_t

---

the rates are much slower (9). This means that the time taken for the binding reaction to come to equilibrium in the presence of an allosteric ligand may be much longer than anticipated from studies on the kinetics of tracer binding alone.

Recruitment of an effector molecule after ligand binding is an extended case of the above mechanism, with the effector formally considered to be an endogenous non-competitive ligand. In general, the effector may react further with added modulators, as in the case of GTP-binding proteins. This can lead to very complex interactions. An example is given in Chapter 11.

2.2.6 Temperature-dependence of kinetic and equilibrium binding constants

The effects of temperature on the equilibrium and rate constants of ligand–receptor interactions have been reviewed by Molinoff *et al.* (4). The temperature dependence of the equilibrium constant is described by the Van't Hoff equation:

$$K = e^{-(\Delta G/RT)} = e^{-(\Delta H/RT)} \cdot e^{(\Delta S/R)}, \tag{14}$$

where ΔG, ΔH, and ΔS are the free energy, enthalpy, and entropy of binding, respectively, and R and T are the gas constant and the absolute temperature.

The temperature-dependence of the first-order rate constant is described by the Arrhenius equation:

$$k_{ij} = kT/h \ e^{-(\Delta G^{\dagger}/kT)} = kT/h \ e^{-(\Delta H^{\dagger}/kT)} \cdot e^{(\Delta S^{\dagger}/k)}, \tag{15}$$

where $\Delta G^{\dagger}$, $\Delta H^{\dagger}$ and $\Delta S^{\dagger}$ refer to the transition state, and k and h are Boltzmann's and Planck's constants, respectively.

The study of the temperature-dependence of rate and equilibrium constants provides information about the standard enthalpy and entropy changes undergone in achieving the transition state of ligand binding, and the final receptor–ligand complex. There often appears to be a large enthalpic contribution to receptor–ligand transition states, which gives marked temperature sensitivity of the kinetic rate constants. In contrast, the equilibrium binding of antagonists tends to be entropy driven, largely it is believed as a result of desolvation of the ligand on binding. Thus the equilibrium constant commonly shows less temperature dependence than the rate constants.

2.2.7 Computer programs

The properties of the foregoing equations are not immediately obvious, except in the simplest of cases. To illustrate them, we have provided short, deliberately simple, BASIC computer programs to accompany the text.

These programs are based on generalizations of the equations in *Tables 1–5*. In most instances, they take tracer ligand depletion into account. They also incorporate a crude model of the non-specific binding of the tracer ligand. It is assumed that non-specific binding is the product of a coefficient, NS, the free ligand concentration L^*, and the protein concentration present in the incubation, expressed in mg/ml. Thus

$$NS \text{ binding} = NS{\cdot}L^*{\cdot}R_t{\cdot}10^9/S, \tag{16}$$

NS is the fraction of tracer ligand non-specifically bound at a protein concentration of 1 mg/ml. R_t is the receptor concentration expressed in molar terms, and S is the receptor specific activity, expressed in pmol ligand bound/mg protein. It should be noted that this simple model does not take into account possible binding of the ligand to components other than the receptor preparation, such as filters, and does not discriminate between ligand entrapment and genuine non-specific binding.

The reader is urged to try out these programs. They were originally written for a BBC B+ micro, but should readily adapt to other machines. They should provide a feel for the nature and behaviour of binding experiments, and a means of simulating the results of specific experiments, and thus checking for, and avoiding some of the commoner experimental artefacts.

2.2.8 Adaptation of models for non-linear, least-squares fitting

Monte-Carlo simulations have shown that determinations of affinity or dissociation constants by non-linear, least-squares fitting yield estimates which are log-normally distributed. Thus, for inclusion in fitting programs, the models of equilibrium binding in *Tables 1, 2, 3, 5* should be recast in terms of log affinity constants. It is also necessary to use log concentrations of ligands (see Section 2.4.3). To take an example, the Langmuir isotherm becomes

$$RL^* = \frac{R_t{\cdot}10^{(p + x)}}{1 + 10^{(p + x)}}, \tag{17}$$

where $p = \log_{10}(K)$ and $x = \log_{10}(L^*)$.

The other functions can be similarly rewritten by putting $pi = \log_{10}(Ki)$ and $xi = \log_{10}(Li)$.

Functions in an appropriate form for fitting to binding data and further discussion is provided in Chapter 11 and Appendix 3 to this volume.

2.3 Ligand association and dissociation

2.3.1 Ligand association

Amongst the earliest experiments performed in the experimental characterization of a receptor–ligand interaction should be exploratory studies of the kinetics of ligand association and dissociation. An important aim is to establish the time needed for equilibrium binding to be achieved under different experimental conditions. An example is shown in the upper curve in *Figure 1a*. This was simulated by running the computer program INTRATE given in

Protocol 2. The program is based on the integrated rate equation for tracer association (*Table 2*), and, therefore, takes tracer ligand depletion into account. It also contains a term for non-specific binding.

Protocol 2. INTRATE program

```
30  REM THIS PROGRAM CALCULATES THE TIME-COURSE OF LIGAND BINDING TO A
40  REM RECEPTOR ALLOWING BOTH FOR DEPLETION AND FOR NONSPECIFIC BINDING
50  REM PROPORTIONAL TO THE FREE LIGAND CONCENTRATION.
60  REM PARAMETERS ARE DEFINED AS IN TABLE 2. THE EQUATIONS HAVE BEEN
70  REM SLIGHTLY MODIFIED TO TAKE ACCOUNT OF NONSPECIFIC BINDING
80  INPUT "FORWARD AND REVERSE RATE CONSTANTS K12,K21",K12,K21
90  INPUT "TOTAL RECEPTOR CONCENTRATION,RT",RT
100 INPUT "RECEPTOR SPEC.ACT. (PMOL/MG),S",S
110 INPUT "TOTAL (BOUND+FREE) LIGAND CONCENTRATION, L*T",LT
120 INPUT "NONSPECIFIC BINDING COEFFICIENT,NS",NS
130 INPUT "INITIAL CONCENTRATION RL*0 OF RL*",RLO
140 INPUT "TIME INCREMENT(IN SEC)",DT
150 MU=NS*RT*1E9/S
160 REM MU MULTIPLIED BY L* GIVES NONSPECIFIC BINDING
170 REM K21/K12 = KD
180 C1=LT+RT+(1+MU)*K21/K12
190 C2=RT*LT
200 ROOT=SQR(C1*C1-4*C2)
210 REM A AND B ARE ROOTS OF A QUADRATIC EQUATION
220 REM A= RL*EQ (STRICTLY, IF THERE IS NO NONSPECIFIC BINDING)
230 A=(C1-ROOT)/2
240 REM B= RT.L*T/RL*EQ (STRICTLY, IF THERE IS NO NONSPECIFIC BINDING)
250 B=(C1+ROOT)/2
260 T=0
270 PRINT " T"," RL*"," TOTAL"
280 FOR I= 1 TO 20
290 X1=EXP((A-B)*K12*T/(1+MU))
300 X2=RLO-B
310 X3=RLO-A
320 REM RL=SPECIFIC BINDING
330 RL=(A*X2-B*X3*X1)/(X2-X3*X1)
340 REM BTOT = TOTAL BINDING
350 BTOT=RL+MU*(LT-RL)/(1+MU)
355 REM PRINTER CONTROL CHARACTERS FOLLOW. MODIFY AS APPROPRIATE
360 §%=&0001030A
370 PRINT T,RL,BTOT
380 T=T+DT
390 NEXT I
400 INPUT "CONTINUE CALCN Y/N",A$
410 IF A$="Y" THEN GOTO 280 ELSE 420
420 INPUT "NEW STEP LENGTH Y/N,A$
430 IF A$="Y" THEN GOTO 140 ELSE 440
440 INPUT "NEW PARAMETERS Y/N",A$
450 IF A$="Y" THEN GOTO 80 ELSE 460
460 END
```

The parameter values used were the following:

- $k_{12} = 10^7/(M \text{ sec})$
- $k_{21} = 10^{-2}/\text{sec}$
- $R_t = 10^{-10}$ M
- S (receptor specific activity) = 0.1 pmol/mg protein
- $L_t^* = 10^{-9}$ M
- $NS = 10^{-2}/(\text{mg/ml})$
- initial value of $RL^* = 0$ M

To give a feeling for the meaning of these numbers:

(a) R_t would be the concentration in a $1:100$ tissue homogenate of a receptor whose tissue concentration is approximately 10 pmol/g wet weight, corresponding to 0.1 pmol/mg protein. A typical example is M_2 muscarinic acetylcholine receptors in rat myocardium (7, 9). Values for receptor concentrations in tissues are usually between 0.1 and 10 times this value, although even lower concentrations are also quite common. Higher concentrations are rare but occur in specialized tissues, for example nicotinic acetylcholine receptors in *Torpedo electroplax* (10). As a rule of thumb, a $1:100$ tissue homogenate, which corresponds to a protein concentration of about 1 mg/ml, is often a good starting point for exploratory binding studies.

(b) The value chosen for k_{12} is $10^7/(M \text{ sec})$. This is typical of a simple biomolecular reaction in which the rate-limiting step is the collision of a small ligand with a receptor-binding site, the collision-limited rate (11). Such values are characteristic of the association rate constants of antagonists for receptors at near-physiological temperatures (25–40°C). Lower values are often found for the binding of rigid, inflexible molecules, or for agonists. They may reflect the occurrence of isomerization, recruitment of an effector, or other reactions of the RL^* complex. Such reactions often lead to the occurrence of complicated, multiphasic kinetic behaviour (Chapter 10).

(c) The ratio k_{12}/k_{21} yields an affinity constant, K, of 10^9/M ($K_d = 10^{-9}$ M). This means that the receptor is half-saturated with ligand at 10^{-9} M free L^*. Such values are characteristic in receptor work, cf. the binding of atropine to muscarinic receptors, naloxone to μ-opioid receptors, or haloperidol to dopamine-D_2 receptors (12–14, see *Receptor Biochemistry, a Practical Approach*, 7). For work with radiolabelled tracer ligands on non-purified receptor preparations such as tissue homogenates, a value of 10^9/M is roughly in the middle of the useful working range of ligand affinities, expressed on a logarithmic scale. As will be demonstrated later, this runs from about 10^7/M to 10^{11}/M.

(d) Several processes contribute to real or apparent non-specific binding of labelled ligands:

> *i. Occlusion.* Practical laboratory methods for the separation of bound from free ligand usually entail the measurement of a proportion of entrapped or occluded free ligand as well as the ligand bound to the receptor. In the case of a microcentrifugation assay, for instance, ligand is entrapped in the membrane pellet. The volume of entrapped water is approximately $10\,\mu l$ after pelleting $1\,ml$ of a $1:100$ tissue homogenate.

> *ii. Partitioning.* If the ligand is relatively non-polar, for example an aromatic tertiary amine, it will partition into membrane phospholipids, or detergent micelles.

> *iii. 'True' non-specific binding.* Ligand may bind to non-receptor sites, for example catabolic enzymes or transmitter uptake sites. Such non-receptor binding sites have been called acceptor sites (15, 16). Under favourable circumstances, this binding will occur with much lower affinity than binding to the receptor, and will appear to be non-saturable with respect to ligand concentration, within the working range.

In addition, partitioning and occlusion are intrinsically non-saturable. Thus, as we have assumed here, the processes of entrapment, partitioning, and non-specific binding can often be lumped together in the form of a simple proportionality constant, the coefficient of non-specific binding, *NS*.

Values of *NS* of approximately 0.01/mg membrane protein/ml are typical for highly polar ligands in microcentrifugation assays. Higher values, for example 0.1/mg/ml membrane protein are often obtained for tertiary amines, which may partition into sealed vesicles, or whole cells, and for peptides, which may have more acceptor sites than simple organic ligands. Lower values may be obtained in filtration assays, which permit more complete removal of occluded and partitioned ligand, as a result of extensive washing, but at the risk of losing specifically bound ligand as well (see Chapter 6). This model of non-specific binding can easily be modified to take into account binding of the ligand to components of the apparatus.

Note: a rather subtle point is that whilst true non-specific binding and partitioning of the labelled ligand into membrane phospholipids will reduce the free concentration of the labelled ligand available in the aqueous phase for receptor binding, entrapment of ligand generated by the separation procedure itself will not necessarily do so. The computer programs used for illustration in this chapter assume the occurrence of 'true' non-specific binding.

The upper curve in *Figure 1a* is an association time-course based on the parameter values listed above. It shows a rapid (effectively instantaneous) transient, corresponding to the non-specific binding, followed by a nearly mono-exponential approach to equilibrium. The bound ligand has reached 97% of its final equilibrium value within approximately 300 sec. This illustrates the first of our criteria for a successful receptor binding assay:

THE BINDING OF THE TRACER LIGAND SHOULD BE RAPID ENOUGH TO ALLOW EQUILIBRATION WITHIN THE PERIOD OF THE ASSAY.

By running INTRATE with different values of L_t^* the reader can verify that for values of L_t^* substantially less than K_d, the time to equilibrium is largely independent of ligand concentration. In contrast, as L_t^* becomes greater than K_d, the equilibration time becomes shorter. This phenomenon reflects the fact that the exponent in eqn (1) of *Table 1* is of the form $k_{12} \cdot L^* + k_{21}$, so that its value is independent of L^* for $L^* \ll k_{21}/k_{12} (= K_d)$, but directly proportional to L^* for $L^* \gg k_{21}/k_{12}$. (NB L^* = free, L_t^* = total ligand concentration.)

2.3.2 Time-course of non-specific binding

The behaviour of the non-specific binding alone can be simulated (trivially) by making k_{12} a relatively small number, for example $10^3/(\text{M sec})$ (*Figure 1*). Experimentally, this effect is achieved by performing the incubation in the presence of a concentration of an unlabelled competitor ligand sufficient fully

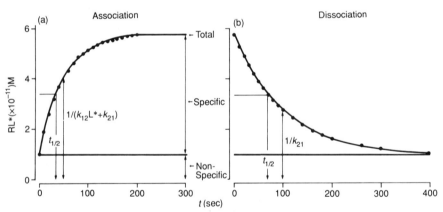

Figure 1. (a) Tracer ligand association time-course, showing total, specific, and non-specific binding. On-rate $k_{12} = 10^7/(\text{M sec})$; off-rate $k_{21} = 10^{-2}/\text{sec}$; $L_t^* = 10^{-9}\,\text{M}$. The effective association rate constant $(k_{12} \cdot L^* + k_{21})$ is $2 \times 10^{-2}/\text{sec}$ and the half-time $(t_{1/2} = \log_e(2)/(k_{12} \cdot L^* + k_{21})) = 0.693/0.02 = 35$ sec. (b) Tracer ligand dissociation time-course initiated by addition of an unlabelled competitor, showing reversal of specific but not of non-specific binding. The half-time is $t_{1/2} = \log_e(2)/k_{21} = 0.693/0.01 = 69.3$ sec. The overall tracer ligand equilibration time is $5 \times 69.3 = 350$ sec, at limiting tracer concentrations. See text for further description.

to occupy the receptor, even in the presence of L*. When using a tracer ligand at a concentration less than its K_d, a concentration of competitor equal to 1000 times the K_d of the receptor–competitor interaction suffices. Under these conditions specific binding of the tracer is blocked. This illustrates the second empirical criterion for the establishment of a valid binding assay:

A COMPONENT OF THE BINDING OF L* SHOULD BE INHIBITED BY PHARMACOLOGICALLY APPROPRIATE CONCENTRATIONS OF UNLABELLED LIGANDS.

By 'pharmacologically appropriate concentrations', we mean concentrations of the drugs which are known from pharmacological assays to bind to the receptor. In contrast, drugs which are *not* receptor-specific, by this test, should *not* inhibit binding. This qualitative statement of the specificity criterion will be developed in more quantitative terms below.

2.3.3 Time-course of specific binding

Provided that ligand depletion is minimal, i.e. that specific and non-specific binding combined do not reduce the free concentration of the tracer ligand by more than 10%, the time-course of the receptor-specific binding of L* can be calculated by subtracting the non-specific from the total binding for each time point (*Figure 1*). When this is done, the resultant mono-exponential time-course can be analysed using the linear plot of *Table 1*, to extract values of the rate constant. Alternatively, the time-course can be analysed by curve-fitting. The appropriate function to use is that given in *Table 2* which allows for tracer ligand depletion. In addition, it is best to incorporate a model of non-specific binding within the fitted function, as is done in the program INTRATE (*Protocol 1*). The analysis of ligand association time curves is discussed further in Chapter 10. Repetition of these measurements at different values of L_t^*, in principle, allows both the forward rate constant and the reverse rate constant to be estimated.

2.3.4 Irreversible ligands

Criteria of specificity similar to those applying to reversible ligand binding may also be applied to ligands which first bind reversibly, and then react chemically, or can be activated (for example photochemically) to generate a reactive species (see Chapters 2 and 5 of reference 7). In the case of such site-directed affinity labels, the kinetics of binding may be more complex, depending on:

(a) the initial reversible binding step;

(b) the rate of the subsequent chemical reaction;

(c) the rate of breakdown of the reaction product.

In many instances, it is the chemical reaction and not the binding step which is rate-limiting, so that the apparent rate constant of the forward reaction is not diffusion or collision-limited. In the case of irreversible agents, it is of course meaningless to talk about 'equilibrium' binding. It is also worth noting that non-specific binding of irreversible ligands often shows a linear dependence on reaction time. A classical example of the use of a specific affinity ligand to label a receptor active site is the alkylation of muscarinic acetylcholine receptors by ^{3}H-propylbenzilylcholine mustard aziridinium ion (17).

2.3.5 Ligand dissociation

The dissociation of L* from the receptor–ligand complex can be studied by preincubation of the receptor with L* until some chosen level of binding has been achieved, followed by (a) the sudden reduction of the concentration of L* by dilution, or (b) by the reduction of the effective forward rate, $k_{12} \cdot L^*$, to a small value by the addition of a receptor-saturating concentration of an unlabelled competing ligand. In the latter case, the cold ligand should have a chemical structure distinct from that of the labelled ligand itself, to avoid displacement of non-specific as well as specific binding.

Using the endpoint of the association time-course calculation as a starting point, the effect of this manoeuvre may be simulated using INTRATE with the following set of parameters:

- $k_{12} = 10^3/(\text{M sec})$; $k_{21} = 10^{-2}/\text{sec}$
- $R_t = 10^{-10}\,\text{M}$
- $S = 0.1\,\text{pmol/mg}$
- $L_t^* = 10^{-9}\,\text{M}$
- $NS = 0.01/(\text{mg/ml})$
- $RL_0^* = 4.85 \times 10^{-11}\,\text{M}$ (receptor initially 49% saturated)

As shown in *Figure 1b*, RL* breaks down with a half-time of around 70 sec $(t_{1/2} = \log_e(2)/k_{21} = 0.693/0.01\,\text{sec})$. The half-time of dissociation is independent of the initial value of RL^*. This is because the exponent which governs the rate of breakdown depends only on k_{21}, the dissociation rate constant (*Table 1*). The eventual level of total binding attained is equal to the level of non-specific binding expected for the total tracer ligand concentration. In the simplest case, the magnitude of k_{21} is given by the slope of the first-order plot of $\log_e RL^*$ vs t (*Table 1*). The half-time of the dissociation reaction is equal to the half-time of the association reaction for values of $L_t^* \ll K_d$.

These considerations lead to a third criterion for the establishment of a valid binding assay employing a reversible tracer ligand:

THE BINDING OF L* SHOULD BE FULLY REVERSIBLE, ON THE TIME-SCALE OF THE ASSAY.

2.3.6 Estimating equilibration time

The above calculations lead to a rule for estimating the time needed for a set of simple bimolecular association reactions to come to equilibrium, even when the ligand concentration is below the lowest K_d in the system. The rule is:

MEASURE THE HALF-TIME OF THE SLOWEST LIGAND DIS-SOCIATION PROCESS IN THE SYSTEM, AND MULTIPLY THIS VALUE BY FIVE.

Whatever the starting point, binding will have reached a minimum of 97% of its equilibrium value after this length of time.

In many instances, straightforward pharmacological antagonists display collision-limited association kinetics ($k_{12} = c.\ 10^7/(\text{M sec})$) at near physiological temperatures. An initial estimate of the incubation time, t_{eq}, needed to allow a low concentration of an uncharacterized labelled antagonist to achieve binding equilibrium is $3.5 \times 10^{-7} \cdot K$ sec, where K is the affinity constant of the ligand; this may be known from pharmacological data. Typical relationships between ligand affinity and equilibration time are shown in *Table 6*. These values are only a rough guide; it must be emphasized that conformationally-restricted ligands may have much slower kinetics. It is evident that lowering the temperature from 30°C to 0°C can increase the equilibration times of a given ligand by 20–50-fold, depending on the initial value of the dissociation rate constant and the thermodynamics and mechanism of the binding process (cf. eqn 15).

2.3.7 The ligand off-rate may determine the assay method

The compatibility of a given receptor–ligand interaction with a particular assay method is determined primarily by the off-rate of the ligand. Filtration, gel-filtration, or ligand adsorption assays all reduce the concentration of free ligand in the vicinity of the receptor-binding site during the separation process, and will, therefore, initiate ligand dissociation. Whether or not the amount of dissociation that ensues is acceptable depends on:

- the separation time, and
- the dissociation rate constant.

If no more than 10% of the specifically-bound ligand is to be permitted to dissociate during the separation, then k_{21} must be less than $0.105/t$, where t is the separation time. The separation time depends critically on the assay method. It is approximately 10 sec for filtration assays (if the filters are washed (Chapter 6)), approximately 30 sec for charcoal adsorption assays (Chapter 8), and approximately 100 sec for gel-filtration assays (Chapter 9). Some rough guidelines are given in *Table 7*.

From *Table 7* it is evident that performing filtration, adsorption, or gel-filtration steps at low temperature often improves the range of these assay methods by slowing the effective ligand off-rate for a given equilibrium affinity constant (Chapter 6). This can be achieved by quenching the reaction

Table 6. Estimated relationship between affinity constant, K of the ligand, and incubation time for full achievement of binding equilibrium, t_{eq}

K (M^{-1})	t_{eq} (sec) 30°C		t_{eq} (sec) 0°C	
10^7	3.5		100	
10^8	35		1250	(21 min)
10^9	350	(5.8 min)	16 000	(4.6 h)
10^{10}	3500	(58 min)	210 000	(58 h)
10^{11}	350 000	(9.7 h)	2 760 000	(766 h)

Table 7. Limiting values of kinetic and equilibrium binding constants compatible with non-equilibrium assay methods

	K_{21} (sec^{-1})	K (M^{-1}) 30°C	K (M^{-1}) 0°C
Filtration assay	0.01	10^9	3×10^7
Adsorption assay	0.003	3×10^9	10^8
Gel filtration assay	0.001	10^{10}	2.5×10^8

k_{21} represents the maximum dissociation rate constant and K the minimum affinity constant compatible with a given assay method, under normal circumstances.

by dilution or washing with cold buffer before or during the separation. Even so, there may be complete failure to detect a rapidly-dissociating component of binding by using these methods. If this is suspected, it is advisable to perform centrifugation assays, in which, in principle, the binding equilibrium is minimally disturbed by the separation process (Chapter 7), and to compare these results with those from the filtration or other assays.

Unfortunately, centrifugation assays suffer from limitations imposed by the relatively unfavourable ratio of specific:non-specific binding of this assay method. If a minimum requirement is that there should be equal amounts of specific and non-specific binding at a receptor occupancy of 10%, we find that the minimum ligand affinity which is usable in a centrifugation assay is

$$K = 10^9 \times NS/S, \qquad (18)$$

where NS is the coefficient of non-specific binding, and S the receptor specific activity. For typical values of these parameters, we find that K should exceed 10^8/M. Under favourable circumstances, i.e. high receptor specific activity, high receptor concentration and low non-specific binding, an affinity of 10^7/M is the lower limit compatible with a centrifugation binding assay. However, there are only a few credible examples of this (see also Chapters 6 and 7).

2.3.8 Initial survey of potential binding artefacts

Violations of the assumptions embodied in INTRATE provide a foretaste of some of the artefacts which are possible in binding studies. These are as follows:

(a) The program attempts to allow for depletion of the free tracer ligand concentration as the binding reaction proceeds, both as a result of receptor-specific and non-specific binding. It is assumed that the free ligand concentration can be calculated from the difference between the measured total ligand concentration and the sum of specifically and non-specifically bound ligand concentrations, i.e. $L^* = L_t^* - RL^* - NS$. This will not be true if the tracer ligand is chemically or radiochemically impure. A common situation is that the tracer ligand is enantiomeric but unresolved, so that both isomers are labelled but only one isomer binds strongly to the receptor. Another possibility is that the ligand may be chemically unstable or subject to enzymatic action. Bound and free ligand pools commonly show different susceptibilities to breakdown, the bound ligand being protected. As previously stated, INTRATE does not distinguish between 'true' non-specific binding, which reduces the free ligand concentration, and entrapment of ligand, which does not necessarily do so.

(b) The assumption is made that the receptor is completely stable throughout the binding reaction, so that the conservation condition $R_t = R + RL^*$ applies at all times. This will not hold if there is thermal or enzymatic degradation of the receptor.

(c) It is assumed that the total receptor population is freely accessible to the tracer ligand. This may not be true if some of the receptors are present in compartments, such as sealed vesicles, or, in studies on whole cells, in intracellular organelles, such as endosomes.

(d) It is assumed that the assay method permits complete separation of bound from free ligand, so that none of the bound ligand is recovered in the free ligand fraction. If this is not true, receptor-specific binding will be underestimated. This error may also arise if the assay method permits measurable dissociation of the bound ligand. The opposite problem, recovery of free ligand in the bound ligand compartment, may also occur. This will entail over-estimation of specific binding, particularly at high ligand concentrations.

(e) It is assumed that the kinetics of binding are described by the simple bimolecular reaction scheme of eqn 1 in Section 2.2.1. This is frequently not the case. Although complex kinetic behaviour is not in itself an artefact, failure to detect complexity may lead to the misinterpretation of experimental results.

Tactics for detecting, and avoiding, these artefacts will be discussed later (Section 4).

Undetected ligand depletion, and non-attainment of equilibrium are two of the commonest avoidable systematic artefacts encountered in receptor binding studies. Both kinds of artefact cause systematic distortion of kinetic and equilibrium binding curves. Quantitative examples are given in Sections 4.10 and 4.11. As a working rule, LIGAND DEPLETION SHOULD PREFERABLY BE HELD BELOW 10%. DEPLETION LEVELS BETWEEN 10 AND 50% MAY BE CORRECTABLE, BUT WILL INCREASE THE ERROR OF THE RESULTS. Depletion of more than 50% will probably invalidate the results.

The detection and correction of ligand depletion, necessitates measurements of free as well as bound ligand concentrations. This fits naturally into a centrifugation assay, where the supernatant fractions may be sampled after pelleting the membranes. It is more difficult in membrane filtration assays, where collection and sampling of the filtrate before washing the filter is cumbersome, and binding of the ligand to the filter may be difficult to take into account. This being so, it is advisable to perform supplementary centrifugation assays to assess the free ligand concentration and the true level of non-specific binding. This is most important as in many filtration assays non-specific binding normally refers to binding to the filters plus that component of non-specific binding or partitioning which cannot be eliminated by the washing protocol.

Adsorption assays suffer from difficulties similar to those associated with filtration assays. In gel-filtration assays, the true free ligand concentration is also not easy to assess, and may have to be established independently by equilibrium dialysis or equilibrium gel-filtration experiments (Chapter 9).

2.4 Equilibrium binding

2.4.1 Concentration-dependence of equilibrium binding

The typical concentration-dependence of binding of a labelled ligand to a receptor is shown in *Figure 2a*. Total binding is defined as binding measured in the absence of any competing ligand. Non-specific binding is defined as binding in the presence of a concentration of an unlabelled competing ligand sufficient to occupy at least 99% of the binding sites even at the highest concentration of L^* used, i.e. an unlabelled ligand concentration greater than $100 \times K_d \cdot (1 + K \cdot L^*)$, where K_d is the dissociation constant of the competitor, K is the affinity constant of the tracer, and L^* is the highest tracer concentration used.

The values shown in *Figure 2a* were obtained by running the computer program EQBIND (*Protocol 3*) which is based on the equation for equilibrium binding given in *Table 2*.

Protocol 3. EQBIND program

```
10 REM THIS PROGRAM CALCULATES THE EQUILIBRIUM BINDING OF A LIGAND L* TO
20 REM A RECEPTOR R WITH AFFINITY CONSTANT K TAKING INTO ACCOUNT
30 REM NONSPECIFIC BINDING AND LIGAND DEPLETION.SEE TABLE 2.
40 INPUT "ASSOCIATION CONSTANT,K",K
50 INPUT "TOTAL RECEPTOR CONCENTRATION,RT",RT
60 INPUT "RECEPTOR SPEC.ACT. (PMOL/MG),S",S
70 INPUT "COEFFICIENT OF NONSPECIFIC BINDING,NS",NS
80 INPUT "INITIAL AND FINAL CONCNS OF L*T,LLO,LHI", LLO,LHI
90 INPUT "NUMBER OF LIGAND CONCENTRATIONS PER DECADE",N
95 REM CALCULATE PROPORTIONALITY CONSTANT FOR NS BINDING.
100 MU=NS*RT*1E9/S
110 M=10^(1/N)
120 LT=LLO
130 PRINT" L*T"," L*"," RL*"," TOTAL"
140 REPEAT
150 C1=LT+RT+(1+MU)/K
160 C2=RT*LT
170 ROOT=SQR(C1*C1-4*C2)
180 REM RL = RECEPTOR-BOUND LIGAND
190 RL=(C1-ROOT)/2
200 REM L=FREE LIGAND
210 L=(LT-RL)/(1+MU)
220 REM BTOT=TOTAL BINDING
230 BTOT=RL+MU*L
240 REM PRINTER CONTROL CHARACTERS: MODIFY AS APPROPRIATE
250 §% =&0001030A
260 PRINT LT,L,RL,BTOT
270 LT=LT*M
280 UNTIL LT>LHI
290 INPUT "NEW VALUES OF L*T Y/N",A$
300 IF A$="Y" THEN GOTO 80 ELSE 310
310 INPUT "NEW COEFF OF NON-SPECIFIC BINDING Y/N",A$
320 IF A$="Y" THEN GOTO 70 ELSE 330
330 INPUT "CHANGE TOTAL RECEPTOR CONCN Y/N",A$
340 IF A$="Y" THEN GOTO 50 ELSE 350
350 INPUT "NEW ASSOCIATION CONSTANT Y/N",A$
360 IF A$="Y" THEN GOTO 40 ELSE 370
370 END
```

The parameters used were the following:

- $K = 10^9/M$ (total binding) or $10^3/M$ (non-specific binding)
- $R_t = 10^{-10}\,M$
- $S = 0.1\,pmol/mg$

- $NS = 0.01/(mg/ml)$
- $L^* = 10^{-10}-10^{-8}\,M$
- points per decade $= 5$

The program explicitly takes into account the effects of ligand depletion due to specific and non-specific binding, and calculates both bound and free ligand concentrations.

This illustrative case involves only limited ligand depletion. Thus the specific component of binding can be estimated to a good approximation by subtracting the lower from the upper curve in *Figure 2a*. The form of the specific binding curve is described by the simple Langmuir isotherm (*Table 1*).

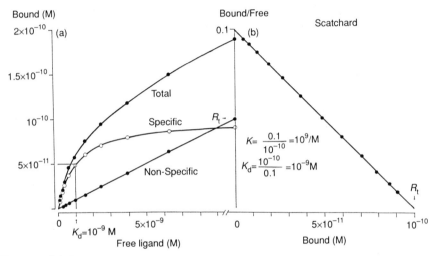

Figure 2. (a) Tracer ligand saturation curve showing total, specific and non-specific binding as a function of free ligand concentration. (b) Scatchard plot of specific binding, showing calculation of initial estimates of K and R_t. See text for further description.

At low concentrations, RL^* shows a linear dependence on L^*. As L^* increases, the slope of the curve decreases, until eventually, RL^* becomes asymptotic to R_t.

On this basis, we can propose a fourth criterion for receptor-specific binding:

SPECIFIC BINDING SHOULD BE SATURABLE, SO DEFINING A FINITE CONCENTRATION OF BINDING SITES.

A similar requirement that a finite number of sites be labelled also applies to the specific binding of irreversible ligands.

It is worth noting that ligand depletion will affect the estimation of non-specific binding, i.e. that non-specific binding must be calculated from the free rather than from the total ligand concentration. Because of this, the most statistically satisfactory method of data analysis requires fitting of the values for total binding to a function incorporating terms for both specific and non-specific binding, by a non-linear, least-squares technique. (Chapter 11.)

2.4.2 Estimating ligand affinity

The concentration-dependence of the specific binding curve evidently provides direct information about the affinity of the tracer ligand for the receptor binding site.

In the simplest case described here, an initial estimate of the dissociation constant of L^* is provided by the equilibrium concentration of free ligand needed to give 50% occupancy of the binding site (*Figure 2a*). A better estimate of this parameter can be obtained by a linear transformation of the Langmuir isotherm (*Table 1*). The most popular of these transformations is the Scatchard plot of RL^*/L^* vs RL^* (bound/free vs bound). This is illustrated in *Figure 2b*. Extrapolation of the Scatchard plot to the x-axis yields an estimate of R_t, whilst the slope provides an estimate of K, or $1/K_d$ (see *Table 1*).

In order to obtain valid parameter estimates from saturation binding curves by Scatchard, or other forms of analysis, it is necessary to measure binding over at least the 10–90% saturation range. The hazards of estimating total concentrations of binding sites by extrapolating Scatchard plots based on measurements made at inadequate levels of occupancy are well known (18). While the Scatchard plot is a useful way of visualizing, or presenting binding data, IT CANNOT BE EMPHASIZED TOO STRONGLY THAT THE LEAST BIASED WAY TO ESTIMATE K AND R_t IS TO FIT A MODEL OF BINDING, SUCH AS THAT EMBODIED IN *EQBIND*, DIRECTLY TO THE UNTRANSFORMED BINDING DATA.

In more complex cases multiple populations of binding sites may be present. In such cases, accurate values of K and R_t for the constituent populations can only be extracted by non-linear, least-squares fitting procedures.

2.4.3 Displaying the Langmuir isotherm: a useful practical aid

The most convenient way to display the simple Langmuir isotherm is to plot binding normalized to the total concentration of binding sites against $\log_{10}$ free ligand concentration, i.e. to plot RL^*/R_t vs $\log_{10} L^*$. The effect of this is to generate a sigmoidal curve which is symmetrical about the value of log K. The shape of this curve,

$$RL^*/R_t = K \cdot L^*/(1 + K \cdot L^*) = L^*/(K_d + L^*), \qquad (19)$$

is independent of the value of K. For example, $RL^*/R_t = 0.091$ for $L^* = 0.1/K = 0.1 \cdot K_d$, 0.5 for $L^* = 1/K = K_d$, and 0.91 for $L^* = 10/K = 10 \cdot K_d$. The reader can easily fill in the intervening values. The effect of changing K is simply to move the curve in a parallel fashion along the x-axis (*Figure 3*).

A useful instant practical aid can be obtained by plotting out this curve on a suitable scale for values of log L^*, from two orders of magnitude below to two orders of magnitude above some suitable K_d value (for example plot $X/(1 + X)$ vs log X for $X = 0.01$ to 100) and then making a transparent template with this shape, preferably machined in perspex. Mark on the template the positions corresponding to 9.1%, 50% and 91% saturation of the binding site. A series of binding curves plotted on the same scale can then be rapidly and efficiently examined, and 'fitted by eye' simply by sliding the template along the x-axis until it best fits the points (*Figure 3*).

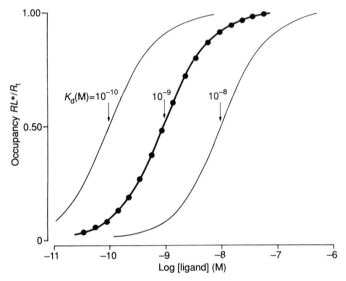

Figure 3. Langmuir isotherms on a semi-logarithmic scale. Occupancy $= RL^*/R_t = K \cdot L^*/(1 + K \cdot L^*) = L^*/(K_d + L^*)$ is plotted vs $\log_{10}(L^*)$ for different values of K (K_d) to illustrate the parallel shift. Initial fits to such curves can be made with the aid of a plastic template.

We have found such templates to be extremely useful as manual laboratory aids for the analysis and plotting of binding curves. If the data deviate from the simple Langmuir isotherm, superimposition of the template makes it instantly apparent. If they fit, the estimated K is usually within the error limit of the value obtained from computerized curve-fitting of the data.

2.4.4 Concentration-dependence of non-specific binding

In *Figure 2a*, it will be noted that non-specific binding shows a linear dependence on free ligand concentration. This is frequently, although not always the case. There have now been numerous reported instances of saturable non-specific binding not only to acceptor sites in biological membranes (15) but also to tubes, filters, or even extraneous materials such as talc (1). Deceptively, such binding can even mimic properties of authentic receptor binding, such as stereoselectivity.

2.4.5 Working range of labelled ligand concentration

Non-specific binding and ligand depletion between them place a limit on the range of affinities which are practical for measuring a tracer ligand saturation curve.

Accurate determination of the saturation curve minimally requires measurements between 10 and 90% receptor saturation. Precise measurement of bound ligand at 90% saturation necessitates at least equal magnitudes of specific and non-specific binding. Thus, we require

$$0.9 \geq 10 \cdot NS \cdot 10^9 \cdot K_d / S, \tag{20}$$

(since at 90% saturation, $L^* = 10 \cdot K_d$),

$$\text{i.e. } K_d < 10^{-10} S/NS \tag{21}$$

Thus, if $S = 0.1–1 \, \text{pmol/mg}$ and $NS = 0.01/(\text{mg/ml})$ (a value typical of centrifugation assays) K_d values of $10^{-9}–10^{-8} \, \text{M}$ represent the maximum ligand dissociation constants (minimum affinity) which will allow measurement of a reasonably complete saturation curve.

Ligands with higher affinities, for example $K_d < 10^{-10} \, \text{M}$ are well suited to the accurate measurement of R_t by binding assays at near receptor-saturating tracer ligand concentrations. However, severe problems of ligand depletion may be encountered when using the lower concentrations needed to define the initial part of the binding curve. At low tracer concentrations ligand depletion of 10% occurs if:

$$K \cdot R_t = R_t / K_d = 0.1. \tag{22}$$

Ideally, this value should not be exceeded. In practice, it frequently is. Usually, it is possible to correct for ligand depletion up to about 50%. This demands accurate measurement of bound and free ligand concentrations which may be possible, provided that the tracer ligand is chemically and radiochemically pure and separation of bound from free is efficient. Ligand depletion can be considered to present a serious and probably irremediable problem when

$$K \cdot R_t = R_t / K_d > 1. \tag{23}$$

Thus, for depletion problems to be containable,

$$K_d > R_t. \tag{24}$$

The window of usable K_d values is, therefore,

$$R_t < K_d < 10^{-10} S/NS. \tag{25}$$

S/NS depends on the assay method, and will usually be larger for filtration (both membrane and gel) than for centrifugation assays, although the warning must be reiterated that filtration assays disturb the binding equilibrium, since they incorporate a buffer wash, and may underestimate specific binding. The relationships between the maximum permissible values of K_d, receptor specific activity, and ligand non-specific binding for different assay methods are summarized in *Figure 4*.

R_t depends on the dilution of the preparation, and can thus be controlled. However, the degree of dilution necessary may actually determine the mode of assay used. For instance, microcentrifugation assays may not be practical for membrane homogenates containing less than $100\,\mu g/ml$ protein because the small amounts of protein may not pellet properly. If high dilutions are needed, filtration assays will be necessary. These allow harvesting of membranes from large volumes of dilute suspensions. However, large additional

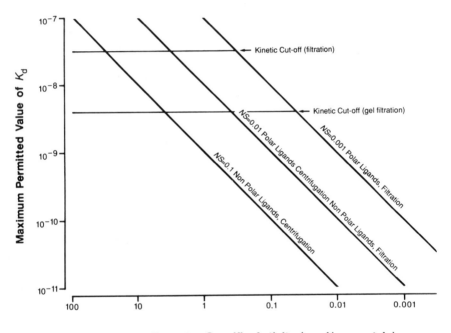

Receptor Specific Activity (pmol/mg protein)

Figure 4. Guide to choosing a ligand and an assay method to suit a particular receptor system. For any particular assay method, the working area lies within the trapezium defined by the ordinate (K_d (M)), abscissa (receptor specific activity (pmol/mg protein)), the diagonal line showing the maximum permitted K_d (minimum ligand affinity) for a given receptor specific activity, and the kinetic cut-off, if applicable, for example if specific activity = 0.1 pmol/mg, a $K_d < 10^{-10}$ M will be necessary for a centrifugation assay if the ligand is non-polar, a $K_d < 10^{-9}$ M if it is polar. Use of filtration with an ice-cold buffer wash (*Table 7*) extends the working range of ligand affinities, but rapidly-reversible components of binding may be lost. *NS* = coefficient of non-specific binding.

dilutions are often not practicable for solubilized preparations, which may have to be assayed by gel filtration, precipitation, or ligand adsorption.

2.4.6 Minimum receptor concentration

In the case of radioligands, it is simple to see how the specific activity of the ligand determines the minimum receptor concentration which can be used in a binding experiment. Consider a direct saturation experiment using 1 ml aliquots of a receptor preparation, and a ^{3}H-labelled ligand with specific activity 80 Ci/mmole (Chapter 1). Assume that 1000 d.p.m. of specific binding at 10% saturation is the minimum needed for accurate measurement of the binding curve (this is a rather stringent requirement). Then the total concentration of binding sites will need to be at least:

$$R_t = 10\,000/(2220 \times 80) = 0.056\,\text{pmoles/ml} = 5.6 \times 10^{-11}\,\text{M}. \quad (26)$$

For ^{125}I-labelled ligands, whose specific activities are often 10–20-fold higher than those of ^{3}H-labelled ligands, R_t may be as low as 2×10^{-12} M. For simple estimates of concentrations of binding sites, the minimum values of R_t are around 5×10^{-12} M for ^{3}H-labelled ligands, and 2×10^{-13} M for ^{125}I-labelled ligands. In summary, we may conclude as a result of these considerations that the OPERATIONAL RANGE OF R_t IS USUALLY BETWEEN 2×10^{-11} M and 10^{-9} M, AND THAT OF K IS BETWEEN 10^8 AND 10^{11}/M. The most valuable range for K is 10^8–10^{10}/M. It is difficult to obtain equilibrium binding of ligands with higher affinity whilst maintaining receptor stability during the long incubations necessary for equilibrium to be achieved.

2.4.7 Correction for ligand depletion

The analysis of a receptor saturation curve depends on knowledge of both the bound and the free ligand concentrations. Measurement of the free ligand concentration is necessary both for the analysis of the direct binding curve using the simple Langmuir isotherm, and for the development of a model of non-specific binding for use in non-linear, least-squares fitting of binding as a function of total ligand concentration, the preferred method of analysis (Chapter 11).

As pointed out previously, determination of the free concentration fits naturally into the process of a centrifugation assay, since the supernatant is accessible for sampling after separation of bound from free ligand.

In general, this is not the case with filtration assays; special arrangements need to be made to collect the filtrates separately from the washings. Thus, if a membrane filtration assay is used, supplementary centrifugation assays should be performed in order to check the level of free ligand, and measure the true level of non-specific binding. IT CANNOT BE ASSUMED THAT THE FREE LIGAND CONCENTRATION CAN BE CALCULATED FROM THE DIFFERENCE BETWEEN TOTAL AND BOUND LIGAND

CONCENTRATIONS WITHOUT AN EMPIRICAL CHECK BEING MADE. The loss of rapidly reversible components of both specific and non-specific binding during washing of the filters, or incomplete retention of the receptor preparation may invalidate such a calculation.

Note that where partitioning of a lipophilic ligand into the membrane contributes appreciably to depletion, a rather subtle artefact can occur, in the sense that the apparent dissociation constant of the ligand can be multiplied by a factor of $(1 + mu)$ where *mu* is the proportionality constant for non-specific binding in the actual experiment (see *Protocol 3* line 150 and Chapter 11). If an inappropriate expression is used to analyse the binding data, the apparent affinity of the ligand will be an underestimate of its true affinity. This phenomenon may account for a noticeable tendency in the literature for the measured affinities of lipophilic ligands to creep up as their specific radio-activity increases as a result of improved methods of synthesis and purification, which encourages the use of lower and lower protein concentrations in binding assays. This emphasizes the desirability of working at lower rather than higher membrane concentrations.

2.5 Effect of unlabelled ligands on the binding of the tracer

2.5.1 Effect of competing ligand on tracer kinetics

Inhibition of the binding of a tracer ligand by an unlabelled ligand yields quantitative information about the kinetics and equilibrium binding of the unlabelled ligand. This information can be extracted, provided that the binding of the tracer ligand is well-enough understood. The computer program COMPKIN (*Protocol 4*) simulates the binding of a tracer ligand in the presence of an unlabelled competitor, which binds to the same site on the receptor, with simple bimolecular kinetics. The program is based on eqn 1 of *Table 3*. To keep the mathematics reasonably simple, COMPKIN does not allow for depletion of the tracer ligand or its unlabelled competitor.

THE EFFECT OF THE PRESENCE OF A COMPETING LIGAND IS ALWAYS TO SLOW THE APPROACH TO BINDING EQUILIBRIUM.

A range of complex behaviour is possible, including overshoots in the binding of L^*. This is illustrated using the following parameters, chosen so that the tracer binds fast, but the inhibitor binds slowly:

- $k_{12} = 10^7/(\text{M sec})$, $k_{21} = 10^{-1}/\text{sec}$, $L^* = 10^{-9}\,\text{M}$ (tracer ligand)
- $k_{13} = 10^7/(\text{M sec})$, $k_{31} = 10^{-3}/\text{sec}$, $A = 10^{-10}\,\text{M}$ (inhibitor)
- $R_t = 10^{-10}\,\text{M}$

The resulting time-course is shown in *Figure 5a*. The bound tracer concentration overshoots substantially, before gradually subsiding towards the final

Protocol 4. COMPKIN program

```
 30 REM THIS PROGRAM COMPUTES THE TIME COURSE OF BINDING OF A TRACER LIGAND
 40 REM L* IN THE PRESENCE OF A COMPETITOR LIGAND A. DEPLETION IS NOT ALLOWED
 50 REM FOR. NON-SPECIFIC BINDING CAN BE ALLOWED FOR BY ADDING A CONSTANT
 60 REM AMOUNT OF BINDING TO THE BOUND LIGAND CONCENTRATION.SEE TABLE 3
 70 CLS
 80 REM READ INITIAL VALUES
 90 INPUT "RATE CONSTANTS FOR TRACER,K12,K21",K12,K21
100 INPUT "RATE CONSTANTS FOR COMPETITOR,K13,K31".K13,K31
110 INPUT "INITIAL CONCNS OF RECEPTOR COMPLEXES R,RL*,RA",R0,RL0,RA0
120 INPUT "CONCNS. OF TRACER AND INHIBITOR L*,A",L,A
130 REM INITIALISE THE CALCULATION
140 K12L = K12*L:K13A=K13*A
150 KL=K12L/K21:KA=K13A/K31
160 RT=R0+RL0+RA0
170 RLE=KL*RT/(KL+KA+1)
180 C1=K12L+K21+K13A+K31
190 C2=K21*K13A+K31*K12L+K31*K21
200 AA=(-C1+SQR(C1*C1-4*C2))/2
210 BB=(-C1-SQR(C1*C1-4*C2))/2
220 DRL0=K12L*R0-K21*RL0
230 PA=(DRL0+BB*(RLE-RL0))/(AA-BB)
240 PB=(AA*(RL0-RLE)-DRL0)/(AA-BB)
250 REM PRINT THE EQUILIBRIUM VALUE OF BINDING
260 PRINT "THE EQUILIBRIUM LEVEL OF BINDING IS ",RLE
270 REM PRINT THE COEFFICIENTS AND EXPONENTS
280 PRINT "THE SOLUTION HAS THE FORM PA*exp(at) + PB*exp(bt) + RL*EQ"
290 PRINT "a = ",AA
300 PRINT "PA = ",PA
310 PRINT "b = ",BB
320 PRINT "PB = ",PB
330 REM INPUT THE STEP LENGTH
340 INPUT "STEP LENGTH,SEC",DT
350 PRINT:PRINT
360 REM PERFORM CALCULATION IN BATCHES OF 20
370 T=0.0
380 PRINT TAB(8);"TIME";TAB(20);"RL*"
390 FOR J=1 TO 20
400 RL=PA*EXP(AA*T)+PB*EXP(BB*T)+RLE
410 PRINT T;TAB(20);RL
420 T=T+DT
430 NEXT J
440 INPUT "CONTINUE CALCN,Y/N",A$
450 IF A$="Y" THEN GOTO 380 ELSE 460
460 INPUT "DEFINE NEW STEP LENGTH",A$
470 IF A$="Y" THEN GOTO 340 ELSE 480
480 INPUT "DEFINE NEW L*,A",A$
490 IF A$="Y" THEN GOTO 120 ELSE 500
500 INPUT "DEFINE NEW R0,RL0,RA0",A$
510 IF A$="Y" THEN GOTO 110 ELSE 520
520 INPUT "DEFINE NEW RATE CONSTANTS",A$
530 IF A$="Y" THEN GOTO 90 ELSE 540
540 END
```

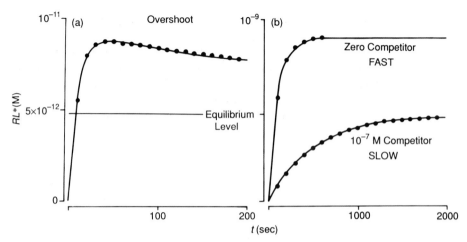

Figure 5. (a) Overshoot in tracer binding. See text for kinetic parameters. The rate of approach to equilibrium is governed by the off-rate of the competitor. (b) The presence of a competitor slows the tracer kinetics. See text for kinetic parameters.

equilibrium level of binding. The timing of the overshoot is governed by the dissociation rate constant of the tracer, and that of the declining phase by the dissociation rate constant of the competitor. Overshoots of this kind are possible when $k_{21} > k_{31}$. In effect the tracer acts as a monitor of the free receptor concentration, which diminishes only slowly as the competitor binds.

Another possible pitfall is a transition from 'on'-rate limited tracer kinetics to 'off'-rate limited kinetics as the concentration of competitor increases. This may be illustrated by the following set of parameters chosen so that the tracer binds slowly, but the competitor binds fast.

- $k_{12} = 10^7/(\text{M sec})$, $k_{21} = 10^{-3}/\text{sec}$, $L^* = 10^{-9}\,\text{M}$ (tracer ligand)
- $k_{13} = 10^7/(\text{M sec})$, $k_{31} = 10^{-1}/\text{sec}$, $A = 0$ or $10^{-7}\,\text{M}$ (competitor)
- $R_1 = 10^{-9}\,\text{M}$

(See *Figure 5b*.)

Binding of the competitor rapidly reduces the free concentration of receptor, thus diminishing the effective association rate constant of the tracer ligand. In the above case, the equilibration time is increased by a factor of approximately six as a result. This can result in a considerable overestimation of the apparent percentage inhibition of tracer binding by competitor at the higher concentrations of competing ligand unless equilibrium has been reached. For example, the percentage inhibition by the ligand is 68% at 500 sec but 50% at equilibrium. Lack of awareness that such complex kinetic behaviour is possible can lead to the assumption that binding reactions have

96

reached equilibrium when, in fact, they have not. The resulting artefacts are discussed in Section 4.11 of this chapter and Chapter 11.

If the receptor has been pre-equilibrated with the competing ligand, the resulting reduction in the initial rate of binding of either a reversible, or an irreversible ligand can be used to deduce the affinity constant of the competitor. In addition, the kinetic time-courses can be fitted to the equations of *Table 3* to provide estimates of the kinetic constants of the competitor as well as the tracer ligand (19). The extraction of useful information of this nature depends on the tracer kinetics being faster than the competitor kinetics, so that the rate-limiting steps are due to the binding or dissociation of the competitor, not the tracer. The competitor dissociation rate constant can often be estimated by these techniques. An example is given in Section 4.11.2.

In the exploratory characterization of a binding assay, WE ARE MOST INTERESTED IN DEFINING THE TIME NECESSARY FOR THE ATTAINMENT OF EQUILIBRIUM. As stated previously, THE DETERMINING FACTOR IS THE SLOWEST RATE-CONSTANT IN THE SYSTEM. As we have seen, this may be contributed by the tracer ligand, OR the competitor. In simple systems, the dissociation rate constants are the rate-limiting factors.

In defining the equilibration time, it is important not to confuse the maximum of a prolonged overshoot (cf. *Figure 5a*) with the final plateau which indicates the attainment of equilibrium. It is also necessary to bear in mind that instability of either the tracer or the receptor itself may lead to premature cut-off of the rising phase of the binding reaction, and so to underestimation of both the equilibration time, and the level of equilibrium binding.

The effect of receptor instability may be simulated using COMPKIN by setting $k_{13} \cdot A$ to some appropriate value, for example 10^{-4}/sec and k_{31} to a small value, for example 10^{-16}/sec (not zero, otherwise the program will fail). The effect of this slow rate of receptor breakdown is to cause substantial undershoot of the true level of equilibrium binding. Binding may falsely appear to be in equilibrium if incubations are not carried out for a sufficiently long time to allow receptor breakdown to be manifest as a clear decrease in binding with time.

Not only the tracer and the receptor, but also the competitor ligand may be unstable under the conditions of incubation. The detection and avoidance of artefacts due to these problems is discussed in Sections 4.4, 4.7, and 4.11. The golden rules are:

(a) To study the binding kinetics of the tracer–receptor interactions alone.

(b) To determine the tracer kinetics in the presence of a *range* of concentrations of the competitor ligand, in order to ensure that equilibrium is attained under *all* experimental conditions. It may be necessary to continue incubations for times considerably longer than are required for apparent equilibrium (in the sense of steady-state binding) to be attained, to detect either receptor instability, or post-binding isomerization of the receptor–ligand complex.

2.5.2 Inhibition of equilibrium binding by an unlabelled competitor

Measurement of the equilibrium binding of the tracer ligand in the presence of different concentrations of unlabelled ligands provides readily-interpretable information about the affinities of the latter. The concentration-dependence of binding is described by the equilibrium binding equations of *Table 3*. The ligand concentrations which appear in the latter expression are free concentrations. The extended version of this expression (*Table 3*), written in terms of total ligand concentrations rather than free concentrations, is implemented in the computer program COMPDEP (*Protocol 5*). For the sake of versatility, this program assumes the presence of two rather than one receptor population ('multiple sites'), and allows both tracer and competitor to exhibit different affinities for either population. Depletion of the tracer, but not the competitor ligand, is taken into account.

Protocol 5. COMPDEP program

```
 30 DIM A(100)
 40 REM THIS PROGRAM ASSUMES THE PRESENCE OF TWO POPULATIONS OF BINDING
 50 REM SITES R1 AND R2 WHICH MANIFEST DIFFERENT AFFINITIES FOR THE TRACER
 60 REM LIGAND L AND THE COMPETITOR A. THE PROGRAM COMPUTES THE TOTAL
 70 REM CONCENTRATION OF BOUND LIGAND AS A FUNCTION OF COMPETITOR CONCN.
 80 REM THERE IS ALLOWANCE FOR DEPLETION OF THE TRACER BUT NOT THE COMPETITOR.
 90 REM THERE IS ALSO ALLOWANCE FOR NON-SPECIFIC BINDING OF THE TRACER.
 95 REM NOTE THAT PARAMETERS ARE DEFINED AS IN TABLE 3
100 REM INPUT PARAMETERS
110 INPUT "AFFINITY CONSTANTS K1,K2,K1A,K2A",K1L,K2L,K1A,K2A
120 INPUT "CONCNS OF SITES R1T,R2T",R1T,R2T
130 INPUT "RECEPTOR SPEC ACT. (PMOL/MG)",S
140 INPUT "CONCN OF TRACER L*T",LT
150 INPUT "NONSPECIFIC BINDING COEFF NS",NS
160 MU=NS*(R1T+R2T)*1E9/S
170 INPUT "INITIAL AND FINAL CONCNS OF COMPETITOR A",ATL,ATH
180 INPUT "POINTS PER DECADE",N
190 GOSUB 530
200 Q=0
210 GOSUB 420
220 RLO=RL
230 PRINT "RL* INCLUDES NS BINDING IN THIS OUTPUT"
240 PRINT "COMPETITOR(M)","  RL*(M)","  L*(M) "," RL*/RL*O(%)"
250 I=0
260 REPEAT
270 I=I+1
275 REM Q IS THE CURRENT VALUE OF THE COMPETITOR CONCN
280 Q=A(I)
290 GOSUB 420
```

```
300 RLP=RL/RLO
301 REM PRINTER CONTROL CHARACTERS FOLLOW. MODIFY AS APPROPRIATE
305 §%=&0001030A
310 PRINT A(I)," ",RL,LT-RL,RLP
320 UNTIL A(I)>ATH
330 INPUT "NEW AFFINITY CONSTANTS Y/N",A$
340 IF A$="Y" THEN GOTO 110 ELSE 350
350 INPUT "NEW R1T,R2T",A$
360 IF A$="Y" THEN GOTO 120 ELSE 370
370 INPUT "NEW L*T",A$
380 IF A$="Y" THEN GOTO 140 ELSE 390
390 INPUT "NEW INITIAL AND FINAL A",A$
400 IF A$="Y" THEN GOTO 170 ELSE 410
410 END
420 REM CALCULATE CONCENTRATION OF RECEPTOR-TRACER COMPLEX, RL PLUS NS
425 REM FIRST CALCULATE AN INITIAL ESTIMATE. CALL IT P
430 P=R1T*K1L*LT/(1+K1A*Q+K1L*LT)+R2T*K2L*LT/(1+K2A*Q+K2L*LT)+MU*LT
435 REM IF THIS IS OUT OF RANGE, USE LT/2
440 IF P>LT THEN P=LT/2
450 PP=P
455 REM IMPROVE THE ESTIMATE BY NEWTON-RAPHSON ITERATION.SEE CHAPTER 11.
456 REM NOTE IF YI IS AN ESTIMATE OF THE ROOT OF THE EQUATION
457 REM Y-F(Y)=0, AN IMPROVED ESTIMATE IS YI+H WHERE
458 REM H= (YI-F(YI))/(F'(YI) - 1) WHERE F' IS THE FIRST DERIVATIVE OF F
460 TEMP1=R1T*K1L*(LT-P)/(1+K1A*Q+K1L*(LT-P))+R2T*K2L*(LT-P)/(1+K2A*Q+K2L*
    (LT-P))+MU*(LT-P)
470 TEMP2=-R1T*K1L*(1+K1A*Q)/(1+K1A*Q+K1L*(LT-P))^2-R2T*K2L*(1+K2A*Q)/
    (1+K2A*Q+K2L*(LT-P))^2-MU
480 H=(P-TEMP1)/(TEMP2-1)
490 P=PP+H
495 REM CONTINUE TO ITERATE UNTIL THE ESTIMATE CHANGES BY LESS THAN 1%
500 IF ABS(H/P)<0.01 THEN 510 ELSE 450
510 RL=P
520 RETURN
530 REM CALCULATE CONCENTRATIONS OF COMPETITOR
540 I=0
550 REPEAT
560 I=I+1
570 A(I)=10^(LOG(ATL)+(I-1)*(1/N))
580 UNTIL A(I)>ATH
590 RETURN
```

For illustrative purposes, we assume that the tracer ligand binds with equal affinity to (i.e. is non-selective between) the two populations of binding sites. The parameters used are the following:

- $K1 = K2 = 10^8/M$
- $R1_t = R2_t = 5 \times 10^{-10} M$
- Receptor specific activity = 1 pmol/mg
- Coefficient of non-specific binding = 0.01/(mg/ml)
- $L_t^* = 10^{-9} M$

In *Figure 6* a variety of different values of $K1A$ and $K2A$ have been used, to show:

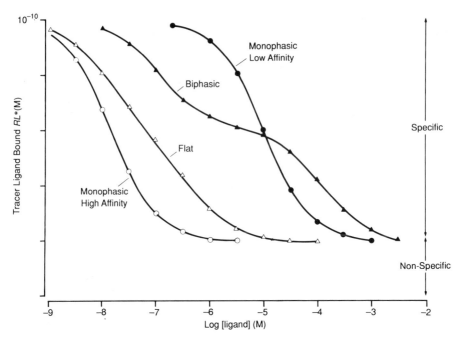

Figure 6. Different forms of ligand displacement curves. Values of $K1A$ and $K2A$ are (○) 10^8, 10^8; (△) 10^8, 3×10^6; (▲) 10^7, 10^4; (●) 10^5, 10^5/M. After subtraction of non-specific binding, monophasic curves can be fitted to a simple Langmuir isotherm.

(a) monophasic inhibition curves, typical of non-selective competitors, which bind with equal affinity to the two populations of sites

(b) flat and biphasic inhibition curves, typical of selective competitors which bind to the two populations of sites with different affinities. Despite their differences in shape, these inhibition curves demonstrate another feature which ideally should become apparent in the exploratory phase of binding studies, namely, that there should exist a generally valid definition of non-specific binding, in the sense that:

A SET OF DIFFERENT AND STRUCTURALLY DISTINCT LIGANDS SHOULD CONSISTENTLY YIELD THE SAME ESTIMATE OF NON-SPECIFIC BINDING.

Although agonists, and selective antagonists, may yield multiphasic in-hibition curves, they should, if competitive with the tracer, be capable of inhibiting the *whole* of the specific element of binding at concentrations in excess of some value. This type of experiment allows a choice to be made of the most appropriate ligand for definition of non-specific binding. In general, it should have high affinity and specificity. Often, high polarity is also a useful attribute since it minimizes partitioning into membranes. The chosen ligand should be chemically stable under the incubation conditions.

2.5.3 Anomalous displacement of tracer ligand binding

Given that a set of pharmacologically appropriate ligands can be found which do in fact give a coherent definition of non-specific binding, one may suspect that an exception to this rule, i.e. a ligand which fails fully to inhibit specific binding but instead produces a plateau at a higher level of binding, may interact non-competitively with the primary receptor-binding site. This may reflect:

(a) binding to a second site on the ligand-binding subunit of the receptor (9)

(b) binding to a site on a separate receptor subunit, which together with the ligand-binding subunit participates in an oligomeric structure. (A good example is provided by benzodiazepine modulation of $GABA_A$ receptors, ref. 20)

(c) interaction with a site on an effector subunit which transduces the effect of ligand binding to the receptor itself. The best example of this is provided by the class of receptors which operate by activation of GTP-binding proteins, (G-proteins) where binding of GTP to the G-protein destabilizes the agonist receptor complex, reducing agonist affinity (21, 26).

In general, such allosteric ligands will have structures noticeably different from those of the 'classical' ligands for a given receptor. Their binding mechanisms must be investigated in detail if they are to be understood.

More trivially, it is often found that a ligand inhibits a component of non-specific binding. This usually reflects a non-specific membrane perturbant effect, a common phenomenon in membrane binding studies. Most pharmacological agents are amphipathic, and can exert local anaesthetic effects at high enough concentrations (for example $> 10^{-4}$ M).

The accurate definition of non-specific binding is essential. It is not sufficient to use the unlabelled tracer ligand for this purpose without further validation. THE UNLABELLED TRACER WILL, BY DEFINITION, BE CAPABLE OF INHIBITING BOTH SPECIFIC AND NON-SPECIFIC BINDING OF THE TRACER LIGAND. Moreover, these two processes may not be readily distinguishable.

2.5.4 Interpretation of inhibition curves

Inhibition curves such as those shown in *Figure 6* clearly provide quantitative information about the affinities of the unlabelled ligands for the receptor.

The simplest case is competition between the unlabelled and labelled ligands for a single, uniform population of binding sites. In this case, binding of both ligands is described by the simple Langmuir isotherm. Under ideal conditions, there will be no significant depletion of either the tracer or the competitor ligand. For this to be true, we require that:

- $K \cdot R_t < 0.1$; $KA \cdot R_t < 0.1$, or if this condition is not fulfilled
- $L_t^* > 10 \cdot R_t$; $KA \cdot R_t / (1 + K \cdot L_t^*) < 0.1$.

Under such conditions there is a simple relationship between the concentration of A which inhibits 50% of the binding of L^*, the IC_{50} value, and the value of KA

$$1/IC_{50} = KA/(1 + K \cdot L_t^*), \tag{27}$$

i.e. $$KA = (1 + K \cdot L_t^*)/IC_{50}, \tag{28}$$

or, in terms of K_d $$K_d A = IC_{50}/(1 + L_t^*/K_d); \tag{29}$$

where K (and K_d) refer to the tracer and KA (and $K_d A$) to the competitor.

The IC_{50} value can be determined by manual or by curve-fitting methods, as discussed in Chapter 11. The correction of the IC_{50} value by dividing it by $(1 + K \cdot L_t^*)$ is known after its popularizers as the Cheng–Prusoff correction (22). A TEST FOR COMPETITIVE BEHAVIOUR IS THAT VALUES OF KA MEASURED USING A RANGE OF CONCENTRATIONS OF L_t^* SHOULD BE INTERNALLY CONSISTENT. If they are not, non-competitive inhibition may be suspected. If ligand depletion occurs, a more complicated correction must be applied. The form taken by this correction is discussed under the heading of depletion artefacts in Section 4.10.3.

If the competing ligand binds to more than one population of binding sites, a Cheng–Prusoff type correction can still be applied to the values of KA determined for the separate populations. In fact, the ENTIRE competition curve is shifted in a parallel fashion along the log ligand axis towards higher competitor concentrations in the manner of *Figure 3*. Correction for the shift regenerates the 'true' competitor binding curve. The magnitude of the shift is the (log of the) Cheng–Prusoff factor $(1 + K \cdot L_t^*)$. However, if the tracer ligand is itself selective, a simple parallel shift will not occur. Increasing the tracer concentration will increase the proportion of low affinity receptor sites labelled. This will also contribute to the observed shift. Multiple-sites binding requires analysis by non-linear, least-squares fitting. There are no really satisfactory or convenient graphical methods for performing such analyses.

If the interaction between tracer and inhibitor ligands is non-competitive, either inhibition or potentiation of tracer binding may be observed, depending on whether the interaction is positively or negatively cooperative.

An example is shown in *Figure 7*. This illustrates a negatively co-operative interaction between ligand and tracer. A diagnostic feature is that the INHIBITION OF SPECIFIC BINDING IS INCOMPLETE. At a low concentration of the tracer ($L_t^* < 0.1 \times K_d$) the maximum degree of inhibition

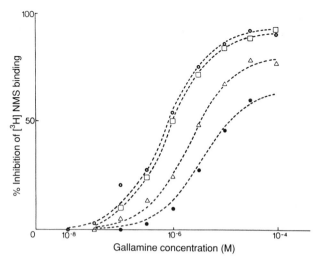

Figure 7. Allosteric inhibition by gallamine of [^{3}H]-*N*-methylscopolamine binding to heart muscarinic receptors. Concentrations of [^{3}H]NMS ranged from 2.5×10^{-11} (○) to 6.8×10^{-9}/M (●) (data from reference 9).

achieved is related to the negative cooperativity of the tracer–ligand interaction:

$$RL^* \, (A = 0)/RL^* \, (A = \infty) = K/K' \text{ (see Section 2.2.5)}. \qquad (30)$$

A characteristic feature is that as the concentration of tracer ligand is increased the inhibition curve flattens, as well as shifting to the right, so that the degree of inhibition attained becomes smaller. Eventually, the entire receptor population is driven into the ternary complex ARL^* (Section 2.2.5). It should be noted that non-competitive interactions may be accompanied by a dramatic slowing of tracer binding which is strongly dependent on the concentration of A. It is, therefore, quite possible for such an inhibition curve to be in equilibrium at low $[A]$, but dramatically far from equilibrium at higher concentrations. This emphasizes the need to check for the attainment of equilibrium, preferably through measurement of the off-rate of the tracer from the ARL^* complex.

A variant of this behaviour is seen in the effect of guanine nucleotides on agonist binding to receptor–G-protein complexes (see also Chapter 11). GTP exerts little or no effect on antagonist binding under normal conditions, but selectively inhibits agonist binding to a high-affinity population of sites. This effect can be studied by agonist/antagonist competition in the presence of different GTP concentrations, as well as by using a labelled agonist to bind directly to the high-affinity sites. Both approaches enable the negatively-co-operative interactions between GTP and agonists to be characterized. These questions are discussed by Haga and Haga in reference 23.

In summary, the inhibition of tracer binding by unlabelled ligands may provide information about:

- The MECHANISM of the interaction: competitive or non-competitive.
- The KINETICS of the unlabelled ligand.
- The BINDING CONSTANT(S) of the unlabelled ligand.
- In the case of non-competitive interaction, the DEGREE OF CO-OPERATIVITY (negative or positive) between the labelled and unlabelled ligands.

2.6 Comparison of binding with pharmacological data

So far, we have been concerned with criteria which establish the internal consistency of binding assays. We have demanded only semi-quantitative agreement between the binding constants obtained by binding studies and those determined by pharmacological measurements. However, the final, and crucial evidence that a set of binding sites defined by *in vitro* ligand binding assays represents bona fide receptor binding sites is the following:

THERE SHOULD BE A QUANTITATIVE CORRELATION BE-TWEEN THE AFFINITIES DETERMINED BY BINDING ASSAYS AND BY PHARMACOLOGICAL OR FUNCTIONAL STUDIES ON THE SAME CELL OR TISSUE TYPE.

In some instances, good agreement is found between receptors in different tissues, as well as within the same tissue. An example, taken from the authors' own work (12) is shown in *Figure 8*. This shows the existence of essentially quantitative agreement between two sets of estimates of affinity constants for a series of muscarinic antagonists, one derived from binding studies on rat cerebral cortex muscarinic receptors in a membrane preparation, and the other calculated from the concentration-dependence of their ability to inhibit the acetylcholine-induced contraction of intact ileal smooth muscle.

Such a quantitative correlation between binding and pharmacological/biochemical estimates of affinity is absolutely necessary for establishing the validity of a binding assay. To be credible, it should:

(a) encompass compounds extending over a wide range of potency, with a reasonably uniform distribution within that range;
(b) embrace a diversity of ligand structures;
(c) preserve characteristic features such as stereospecificity;
(d) include enough compounds to allow evaluation of the statistical significance of the correlation coefficient.

In the case of antagonist binding, under near physiological conditions, it is reasonable to expect near-identity (< 3-fold discrepancies) between binding

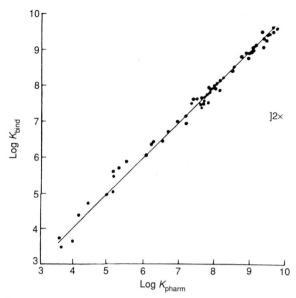

Figure 8. Correlation between the binding affinities of 60 muscarinic antagonists for rat cerebral cortical muscarinic receptors and their pharmacological potencies in antagonizing muscarinic contraction of the longitudinal muscle of the guinea-pig ileum. Subtype-selective antagonists deviate significantly from this correlation.

and pharmacological estimates of affinity, for classical non-selective antagonists. If receptor subclasses are present, selectivity may first show itself as a significant deviation from such a correlation. An example is provided by the binding of the selective antagonist pirenzepine to muscarinic receptors (24).

In the case of agonist binding, actual identity between binding and effect is found less frequently. This is because both binding and response are sensitive to factors such as local concentrations of receptors with respect to effectors (spare receptors) and other essentially environmental factors. These affect the energetics of agonist binding, and hence the binding constants. They also influence the efficacy of receptor-response coupling, and hence the apparent potency of the agonist in the functional response (25).

The persistence of receptor–effector coupling in broken-cell preparations is heralded by:

(a) complex agonist binding curves, which deviate from the simple Langmuir isotherm; and

(b) sensitivity of agonist binding to intracellular modulators, such as guanine nucleotides and divalent cations (26).

The extraction of meaningful binding constants from such data is predicated on the establishment of a valid model encompassing all of the reactions participating in the binding process. This is usually the biggest stumbling block. Sometimes, one has to be content with a simplified model of the binding interactions, such as the multiple sites model. Even so, it should still be possible to demonstrate a clear and unequivocal correlation between binding and response.

2.7 Tissue distribution of binding sites

A corollary of the criterion of pharmacological relevance is the following:

THE BINDING SITES SHOULD BE FOUND IN TISSUES AND CELL TYPES KNOWN BY PHARMACOLOGICAL AND PHYSIOLOGICAL STUDIES TO CONTAIN THE RECEPTOR.

This distribution criterion is obviously difficult to apply as a primary test if pharmacological data are limited. In other cases, it may give useful confirmation of ligand specificity. It provides a basis for the now widespread use of ligand binding techniques in receptor localization studies, both *in vitro*, by autoradiography (27) and now also *in vivo*, by positron-emission tomography of for example fluorine-18-labelled ligands (28).

3. Experimental procedures

In the following section, some experimental procedures which we have found useful in the performance of binding studies are described. The theoretical and experimental basis of these procedures has been outlined in Section 2, and the reader is urged to refer back as necessary. Testing for, and avoiding experimental artefacts is discussed in Section 4. The basic protocols described in this section are extended to cover more complex cases in Chapter 11. Membrane preparations are covered in Chapter 5. Protocols for solubilizing receptors are given in *Receptor Biochemistry, a Practical Approach* (7). Technical details of microcentrifugation and filtration assays for membrane-bound receptors, and of gel-filtration, PEG-precipitation, and charcoal adsorption assays for solubilized receptors are given in Chapters 6–9. The quantitative analysis of binding curves is discussed in Chapter 11. Analysis and interpretation of the kinetics of receptor binding is covered in more detail in Chapter 10.

Throughout this section, emphasis will be placed on the minimization of errors, namely:

(a) pipetting errors, which are a major source of experimental variation;
(b) harvesting errors will be mentioned, but will be covered in detail under the relevant technique;
(c) counting errors;
(d) experimental artefacts will be mentioned as appropriate.

Finally, the reader will be aware of an element of repetition. This is deliberate. Important points are reiterated in *each* protocol, to minimize the need for cross-referencing at the bench.

3.1 Common laboratory equipment and reagents for binding studies

3.1.1 Equipment

We have found that a range of common laboratory equipment is particularly useful for binding studies. This comprises:

i. For setting up and incubating binding assays

(a) 5 ml, 1 ml, 250 µl adjustable pipettes taking disposable plastic tips (for example Gilson, Eppendorf, Finn, Oxford).

(b) A high accuracy, low volume adjustable micropipette, for example Gilson P20. This may be automated for greater reproducibility, e.g. Gilson EDP.

(c) Eppendorf or similar repetitive pipette, taking a range of combitips (12.5 ml, 2.5 ml, 250 µl); these give excellent accuracy (about 1%).

(d) Eppendorf, or similar plastic microfuge tubes, 1.5 ml is the most useful size, though other smaller sizes are also usable.

(e) Racks for holding microfuge tubes. It is best if these actually grip the tubes, allowing the rack to be inverted if necessary, and preventing the tubes from floating during incubation in a water bath. We find a convenient design to be 2 × approx. 12 rows of 4 tubes (*Figure 9*). It is easy to make such racks out of Perspex. It is desirable for the space between at least every other row to be large enough to accommodate a microfuge tube with the cap open. In setting up the incubations, the tubes can then be placed in alternate rows, while later on during processing the tubes can be close-packed, if necessary. Commercially available racks with lids or grips may be suitable, for example those from Scotlab.

(f) Test-tube racks, again laid out in rows of 4, which permit placement of microfuge tubes in an inverted position, allowing draining of the tubes after a centrifugation assay.

(g) Incubations for filtration assays can be performed in plastic or glass culture tubes, with incubation volumes of up to 2 ml. The tubes can then be rinsed out with up to 5 ml of ice-cold buffer (see Chapter 6).

(h) Thermostatted water baths with heating and cooling facilities. Shaking water baths are desirable for some applications.

(i) Magnetic stirrers.

(j) Vortex mixers.

(k) Digital timer.

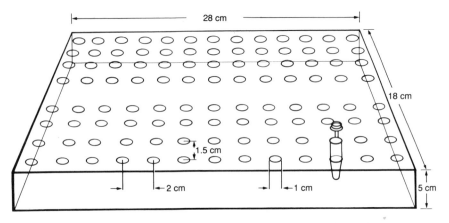

Figure 9. Dimensions of rack suitable for incubating 1.5 ml microfuge tubes.

ii. For preparing membranes and making solubilized preparations

(a) Potter–Elvehjem homogenizer (clearance 0.1/0.2 mm) for homogenization of soft tissue, and resuspension of membrane pellets. We mount the pestle in the overhead chuck of a power drill, rotating at 300 r.p.m., but a range of custom-made drives is also available (e.g. Braun Potter S). 10 and 50 ml sizes are convenient for the glass homogenizing vessel (see Chapter 5).

(b) Homogenizing device for tougher tissues, for example Polytron (Brinkman, Kinematica), Ultra-turrax (Kinematica), Omnimixer (Sorvall). Our personal preference is for the Polytron. A range of models and disrupter probes is available. A suitable size for receptor studies is the PT10 micro model (sample size 2–250 ml). A useful list of tissue disrupting devices is given by Evans (29).

(c) Preparative refrigerated centrifuge with rotor capable of at least 30 000 g with 50 ml tubes, for membrane preparations.

(d) Ultracentrifuge with rotors capable of 100 000 g for separating solubilized receptors from membranes. It is convenient to have an angle rotor capable of handling approx. 50 ml for preparative work, and a swing-out rotor (for example 5 ml) for running sucrose gradients, etc.

(e) Refrigerated low-speed centrifuge, for low-speed spins.

Membrane preparations are covered more fully in Chapter 5.

iii. For implementing separation procedures

(a) Microcentrifuges capable of at least 10 000 g, preferably around 15 000 g (Chapter 7).

(b) Filtration manifold (for example Millipore, Schleicher and Schuell) or cell-harvester (for example Brandel, Skatron), and filters (Chapter 6).

(c) Vacuum pump.

(d) Disposable gel-filtration columns (Chapter 8).

iv. For measuring radioligands

An accurately calibrated scintillation counter and/or γ-counter will be needed. The use of plastic scintillation vials reduces background and is preferable. *Note*: the counter should provide output as d.p.m. otherwise individual counts must be corrected for quenching.

In addition to the above items, it is labour-saving to have some form of automatic or semi-automatic liquid-dispensing device, capable of dispensing aliquots of buffer either in a timed manner, or at the push of a button (for example Gilson Dilutor, Hamilton Microlab 1000). Apart from this, a good range of laboratory glassware, accurate balances, a spectrophotometer for protein assays are also necessary.

For analysis of binding experiments, one needs access to a good non-linear, least-squares fitting algorithm, and appropriate computer. This aspect is covered in Chapter 11 and Appendix 3.

3.1.2 Buffers for binding studies

The choice of buffer for binding studies depends on the receptor type and specific purpose of the experiment. Some workers favour the use of physiological media such as Krebs–Henseleit solution, in which the main buffer ion is bicarbonate, so gassing with carbon dioxide is necessary to maintain pH. A comprehensive list of physiological media is given in *Data for Biochemical Research* (30). Their use is essential for *in vivo* work on cells. In general, however, it is difficult to simulate genuinely physiological conditions using membrane preparations *in vitro*, for the simple reason that, because of their vectorial insertion in the cell-surface membrane, the extracellular and intracellular surfaces of receptors are exposed to different ionic conditions. It may be possible to remedy this defect in reconstitution systems in which receptors are inserted into phospholipid vesicles which can be sealed, thereby entrapping medium of a certain composition, provided that 'right-side out' vesicles can be separated from 'inside-out' vesicles, for example by lectin affinity chromatography. Reconstitution techniques are described in reference 23. The effects of transmembrane ionic gradients and potential differences on receptor binding are largely uninvestigated.

For the study of ligand binding, the use of simple buffers usually suffices. The choice is dictated by the desired pH. For studies around neutral pH, Hepes, phosphate, or Tris buffers are often used in the concentration 20–50 mM. The 'Good' buffers (Hepes, Mes, Pipes, etc.) seem, generally, to be fairly innocuous for receptor binding studies. These buffers can be supplemented with the salts of monovalent cations (NaCl, KCl) to produce the desired ionic strength. Because receptor-specific ligands are almost invariably charged, ionic strength often exerts a strong effect on the effective affinity constant of the receptor–ligand interaction, through the screening of charge–charge interactions. In the case of muscarinic receptors, an increase in ionic strength from 0.02 to 0.1 milliosmoles/litre produces an approximate 3-fold decrease in antagonist binding affinity.

Variations in the pH within the physiological range can produce large effects on receptor–ligand interactions. These can be due either to effects on the ionization of groups within the receptor's ligand binding site, or on groups which affect receptor–effector interactions, which may be on the receptor itself, or on the effector if these are separate entities, as in the case of G-protein-coupled receptors (23). Alternatively, the ionization of groups on the ligand may be affected. Such pH-dependent effects may influence both the equilibrium binding constant, and the kinetic rate constants of binding. The interplay of these effects can be very complex. For an example relating to muscarinic receptors, see reference 31. A borate/citrate/phosphate buffer system which has constant ionic strength over the pH range 3–11 has been described, see reference 32. The interactive effects of ionic strength, pH, and temperature on ligand binding mean that for a comparison of binding data both within a given laboratory and between laboratories it is essential to standardize the conditions used.

Buffers are often supplemented with ions which may promote specific interactions. A classic example is that Mg^{2+} ions (and some other divalent cations such as Mn^{2+}, Co^{2+}, or Ni^{2+}), at concentrations of 0.1–10 mM promote agonist binding to many G-protein-coupled receptors in membranes by favouring the formation of the high-affinity agonist–receptor–G-protein complex. Such additives can also dramatically affect the rate constants of ligand binding. Another example is the ability of Na^+ but not K^- ions to inhibit agonist and promote antagonist binding to μ-opioid receptors. It, therefore, should be considered whether the buffer ions used in binding studies may chelate cations, such as Mg^{2+}, added at millimolar concentrations, as phosphate or citrate buffers will do. 'Good' buffers, such as Hepes, are more non-perturbing in this respect. Buffer solutions containing heavy metal ions should be avoided because of the risk of reaction with sulphydryl groups in the receptor, which may completely alter the structure–binding relationships.

Another consideration is whether the buffer ions may mimic or compete with specific receptor–ligand interactions, or with specific interactions of divalent cations or effectors. In the case of cationic amine receptors, amine buffers such as Tris and triethanolamine run this risk. For example, Tris has been reported to mimic the action of Mg in promoting receptor–G-protein interactions. Buffer ions may also specifically block or modulate ion channels or receptor-associated enzyme activities. For these reasons, Tris buffers must be used cautiously in binding studies, although they are often of great utility in purification work.

The addition of chelating agents may strongly perturb ligand interactions, usually indirectly by removing regulatory multivalent metal cations. Buffer and ionic conditions will also usually affect the handling properties of membrane preparations. Low ionic strength media containing chelating agents promote membrane dispersion. Raising the ionic strength or adding divalent cations promote clumping and sedimentation by screening the ionic charges on cell-surface sialic acids, and phospholipid headgroups. This may make it difficult to maintain a homogeneous membrane suspension. Raising the concentration of monovalent cations to around 1 M increases the density of the solutions to such an extent that membranes may become difficult to sediment in a bench-top microfuge, so that higher g-forces have to be used. This topic is discussed further in Chapter 7.

It will be evident from the above that detailed and specific guidance on the use of buffers and media in binding studies is difficult to give. The preference in this laboratory is to use a buffer such as 20 mM Hepes, 100 mM NaCl, adjusted to pH 7.5 with NaOH, at the temperature of the experiment as a starting point, and to explore from there. However, the constituents will vary with the receptor type, and it is necessary to investigate a range of conditions in setting up a binding assay, in order to optimize specific parameters, such as the ratio of specific to non-specific binding. A useful list of buffers and their properties (pKs, temperature coefficients, chelating abilities) is given in *Data for Biochemical Research* (30). Further information can be gleaned from manufacturer's catalogues, for example Calbiochem, Sigma, Boehringer–Mannheim, Pierce, BDH, and Aldrich.

3.2 Measuring the time-course of tracer ligand association

3.2.1 Aims

Following the outline presented in Section 2.3, the aims of a ligand-association experiment may be regarded as the following:

- To establish the time needed for equilibrium to be achieved at a range of different tracer ligand concentrations.

- To demonstrate that a component of the binding is specific.

- To measure the tracer ligand binding accurately at sufficient time points to allow quantitative extraction of kinetic parameters, both association ('on') and dissociation ('off') rate constants.

- To ensure that the measurements are valid, and that artefacts are avoided.

- To extract the appropriate kinetic parameters, accompanied by statistically meaningful error estimates.

The time-course of tracer ligand association may be determined either in the absence or in the presence of unlabelled competitors, or modulators. Analysis of the resulting effects on the kinetics of the tracer can provide information about both the kinetics and the equilibrium binding properties of the added ligands. However, the experimental protocols described below are not affected by this extension of the basic tracer association experiment.

3.2.2 Receptor preparation and concentration of binding sites

---

Protocol 6. Diluting the receptor preparation

1. Dilute the receptor preparation to give a concentration of binding sites R_t approximately equal to $0.1 \times K_d$, where K_d is the dissociation constant of the tracer ligand. If the resulting value of R_t is less than 5×10^{-11} M, the binding may be too low to measure accurately with ^{3}H-labelled ligands. If so, it may be necessary to use a higher receptor concentration, and to attempt to correct for any ligand depletion.

2. If the dissociation constant of the tracer is unknown, dilute the receptor preparation to give a final concentration of sites in the range 10^{-10}–10^{-9} M.

3. If there is no reliable a priori information, try a concentration of the receptor preparation equivalent to a 1:100 dilution of the original tissue.

4. Allow a large enough volume of the receptor preparation to permit the required number (say 50) measurements of binding. Measurements of non-specific as well as specific binding will be required. Store the preparation on ice until needed.

5. Preincubate the receptor preparation in a conical flask or other suitable container until the desired incubation temperature has been attained. Use a shaking water bath to keep the membranes suspended, or stir them slowly on a magnetic stirrer. For obvious reasons, it is easier to work in a thermostatted room (hot-room, cold-room, or at room temperature) if it is important to minimize temperature changes during the incubation or separation procedure. Under some circumstances this may be difficult to achieve completely, for example if using a non-thermostatted microcentrifuge.

3.2.3 Sequential processing of assays

Several receptor assay methods, most obviously the manifold-based filtration assay, but also gel-filtration, receptor precipitation, and ligand adsorption methods are particularly well suited to sequential processing of incubations (*Figure 10a*). If using such an assay method, it is convenient to add the tracer ligand to the bulk receptor preparation. This has the virtue of minimizing variation due to pipetting of the tracer which is necessarily present at uniform concentration throughout. If possible, in the initial experiments the concentration of the tracer should be approximately equal to its K_d. The range should be extended to measurements between $0.1 \times K_d$ and $10 \times K_d$, subsequently. If there are problems in handling the tracer ligand due to adsorption to glass or plasticware, refer to Section 4.3.

Protocol 7. Sequential processing of assays

1. Add tracer ligand. Mix rapidly. Keep membranes in uniform suspension by stirring or shaking.
2. Withdraw and process aliquots (0.25–1.0 ml) at predetermined times, for instance using a filtration manifold (for example Millipore) attached to a water pump, or vacuum pump and reservoir.
3. Immediately wash the filters with one or more washes with ice-cold buffer (2–5 ml).

Typical fast flow filters for use in such assays are Whatman glass-fibre filters GFB, GFC, or GFF (1.0, 1.2, 0.8 μm pore size). The choice of filter porosity depends on the properties of the receptor preparations. A suitable filter diameter is 2.5 cm. The filters should be rinsed well with ice-cold wash buffer before use. Keep filters not in use blocked off, or wet, to avoid drying off and loss of vacuum. A detailed account of filtration procedures and filter properties is given in Chapter 6.

A convenient way to obtain aliquots at the shorter time points, particularly useful if working in a thermostatically controlled environment or at room temperature, is to perform the reaction in a 12.5 ml Eppendorf combitip.

Protocol 8. Aliquotting at short time points

1. Suck up about 1 ml of air to form a small bubble.
2. Take up a small aliquot of buffer, containing the tracer ligand.
3. Suck up about 10 ml of receptor preparation, as rapidly as possible.
4. Mixing (check with a dye) is reasonably thorough. If necessary, invert once or twice so that the bubble completes the mixing process.
5. Start to pipette out and process aliquots—0.5 ml will allow about 20 time points. Remix frequently to keep membranes in suspension.
6. If using a manifold-based filtration assay, follow up immediately with one or more washes with at least 2 ml of ice-cold buffer. The buffer is conveniently dispensed by a manually operated repetitive-pipette, or better, by a push button operated pipetting device (for example Gilson Dilutor). This protocol is best performed by two people, one responsible for pipetting the receptor, and the other for subsequently washing the filter.

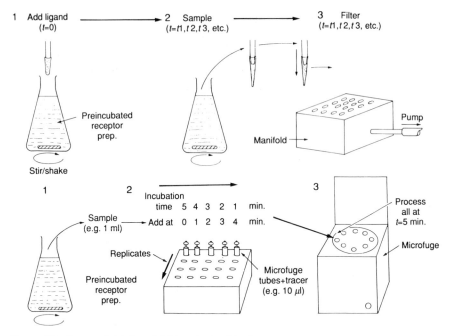

Figure 10. Sequential (a) and (b) batchwise binding assays.

Whichever protocol is used, pipette out and count replicate aliquots (for example 4) of the incubation mixture to determine the total tracer ligand concentration. Measure the counter background using four background samples and subtract the mean value of this parameter from the measurements of total and non-specific binding, and total ligand concentration.

3.2.4 Batchwise processing of assays
Methods other than manifold-based filtration assays, particularly microcentrifugation (Chapter 7) or cell-harvester (Chapter 6) filtration lend themselves better to processing incubations batchwise. In such instances, a reversed time-course strategy is better (*Figure 10b*).

Protocol 9. Batchwise processing of assays

1. Pipette the tracer ligand in small aliquots into individual assay tubes; $10 \, \mu l$ is convenient, and can be dispensed with a repetitive pipette, or with an automatic pipette.

2. Pipette aliquots of the preincubated receptor preparation (for example 1 ml) at the desired temperature into the assay tubes in reverse order, with the longest time points pipetted out first, and the shortest last.

3. Vortex-mix each tube. Transfer to a water bath, if necessary. Shake or remix as necessary to maintain an even suspension.

4. Transfer the tubes to the microfuge, or cell-harvester as time zero approaches. In the case of microfuges with drum rotors, and detachable racks, incubation and centrifugation may possibly be carried out in the centrifuge rack itself.

5. At time zero, initiate centrifugation or batch filtration to terminate the binding reaction.

6. Pipette out replicate (for example four) aliquots of tracer ligand for measurement of the total ligand concentration. Measure counter background using four background samples, and subtract the mean value from the other measurements.

The disadvantage of multiple pipettings of the tracer ligand is the variability introduced due to pipetting error. This can be partly compensated for by replicating the assays. A minimum of four replicates is necessary for calculating a proper standard error, but this is probably not necessary in exploratory experiments. A non-linear, least-squares fit to the time-course will in itself provide a measure of parameter error.

3.2.5 Choice of time intervals

Time-courses of ligand association (and dissociation) are, intrinsically, multi-exponential processes. It is, therefore, natural to schedule measurements to provide an approximately geometric distribution of time intervals. For initial exploration, a sequence such as 5, 10, 20, 45, 60 sec, 2, 5, 10, 20, 40, 60, 90, 120, 240, 360, 480 min should give information both about rapid initial processes, and subsequent slower events. As noted, several of the assay procedures are intrinsically slow, for example gel-filtration (around 120 sec), precipitation (around 10 min), or adsorption (around 60 sec), and may initiate dissociation of the receptor–ligand complex RL*. Timing of intervals in the second time-range is impossible to establish in centrifugation assays. Filtration is the method best suited to kinetic measurements (Chapter 10), provided that rapidly-dissociating components of binding are not lost during the washing of the filters. In favourable instances it is possible to use the filtration assay, *without* washing the filters. In these circumstances, it is possible despite decreased reproducibility of replicates, to determine whether there is a significant loss of rapidly dissociating components.

3.2.6 Measurement of free ligand concentration

As pointed out earlier, it is necessary to measure and check the free ligand concentration, to detect and preferably to avoid ligand depletion. In the case of centrifugation assays, the free ligand concentration can be measured by sampling and assaying aliquots of the supernatants provided that separation of bound from free ligand is complete. In the case of filtration assays, this may not be convenient as a routine procedure, and it is advisable to perform some supplementary centrifugation assays, in order that the true free ligand concentration may be ascertained. In the case of assays on solubilized receptors, the true free ligand concentration is not easy to measure. If there is any doubt about it, perform equilibrium dialysis or equilibrium gel-filtration measurements. As has been emphasized, knowledge of the free ligand concentration is essential to permit statistically valid analysis of the time-course of binding. In simple cases, the free ligand concentration is equal to the difference between the total ligand concentration and the bound ligand concentration, but this needs to be verified in each case.

3.2.7 Time-course of non-specific binding

Repeat the time-course measurements using a receptor preparation preblocked with a competing ligand. This will allow assessment of the time-course of non-specific binding. Pre-blocking can conveniently be carried out during the pre-incubation period. The concentration of competing ligand used should be, initially, 1000 times its K_d for the receptor, i.e. the stock solution should have a concentration of $10^5 \times K_d$ if a 1:100 dilution is used in the final assay. This will provide at least 99% blockade, even if the tracer ligand is present at 10 times its own K_d. The use of a very high affinity ligand to define non-specific binding may require an extended period of preincubation, particularly if measurements are made at relatively low temperature. At 0–4°C, a 15–60 min preincubation is advisable.

3.2.8 Measurement of total receptor concentration

For anything more complicated than a preliminary, semi-quantitative measurement of the time-course of association, it is necessary to know the total receptor concentration (R_t) present in the assays. Thus it is advisable to include several (for example 4) replicate incubations employing a near receptor-saturating concentration of the ligand, together with appropriate measurements of non-specific binding. If possible, these assays should be conducted at 30–100 times the K_d of the tracer ligand, and non-specific binding defined using 3000–10000 times the K_d of the unlabelled ligand. If this is not feasible, it is necessary to perform measurements of the ligand saturation curve at equilibrium over the 10–90% saturation range, in order to define R_t (see Section 3.5). If the affinity constant of the tracer ligand is known, these measurements can be used to check the internal consistency of the time-course measurements, provided that the bound and free concentrations of the tracer ligand are related as described by eqn 3 of Section 2.1. If so, $R_t = RL^*_{eq} \cdot (1 + K \cdot L^*_{eq})/(K \cdot L^*_{eq})$, where RL^*_{eq} is derived from the association time-course should agree with R_t determined from the measurements at high tracer concentration, or from the direct ligand saturation curve. The subscript 'eq' denotes measurement at equilibrium. Data from several association experiments can be pooled for analysis if the corresponding values of R_t are known.

3.2.9 Addition of unlabelled competitors or modulators

Unlabelled ligands can be added either during the preincubation phase, or simultaneously with the labelled ligand, as best suits the aim of the experiment. The presence of competing ligands will slow the rate of approach to equilibrium (see Sections 2.5.1 and 4.11). The presence of non-competitive ligands or modulators may either retard or accelerate equilibration. To avoid artefacts, it is obviously important to measure the time-course of tracer binding under the actual conditions of the proposed experiment.

3.2.10 Preliminary analysis and interpretation

i.

Examine the extent of depletion of the free radioligand at the longest incubation times studied. If depletion is in excess of 50% the experiment should be repeated with a lower value of R_t. Such repetition is desirable unless depletion is less than approximately 10%. Convert measurements of bound and free ligand to molar concentrations. In the case of radiolabelled ligands:

$$RL^* = 10^{-9} \cdot B/(V \cdot SA \cdot 2220) \quad M = B/(V \cdot SA \cdot 2220) \quad \text{nM}, \quad (31)$$

where B is the radioligand bound (d.p.m.) corrected for counter background, V is the volume assayed (ml), SA is the specific activity of the radioligand (Curies/mmole), and 2220 is the conversion factor from d.p.m. to nanocuries.

ii.

Plot bound ligand RL^* vs *time*. Estimate specific binding by subtracting the non-specific binding for each time point from the total binding for that time point. This calculation is only accurate for conditions under which depletion is less than 10%. If replicate assays are available, calculate the mean and standard error of the mean of total and non-specific binding (SEM = $\sigma/\sqrt{r}$, where σ is the standard deviation, and r the number of replicates). Let these values be SE_t and SE_{ns}. The standard error of the specific binding is then $SE_{spec} = (SE_t^2 + SE_{ns}^2)^{1/2}$ for equal numbers of replicates. From the apparent equilibrium value, RL_{eq}^*, estimate the apparent affinity constant:

$$K = 1/K_d = RL_{eq}^*/((R_t - RL_{eq}^*) \cdot L_{eq}^*). \quad (32)$$

Note that the apparent value of RL_{eq}^* may not accurately define the true value. If binding is slow, and has not reached a plateau indicating the attainment of equilibrium, longer incubation times are necessary. If RL^* passes through a maximum and then declines (see Sections 2.5.1, 4.4, and 4.7), degradation of the receptor or of the tracer ligand may be occurring. An increasing value of the non-specific binding may indicate partitioning of the ligand into phospholipids, or diffusion into sealed vesicles.

iii.

If ligand depletion is minimal ($< 10\%$) plot:

$$-\log_e (1 - RL^*/RL^*_{eq}) \text{ against } t \text{ (see } Table\ 1\text{).} \tag{33}$$

In the simplest case, this elementary first-order plot yields a straight line with a slope:

$$k = k_{12}{\cdot}L^* + k_{21}. \tag{34}$$

The slope of the plot can be estimated by eye, or by a linear least-squares fit.

If a straight line is not obtained, a more complex binding process is indicated. Multiple exponential time-courses may be generated by:

- the occurrence of isomerization processes after the formation of the initial RL^* complex, including bimolecular processes involving the formation of a receptor–effector complex,
- the presence of several independent classes of binding sites with different kinetic constants,
- artefacts, such as receptor and ligand instability, or
- the presence of competing or modulating ligands (see Section 4.11.2).

The detection and avoidance of some possible artefacts is covered in Section 4. The analysis of complex association kinetics is discussed in Chapter 10.

iv.

If ligand depletion is more pronounced (10–50%) use the integrated rate equation (*Table 2*). (See also program INTRATE, *Protocol 2.*) Thus, let a, b, be the roots of the quadratic equation:

$$X^2 - X{\cdot}(R_t + L_t^* + K_d) + R_t{\cdot}L_t^* = 0; \tag{35}$$

whose solution determines the level of equilibrium binding. The roots are: $a = RL^*_{eq}$, $b = R_t{\cdot}L_t^*/RL^*_{eq}$. We then find by rearranging the integrated rate equation that:

$$\frac{1}{(a - b)} \cdot \log_e \frac{(b\,(RL^* - a))}{(a\,(RL^* - b))} = k_{12}{\cdot}t. \quad (Table\ 2) \tag{36}$$

Thus a plot of the left hand side of this equation vs time should yield a straight line with slope k_{12}. Again, deviations from linear behaviour may indicate isomerization of the RL^* complex, receptor heterogeneity, or receptor or ligand instability. Note that a non-linear, least-squares fit to the untransformed data is desirable for unbiased extraction of the kinetic constant (see Chapters 10 and 11).

v.

From the slope of the first-order, or integrated rate plot and the estimated K_d value of the ligand, initial estimates of the dissociation as well as the association rate constant can be obtained: thus the association rate constant is obtained from the equation $k_{12} = k/(L^* + K_d)$ where k is the slope of the first-order plot (cf. *iii*) and the dissociation rate constant estimated from the equation $k_{21} = k_{12} \cdot K_d$.

vi.

Repeat the association time-course measurements using different L^* concentrations to span the range $0.1 \cdot K_d$ (off-rate limited kinetics) to $10 \cdot K_d$ (on-rate limited kinetics). As discussed in Section 2.3, the time taken to reach equilibrium under off-rate limited conditions is a critical piece of information for equilibrium binding studies.

Under non-depletion conditions, and provided that the time-courses are mono-exponential, a replot of the slopes of the first-order plots against L_t^* will yield a straight line with a slope of k_{12}, and a y-intercept of k_{21}. Initial estimates of both these parameters can be obtained from a linear least-squares fit to the replot.

vii.

The full quantitative analysis of an association time-course requires that it be fitted to an appropriate theoretical model. The general function is (in the absence of depletion):

$$RL^* = \sum_i Ci \cdot e^{-ki \cdot t} + NS + NS' \cdot t, \tag{37}$$

which incorporates a sum of exponential functions, and both a constant and a time-dependent (sloping baseline) component of non-specific binding.

A specialized form of this function, applicable to multiple independent receptor populations, each exhibiting bimolecular kinetics is:

$$RL^* = \sum_i RiL_{eq}^* \cdot (1 - e^{-ki \cdot t}) + NS + NS' \cdot t, \tag{38}$$

where each ki is the sum of a term such as $k_{12} \cdot L^* + k_{21}$. The dissection of multiexponential time-courses using such functions is described in Chapter 10.

3.3 Measuring the time-course of tracer dissociation

3.3.1 Aims

Following the outline given in Section 2.3.5, the aims of a tracer ligand dissociation experiment are the following:

● To show that specific binding is fully reversible, and to determine the time needed for such reversal to occur (provided that the ligand does not bind covalently).

- To determine the time-course with sufficient precision to allow accurate estimation of the dissociation rate constant(s).
- To ensure the avoidance of experimental artefacts.
- To analyse the dissociation time-course quantitatively, and so extract estimates of the appropriate kinetic parameters, with realistic error estimates.

3.3.2 Prelabelling a batch of receptor

Protocol 10. Bulk prelabelling

Under selected conditions of temperature and buffer composition:

1. Label a batch of receptor preparation with tracer ligand as described in Section 3.1.2. Allow enough for about 25 time-points.

2. Include an aliquot incubated in the presence of an unlabelled competitor at a concentration around 1000 × its K_d value for determination of non-specific binding; *NB* do not dilute the preparation more than 1% as a result of this addition for determination of non-specific binding. Allow enough material for quadruplicate assay of the level of non-specific binding.

3. Sample the prelabelled preparation for determination of L_t^* (4 replicates).

4. Process a minimum of four aliquots of both the total and non-specific binding incubations, before initiating the dissociation reaction for determination of RL_0^*, the initial concentration of the receptor–ligand complex.

5. Centrifuge several (for example four) aliquots and sample the supernatant to measure the free ligand concentration. In centrifugation assays, this may be combined with step 4. Measure the counter background accurately, using four background samples, and subtract the mean value from the other measurements.

To avoid problems which might be associated with variation in the initial level of binding over the time period necessary for the initiation of a number of assays, it is advisable to label to apparent equilibrium. However, note that if the binding mechanism is complex, conformational changes, formation of ternary complexes, or similar processes may occur after the initial binding step. These may affect the time-course of dissociation, which may, therefore, be affected by the length of the preincubation, even though maximal binding has, apparently, been attained (see Section 4.1). The kinetic consequences of complex binding mechanisms are discussed in more detail in Chapter 10.

In an initial experiment, it is advisable to aim for a high level of receptor saturation ($L_t^* =$ approx. $10 \times K_d$), so that both high and low affinity ligand binding sites will be occupied if present, and both rapid and slow dissociation processes seen. Lower degrees of saturation ($L_t^* = 0.1 \times K_d$) should be subsequently investigated, to parallel the range of concentrations employed in investigating the association time-course. As pointed out in Section 2.3.4, from these data it will also be possible to derive an independent estimate of the equilibration time of the association reaction when $L_t^* < K_d$.

3.3.3 Processing the assays

Protocol 11. Sequential processing

1. Initiate dissociation by adding a small volume (preferably not more than 1% of the total incubation volume) of displacing ligand to the bulk pre-labelled preparation. Aim to achieve a final concentration of the displacer of $1000 \times K_d$ in the incubation. It is convenient to use a concentration of displacer which is the same as that used to define non-specific binding. The displacer should NOT be the unlabelled tracer ligand unless it is known *beyond doubt* that the displacement of non-specific binding will not occur. The displacer should be known to interact by a competitive rather than an allosteric mechanism.

2. Mix rapidly.

3. If using a membrane suspension, maintain uniformity by using a shaking water bath, or mix gently on a magnetic stirrer.

4. Withdraw aliquots of convenient size (0.25–1 ml) for filtration, centrifugation, etc. at predetermined times.

5. Carry out the reactions, for convenience as in Section 3.2, *Protocol 8*, in an Eppendorf combitip, using approximately 20×0.5 ml aliquots to define the time-course. To use this method, suck up about 1 ml of air, followed by the unlabelled displacer in a small volume of buffer (50–100 μl). Then rapidly draw up the prelabelled receptor preparation (10 ml). If necessary invert once or twice to allow the air bubble to complete mixing before commencing the pipetting of aliquots. Remix frequently to keep membrane fragments in suspension.

Protocol 12. Batchwise processing of assays (microfuge or cell harvester)

1. Pipette the unlabelled displacer in 10 μl aliquots (using a combitip or automatic pipette) into individual tubes.

Protocol 12. *Continued*

2. Use a reversed order of addition when adding the labelled receptor preparation, as in Section 3.2.4.
3. Mix and incubate as described in Section 3.2.4.
4. Terminate binding by centrifugation, batch filtration, etc.

3.3.4 Selection of incubation times

The considerations given in Section 3.2.5 also apply to the selection of time points for dissociation. A sequence such as 5, 10, 20, 40, 60 sec, 2, 3, 5, 10, 20, 40, 60, 90, 120, 240, 360, 480 min should prove reasonable for initial explorations, and can be filled in or extended later. Again, note that the nature of the assay defines the time-resolution which can be obtained (Section 3.2.5).

3.3.5 Measurement of total receptor concentration

It is advisable to include an estimate of R_t in the experiment, either by replicate measurement of the binding of a near receptor-saturating concentration of L^*, or by independent performance of a saturation experiment. This will allow data from different experiments to be compared and combined for analysis.

3.3.6 Dissociation by dilution

An alternative way to initiate the dissociation process is to dilute the incubation mixture substantially (10–100-fold). This entails prelabelling a concentrated preparation with tracer, to provide scope for dilution, or, alternatively, harvesting (for example by centrifugation) followed by rapid resuspension (Chapter 7). In the simplest case, such a manoeuvre should give a time-course of dissociation compatible with that yielded by the addition of an excess of unlabelled displacing ligand. This may not be so, however, if the displacing ligand exerts a non-competitive effect on the receptor, by binding to a second site, or selectively to one of a number of specific conformations of the receptor (Section 2.5.4; Chapter 11).

More subtly, the presence of high local concentrations of receptors, particularly if they are physically occluded to some degree, may retard dissociation owing to rebinding of the ligand after initial dissociation. This is more likely after dilution than after addition of a competitor. Ligand–ligand interactions are also capable of altering apparent dissociation rates. This makes the interpretation of dilution experiments somewhat more problematical than that of displacement experiments. However, a combination of dilution and addition of a competitor may be used as a practical method of reducing non-specific binding in dissociation experiments. Dilution will also affect non-specific binding, so the same set of manipulations should be carried out on the non-specific binding assays, in order that any time-dependence of non-specific binding may be detected.

3.3.7 Preliminary analysis and interpretation

i.

Convert measurements of bound (and free) ligand to molar concentrations (see Section 3.2.10). Plot bound ligand RL^* vs *time*. Ensure that reversal to the original level of non-specific binding occurs. If necessary, repeat the experiment using longer incubation times. Failure to obtain full reversal may imply uptake and/or metabolism of the tracer ligand (see Sections 4.5, 4.6). If replicate assays are available, calculate the mean and standard error of total and non-specific binding. If ligand depletion has been avoided, subtract non-specific from total binding to yield an estimate of specific binding. Again, the standard error of the mean specific binding is given by $SE_{spec} = (SE_t^2 + SE_{ns}^2)^{1/2}$ for equal numbers of replicates. If significant ($> 10\%$) tracer depletion occurred, the level of non-specific binding will increase as the dissociation reaction proceeds. In this case, it will be necessary to determine non-specific binding as a function of free ligand concentration in order to obtain the appropriate estimates. Even these values will be incorrect if the non-specific binding is time-dependent.

ii.

An initial analysis of the dissociation time-course can be carried out graphically, by plotting $\log_e RL^*$ vs t (*Table 1*). In the case of simple exponential dissociation, this plot yields a straight line with slope k_{21}, and y-intercept RL_0^* (RL_{eq}^* if the binding reaction was initially in equilibrium). The slope and intercept of the line can be determined visually, or by a linear least-squares fit, to provide preliminary estimates. If the binding process is more complex, involving a conformational change after binding, the ligand-induced formation or breakdown of receptor–effector complexes, or the presence of multiple independent receptor populations the above first-order plot will deviate from linearity and must be analysed by least-squares fitting to a multi-exponential model or by 'curve peeling'. The use of the log–linear plot to extract measurements of dissociation rate constants is explained in Chapter 10.

iii.

A suitable multi-exponential model of dissociation takes the form:

$$RL^* = \sum_i RiL_0^* \cdot e^{-ki \cdot t} + NS. \tag{39}$$

The non-specific binding, if it has been defined properly and is well behaved, should be constant, at least under non-depletion conditions where L^* does not alter significantly. Note that the rate constant (or half-time) of the slowest dissociation process detected determines the time taken for the receptor to equilibrate with L^* under off-rate limited conditions, i.e. at concentrations of L^* much less than K_d. The time needed for the slowest binding process to reach 97% of its final equilibrium value is $5 \times t_{1/2}$ for this process (Section 2.3.6).

3.4 Rapid components of association or dissociation

3.4.1 Aims

A common situation in binding studies is that the association or dissociation time-course of a ligand is found to be markedly biphasic. Often, the relative magnitudes of the fast and slow components are found to vary as a binding reaction progresses. A typical example is provided by a ligand which undergoes rapid initial binding, followed by a subsequent slowly-reversible conformational change of the initial receptor–ligand complex. In order to study the evolution of the fast and slow components, it is desirable to allow the binding reaction to progress for a series of increasing times, then add excess of an unlabelled competitor, incubate for a short, fixed period of time, sufficient to allow reversal of the fast but not the slow component, and then assay.

Protocol 13. Simultaneous initiation of a set of incubations

1 Set up an array of microfuge tubes in a rack which grips them tightly or has a lid, allowing simultaneous inversion of the whole set of assays (Section 3.1.1). Allow four tubes per time point.

2. Pipette 10 μl of a solution of the unlabelled displacing ligand into the angle of the inner surface of the open lid of each Eppendorf tube (*Figure 11*).

3. Set up a reversed-order tracer ligand association time-course using 5 times the normal assay volume, for example 5 ml in suitable containers, such as, e.g. glass or plastic scintillation vials each containing 50 μl of the tracer ligand. Cap and mix thoroughly.

4. Pipette 4×1 ml aliquots of the tracer ligand incubation mixture into the corresponding Eppendorf tubes.

5. Carefully cap the tubes. Surface tension will keep the 10 μl of displacing ligand as a 'hanging drop' in the lid, without contacting the incubation mixture in the base of the tube.

6. At time zero invert the rack (and thus at the same time the tubes) several times to add the displacer and initiate the dissociation reaction simultaneously in the whole set of assays.

7. Transfer the tubes to microcentrifuges, or to the cell harvester.

8. Continue the incubation for a time predetermined to give optimum discrimination between the fast and slow components of dissociation.

9 After the elapse of the predetermined period, terminate the reaction by separating bound from free ligand.

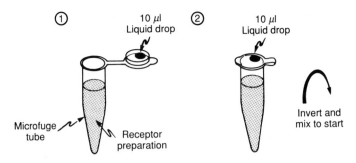

Figure 11. Hanging drop technique for initiating incubations.

Protocol 13 is best performed in a constant-temperature environment. We have found it to be particularly useful for studying ligand binding at cold-room temperatures (4°C) in the context of microfuge assays. With care, $2 \times 10 \mu l$ aliquots can be pipetted into the lid of a microfuge tube in separate positions, and the lid closed without contaminating an incubation in the body of the tube, thus allowing the simultaneous addition of displacers and modulators.

3.5 Tracer ligand saturation curves

3.5.1 Aims

Following the outline given in Section 2.4, the aims of a tracer ligand saturation experiment are the following:

- to define the concentration-dependence of total binding
- to define the concentration-dependence of non-specific binding
- to demonstrate the saturability of specific binding
- to ensure the accumulation of sufficient data to allow extraction of the affinity(s) of the tracer ligand for the sites so defined, and to determine the absolute concentration(s) of these sites
- to ensure that the binding measurements are free of artefacts
- to analyse the binding curve quantitatively, extracting valid estimates of the parameters, and statistically correct estimates of the associated errors

Tracer ligand binding curves ('direct' binding curves) may be determined either in the presence, or in the absence, of competing ligands or modulators. The effect of added unlabelled ligands on the tracer binding curve provides information about the interactions of the former, as well as the latter, with the receptor. However, the experimental protocol described below is not fundamentally affected by this extension of the scope of the basic saturation experiment.

125

3.5.2 Receptor preparation: concentration of binding sites

Dilute the receptor preparation as in Section 3.1.2 to give a concentration of binding sites which, ideally, will be around $0.1 \times K_d$ of the tracer ligand. Concentrations up to $0.5 \times K_d$ may be acceptable, provided that the tracer ligand is chemically and radiochemically pure. Methods of testing the radiochemical and chemical purity of the tracer ligand are given in Section 4.1. The ideal concentration of binding sites for use with ^{3}H-labelled ligands (~ 80 Ci/mmole) is not less than 5×10^{-11} M, and for use with ^{125}I-labelled ligands (~ 2000 Ci/mmole) not less than 2×10^{-12} M, if the saturation curve is to be measured accurately in the 10–90% saturation range using 1 ml aliquots of receptor preparation (Section 2.4.6). Using this concentration of sites it should be possible to obtain 1000 bound d.p.m. at 10% saturation, which will be measurable with $\sim 1\%$ counting error over 10 min. If the receptor preparation is to be diluted further, it is advisable to harvest larger volumes. Otherwise counting and, in practice, other handling errors will be increased. Store the receptor preparation on ice until it is needed.

3.5.3 Setting up the binding curve

Two protocols (*Protocols 14* and *15*) are described to measure direct ligand saturation curves. The first involves making a series of dilutions of the radioligand which are then pipetted out into replicate tubes. The advantage of this method is that SEMs are calculable for each point on the saturation curve. The disadvantage is that errors due to the pipetting of the tracer ligand are not minimized. The second involves making a series of serial dilutions of the tracer ligand in the receptor preparation itself. This method minimizes the errors due to pipetting of the tracer, and allows the determination of a larger number of points per decade. It does, however, demand full reversibility of specific and non-specific binding within the incubation time employed.

Protocol 14. Replicate incubations

1. Dilute the tracer ligand in assay buffer to give a concentration 100× the value desired for the highest final concentration to be assayed. The latter should be at least 10× and preferably 30–100 $\times K_d$. For example, if the tracer ligand has a K_d of 10^{-9} M, the saturation curve should extend at least to 10^{-8} M, and the stock tracer ligand solution should be 10^{-6} M. The stock solution should be sampled for assay. In the case of a radioligand, the concentration present in the stock solution is $d.p.m./(V \times 2220 \times SA)$ nM, where $d.p.m.$ is the d.p.m. corrected for counter background measured in V ml of radioligand of specific activity SA (Ci/mmole).

2. Set up a series of $M \times N$ tubes (microfuge tubes are again convenient), where M is the number of concentrations of radioligand to be used per

decade of concentration, and N is the number of decades of concentration to be covered. N is minimally 2, and preferably up to 4. M should be at least 2 for accurate definition of the saturation curve and can be up to 6.

3. Pipette volumes of buffer into these sets of tubes as shown in *Table 8*. Make up one such set for each decade of concentration.

Table 8. Volumes for making ligand dilutions

Points per decade	Volume/tube (ml)
1	0.900
2	0.216, 0.900
3	0.115, 0.364, 0.900
4	0.0778, 0.216, 0.462, 0.900

4. Pipette 0.1 ml of stock tracer ligand into each of the first set of tubes; for example for 2 points per decade, obtaining 0.1 ml + 0.216 ml (1:3.16-fold dilution) and 0.1 ml + 0.9 ml (1:10-fold dilution). *Discard the pipette tip and take a fresh one*. Cap and vortex mix the tubes.

5. Take 0.1 ml aliquots from the 1:10 dilution and repeat the above process with the next set of tubes. *Discard the pipette tip*. Repeat the process, using the previous 1:10 dilution as the starting point until the dilutions are completed (see *Figure 12*). If the tubes are laid out in a row, it is a good idea to displace them one space to the side as the dilutions are made, to keep count. If your preference is to work from left to right, you will end up with the highest concentration of tracer ligand on the LHS and the lowest on the RHS. Reverse this order, so that the lowest concentration is on the LHS and the highest on the RHS (see *Figure 12*).

6. Take small replicate samples (for example $4 \times 10\,\mu l$) from each tube and count to ascertain that the dilutions have been made correctly. At this stage, any problems with handling the tracer ligand, owing to adsorption on to plasticware or glassware will become apparent. For a discussion of how to avoid handling losses see Section 4.3.

7. Set up two racks of each containing $M \times N$ rows of r tubes so that the number of rows of tubes in each rack corresponds to the number of tracer ligand dilutions made. r is the number of replicates. Into each of the r tubes in a given row, pipette an appropriate volume (for example $10\,\mu l$ to minimize dilution of the assays) of the corresponding concentration of the radioligand, preferably using the same pipetter and setting used to sample the radioligand dilutions. Having completed this process, pipette in any competitor or modulator ligands. Finally, pipette the unlabelled

Protocol 14. *Continued*

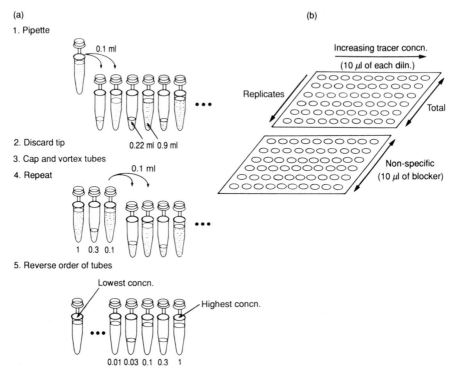

Figure 12. Making dilutions (a), and (b) setting out direct tracer binding or competition assays.

blocking ligand used for definition of non-specific binding into the corresponding set of tubes; $10\,\mu l$ is a convenient volume. Aim for a concentration of $1000 \times K_d$ of the blocking ligand in the final incubation (see Section 3.2.7) to ensure 99% blockade and up to 90% tracer ligand occupancy. For maximum accuracy to equalize volumes pipette $10\,\mu l$ of buffer into the tubes used for total binding, using a fresh tip.

8. Divide the receptor preparation into two separate aliquots, one for measurement of total binding, the other for measurement of non-specific binding. If necessary, preincubate until the appropriate temperature is attained. The minimum total volume of receptor preparation needed will be $2 \times r \times M \times N \times V$, where r is the number of replicates per point (a bare minimum of 2; 4 for a reasonable estimate of the SEM), and the factor of 2 comes from the need to measure the concentration-dependence of non-specific as well as total binding. V is the volume of each assay (typically 0.5–$1.0\,ml$). It is advisable to use a total volume

128

which is at least 2 ml greater than the minimum volume in order to ensure that you do not 'run out' of membranes when pipetting.

9. Dispense aliquots of the preincubated receptor preparations into the assay tubes, starting with those set up for total binding. Work from the lower tracer ligand concentrations towards the higher, to minimize problems due to carry-over of small volumes of tracer. Do this SEPARATELY for the total and non-specific binding measurements. Make sure that the receptor preparation remains uniformly suspended throughout, for example by gentle stirring.

10. Cap and vortex-mix the tubes. Transfer to a water bath if necessary. Incubate for a time sufficient for tracer equilibration to occur under off-rate-limited conditions. *Note* that the presence of competitors or modulators may slow the approach to equilibrium (see Sections 2.5 and 4.11). Shake or remix as necessary to maintain an even suspension.

11. Process the tubes to obtain separation of bound from free ligand. If performing a microcentrifugation assay, sample the supernatants for determination of the free tracer ligand concentration. If performing a filtration assay on membranes, or precipitated soluble receptors, check the free ligand concentration by supplementary centrifugation assays. If necessary, use equilibrium dialysis or equilibrium gel filtration to establish the free ligand concentration. Samples can be taken *before* processing to check total ligand concentrations.

12. Measure the counter background (4 replicates) and subtract the mean value from the other measurements.

The aim of *Protocol 15* is to combine the action of diluting the tracer ligand with that of dispensing the receptor preparation, thus eliminating the variation due to errors in pipetting the radioligand.

Protocol 15. Exponential dilutions

1. Dilute enough of the receptor preparation to give $6 \times N$ assays, where N is the number of decades of concentration to be covered. Ensure that the receptor preparation remains uniform, by stirring gently on a magnetic stirrer. Keep cold, if necessary, for stability (Section 4.4).

The following step will yield six equally-spaced points, on a logarithmic scale, per decade of tracer ligand concentration.

2. Pipette 3.136 ml of receptor preparation into a small container, for example a 20 ml scintillation vial, containing a magnetic stirring bar. Add sufficient tracer ligand to yield the highest concentration to be studied in the saturation experiment. Stir sufficiently to maintain a uniform suspen-

Protocol 15. *Continued*

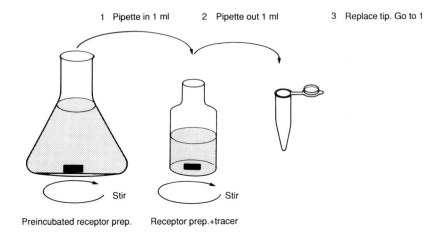

Figure 13. Exponential serial dilution protocol.

sion, and give good mixing, without splashing. Pipette 1 ml of the suspension into an Eppendorf tube. *Discard the pipette tip* (*Figure 13*).

3. With the same pipette but a fresh tip, add 1 ml of fresh receptor preparation into the container. Mix well. Pipette out a further 1 ml of the mixture
into an Eppendorf tube. *Discard the pipette tip.* Using the same pipette for
both procedures ensures that the volume in the mixing chamber remains
constant.

4. Repeat the above dilution and pipetting process until the desired concentration range has been traversed. Each step reduces the tracer concentration to 0.681 of its value before that step. Six steps reduce the concentration 10-fold. However, because of the accumulation of pipetting errors the
actual concentrations obtained will deviate somewhat from the theoretical
values. Thus, it is necessary to measure the free or total ligand concentration, when using this protocol. Total ligand can be measured by sampling
before commencing incubation.

5. Repeat steps 2–4 using receptor preblocked with unlabelled competitor
($1000 \times K_d$) to obtain the concentration-dependence of non-specific binding.

6. Incubate until equilibrium is attained. As used, it is best to assume off-
rate-limited kinetics, and allow five times the half-time of the slowest
dissociation process.

7. Process the incubations.

8. Sample the supernatants for determination of the free ligand concentrations. Measure the counter background (four replicates), and subtract the mean value from all other measurements.

3.5.4 Preliminary analysis and interpretation

i.

Convert measurements of bound and free ligand to molar concentrations as shown previously. In the case of radiolabelled ligands:

$$RL^* = 10^{-9} \cdot B/(V \cdot SA \times 2220) \text{ M (see 3.2.10)} \qquad (40)$$

ii.

Plot total and non-specific binding vs concentration. If replicates are available, calculate means and SEMs (SEM = $\sigma/\sqrt{r}$). If depletion of the tracer ligand was kept to a minimum ($< 10\%$) calculate specific binding by subtraction of non-specific from total binding. As usual, the SEM of the specific binding is given by $SE_{spec} = (SE_t^2 + SE_{ns}^2)^{1/2}$ for equal numbers of replicates. If depletion was more marked it will be necessary to use the non-specific binding curve to read off the non-specific binding appropriate to a given value of total binding from the corresponding free ligand concentration.

Plot RL^* (specific binding) against L^* (free concentration) and against $\log_{10}L^*$, to gain a visual impression of its saturability (see *Figures 2 and 3*).

iii.

Plot RL^*/L^* vs RL^* (Scatchard plot) or RL^* vs RL^*/L^* (Eadie–Hofstee plot) (*Table 1; Figure 2*).

The use of these plots has been described by Zivin and Waud (33) and Munson and Rodbard (34–37, 75). If the data allow it, fit a straight edge through the points to determine R_t (intercept on x-axis in Scatchard plot, on y-axis in Eadie–Hofstee plot) and K (slope = $-K = -1/K_d$). In the opinion of Munson (36) it is better not to attempt linear regression analysis of the Scatchard plot because the distribution of errors is complex, but simply to draw a line by eye, to obtain initial estimates of the binding parameters. Note that the ability of the Scatchard plot to lead the eye on is notorious (18), and may entice one to make a large extrapolation from a limited linear segment of the plot (see *Figure 14a*) to yield an apparent R_t value which the data do not really justify. Thus Scatchard analysis should be regarded as a means of visualizing, or presenting data. It is not a substitute for a non-linear, least-squares fit to a properly justified model of the binding process, which we again emphasize should incorporate terms for both specific and non-specific binding.

A linear Scatchard plot implies a uniform, non-interacting population of binding sites. If this is not the situation, or if certain binding artefacts occur,

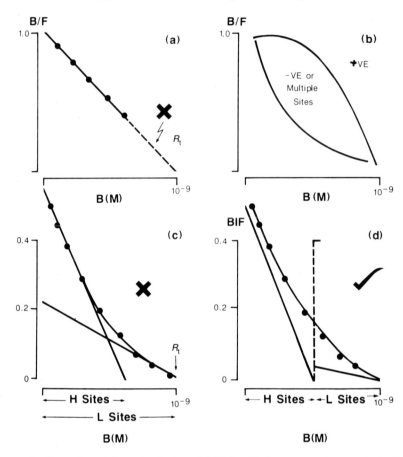

Figure 14. Examples of Scatchard plots. (a) Unjustified extrapolation to obtain R_t. (b) Examples of apparent positive co-operativity (+VE) and negative (−VE) co-operativity or multiple sites. (c) Incorrect analysis of a biphasic Scatchard plot. (d) Correct analysis of a biphasic Scatchard plot.

the plot will deviate from linearity. A Scatchard plot which is convex upwards (*Figure 14b*) may indicate the occurrence of positive cooperativity between binding sites in an oligomeric structure (cf. the binding of acetylcholine to nicotinic acetylcholine receptors (10)). Alternatively, a variety of binding artefacts can give rise to the appearance of positive cooperativity, for instance:

(a) Incomplete equilibration with the lower but not the higher concentrations of the tracer ligand.

(b) Over-estimation of the free tracer ligand concentration. This is a particular hazard under depletion conditions at the low end of the saturation curve if the tracer ligand is impure, if part of the bound ligand is recovered in the free ligand pool, or if non-specific binding is overestimated.

A Scatchard plot which is concave upwards (*Figure 14b*) may have a variety of genuine explanations. It may arise from site–site negative cooperativity within an oligomer, from the presence of multiple independent populations of receptors with different affinities for the tracer ligand (Chapter 11), or from the presence of receptor–effector complexes which manifest different affinities for the ligand, as in the case of receptor–G-protein interactions (see ref. 23).

Alternatively, an upwardly concave Scatchard plot can arise from the presence of an undiagnosed artefact:

(a) from the improper definition of non-specific binding with the recovery of part of the non-specific binding in the apparently specific component

(b) from incomplete equilibration of the ligand with an occluded component of the receptor population, which would lead to an underestimation of its contribution at low, but not at higher ligand concentrations

(c) under certain circumstances, from ligand–ligand interactions or ligand heterogeneity; fortunately, these are relatively infrequent phenomena (5).

A fuller discussion of such artefacts is presented in Section 4.

When a biphasic Scatchard plot is encountered, a common assumption is that the two limbs of the curve represent high and low affinity sites in the manner shown in *Figure 14c*. THIS IS NOT SO. The true situation is as shown in *Figure 14d*. The values of the affinity constants and relative proportions of the two populations of sites must be estimated from the ASYMPTOTES of the Scatchard plot. This is almost impossible to do by eye. Thus, it is much better not to attempt to dissect a Scatchard plot manually, but instead to fit an appropriate model of binding directly to the untransformed data (Chapter 11).

iv.

Using the value of R_t obtained as above, or from a least-squares fit, it is possible to normalize the binding data by dividing each value of RL^* by R_t to get an occupancy (*P*) value. Plot *occupancy* vs $\log_{10} L^*$ on a suitable scale. Examine the plotted curve with the template described in Section 2.4.3 to confirm either that it fits, or deviates significantly from the simple Langmuir isotherm.

v.

The extent of the deviation can be quantified by the calculation of a Hill coefficient ('slope factor') (*Figure 15*). Plot $\log_{10} (P/(1 - P))$ vs $\log_{10} L^*$ in the

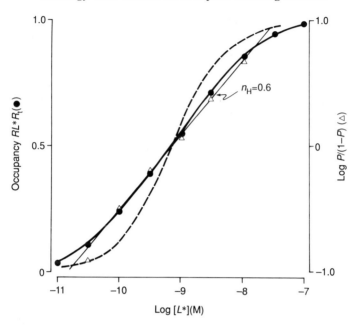

Figure 15. Estimation of a Hill coefficient for a flat binding curve. The dotted curve is a Langmuir isotherm drawn through the IC_{50}. The Hill plot (thin line) has a gradient (n_H) of 0.6. An alternative approach is to fit the Hill equation $RL^* = R_t \cdot (KL^*)^{n_H}/(1 + (KL^*)^{n_H})$ directly to the binding data.

range 10–90% saturation. If the binding curve follows the simple Langmuir isotherm, the slope of the plot, the Hill coefficient n_H, will be 1.0. Apparent positive cooperativity is exemplified by n_H values greater than 1.0, and negative cooperativity or multiple sites by n_H values less than 1.0—the larger the ratio of the affinities of the ligand for the high to the low affinity sites $(K1/K2)$, the lower the value of n_H. However, this definition of n_H presents a problem. Hill plots are usually not completely linear; n_H must, therefore, be taken to be the minimum (or maximum) slope of the plot. The difficulty of obtaining a global measure of n_H reduces its usefulness. It should be regarded as a semi-quantitative measure of the 'flatness' or 'steepness' of a binding curve. The diagnostic use of Hill plots is discussed in reference 38. n_H values in the context of multiple sites and ternary complex models of receptor binding are discussed in Chapter 11.

vi.
Several different models are available for the analysis of ligand binding curves which deviate from the simple Langmuir isotherm. In practice, one of the most useful of them is the multiple sites model:

$$RL^* = \sum_i Ri_t \cdot Ki \cdot L^*/(1 + Ki \cdot L^*) + C \cdot L^* \ (Table\ 3);\qquad(41)$$

according to which the bound ligand, RL^*, is made up of the sum of contributions from sites with different abundances Ri_t exhibiting different affinity constants Ki for the ligand, and a linear term $C \cdot L^*$ describing non-specific binding. The latter can be replaced by a saturable function if required. The use of expressions such as these to fit saturation binding curves is described in Chapter 11. The LIGAND suite of programs (34, 35) is explicitly designed to fit multiple sites models (see Appendix 3) and asks for input in terms of total rather than free ligand concentration, i.e. it corrects for ligand depletion. The model should be rewritten in terms of log affinity constants and log ligand concentrations before being used in fitting programs such as ENZFITTER or GRAFIT (Appendix 3); i.e.

$$RL^* = \sum_i \frac{Ri_t \cdot 10^{(pi+x)}}{(1 + 10^{(pi+x)})} + C \cdot 10^x, \tag{42}$$

where $pi = \log_{10}(Ki)$ and $x = \log_{10}(L^*)$ (see Section 2.2.7). We have found ENZFITTER adequate for most routine binding analyses, although it cannot handle complex non-linear models.

3.6 Inhibition or modulation of tracer ligand binding

3.6.1 Introduction

We have seen that the kinetics and the equilibrium binding of a tracer ligand may be affected by the presence, or addition of a second ligand. An added ligand may:

- bind competitively to the site occupied by the tracer
- bind non-competitively to a second site on the receptor binding subunit
- may interact with another subunit of the receptor
- may interact with an effector macromolecule, and thus indirectly either enhance or diminish the binding of the tracer

The mechanistic consequences of these distinct models are obviously different; some aspects of their diagnosis and interpretation are outlined in Chapter 11. However, their impact on the practical aspect of performing binding studies is minimal; fundamentally, the same protocol suffices for all.

The kinetic effects of added competitors and modulators have been outlined in Section 2.5.1. Further discussion of these matters is given in Chapter 10. Kinetic considerations will not concern us in this section, except in so far as they affect the tracer ligand equilibration time.

3.6.2 Aims

- To define the effect of an added ligand on the equilibrium binding of the tracer ligand. A special case is when the added ligand is the unlabelled tracer ligand—the 'cold vs hot' experiment—see Section 3.6.7 below.

- To ensure that measurements are devoid of artefacts.

- To establish the mechanism of the interaction between the added ligand and the tracer.

- To analyse the data quantitatively and to extract binding parameters characterizing the interaction of the added ligand with the receptor, accompanied by statistically valid error estimates.

3.6.3 Receptor preparation: concentration of binding sites

Dilute the receptor preparation to give a concentration of binding sites equal to around $0.1 \times K_d$ of the tracer ligand, or $0.1 \times K_d$ of the inhibitor ligand, whichever is smaller. Perturbation of the inhibition curve will be sufficient to necessitate corrections if tracer ligand depletion exceeds 10%. This will reduce the precision of parameter estimation. Such corrections are also model-dependent. Double depletion of both tracer and competitor is likely to invalidate the experiment, and must be avoided. If the resultant concentration of receptor binding sites is too low to permit accurate measurements ($< 5 \times 10^{-11}$ M if working with a tritiated ligand), increase the receptor concentration, and use a radioligand concentration of $10 \times R_t$. The resultant competition curves will be competitively right-shifted as a result of the high occupancy of the binding site by the tracer ligand (see Section 2.5.4). The analysis of the data will then require accurate knowledge of the mechanism of binding, and of the tracer ligand affinity constant. Ligand depletion of up to 50% may be correctable, at the expense of some reduction in the accuracy of the parameter estimates (Section 4.10). Store the diluted receptor preparation on ice until it is needed.

3.6.4 Setting up the binding curve

As in Section 3.5.3, two protocols (*Protocols 16, 17*) are described for setting up inhibition or modulation curves. The first involves replicate pipettings of the tracer and the inhibitor/modulator. This is more convenient for screening assays, and in situations in which accurate timing of the exposure of the receptor to the second ligand is necessary. The second is a variation on the serial dilution procedure described in *Protocol 15*. It minimizes pipetting errors, and allows the generation of large numbers of ligand concentrations, at the expense of requiring longer incubations to achieve equilibrium under certain conditions.

Protocol 16. Replicate incubations

1. Dilute the stock tracer ligand in assay buffer to give a concentration 100 times the value required in the incubations. To minimize corrections of competition curves caused by L^*-induced shift (Section 2.5.4), at least in initial experiments it is desirable to keep the tracer concentration at

~0.1 × K_d. However, if the affinity of the tracer ligand for the receptor is such that tracer depletion is unavoidable at such low concentrations, even by further dilution of the receptor preparation, then it is advisable to increase L_t^* to a value of at least $10 × R_t$. This will avoid the problem of tracer ligand depletion, but will necessitate the application of a parallel shift correction to an inhibition curve if the ligand binds competitively. This correction is straightforward (Section 2.5.4) provided that the tracer binds to a uniform set of sites in a manner conforming to the simple Langmuir isotherm. If the tracer–modulator interaction is complex, no simple correction can be applied, and the binding curve must be analysed by fitting to an appropriate model. For a complete investigation, it is necessary to explore a wide range of L_t^* concentrations (see Section 3.6.5 and Chapter 11 for further discussion).

2. Make up a series of dilutions of the competitor/modulator ligand as described in *Protocol 14* (see *Table 8* and *Figure 12*). Allow for the dilution factor involved in setting up the assay. A convenient final dilution is 1:100 (10 µl to 1 ml). If necessary, carry out checks to make sure that the ligand is not subject to handling losses during dilution, etc. The avoidance of handling losses is discussed in Section 4.4.

3. Set up a rack containing $N × M × r$ tubes, where N is the number of decades of competitor concentration to be studied, M is the number of competitor concentrations per decade, and r is the number of replicates to be studied. Two replicates are adequate for screening assays, for example for an initial survey of the effects of a large number of different drugs, but four replicates is the norm for more accurate measurements, for which it is desirable to compute an SEM for each data point. It is also desirable to include $2 × r$ replicates of measurements of the binding of the tracer ligand alone, and of non-specific binding defined in the standard manner for the receptor in question (see *Protocol 14* and Section 3.2.6). Any effect of the competitor/modulator on non-specific binding should be tested directly.

   Inclusion of measurements of total and non-specific binding in each set of competition/modulation dose-effect measurements provides the basis for a check on internal consistency, and for the comparison and pooling of data from different experiments. The use of a larger-than-normal number of replicates allows more accurate definition of the top and bottom of the inhibition curve. This is important in subsequent computerized data analysis.

   The range of competitor concentrations to be studied should extend from 3 decades below to 3 decades above the K_d, i.e. from 0.1% to 99.9% inhibition of tracer binding. This range will have to be extended if the competitor/modulator binds in a complex manner, for example to multiple sites with widely different binding constants.

Protocol 16. *Continued*

4. Pipette the tracer ligand (for example 10 µl aliquots using a repeating pipette) into the tubes. Pipette out each concentration of the competitor/modulator into the appropriate set of *r* tubes. Start pipetting with the lowest concentration of competitor/modulator, and finish with the highest, to minimize problems of carryover. Set up the zero, and non-specific binding measurements, using for example 10 µl of buffer or the standard blocking ligand as appropriate.

5. Include replicate assays for the determination of R_t (cf. Section 3.2.8). These are essential for interpretation of the competition curve if there is any ligand depletion. They also allow a single-point estimate of the tracer ligand affinity, and thus permit an internal check on the assays. Finally, they are useful in normalizing different sets of experimental data. Preincubate the receptor preparation as required. Pipette aliquots (0.5–1 ml) into the assay tubes, working from low competitor/modulator concentration towards high to minimize the chances of cross-contamination. For this reason, use a separate aliquot of receptor preparation for each separate competition curve in a series. Also set up the preincubation for determination of R_t separately, although using the same receptor dilution, to avoid all risks of cross-contamination of one incubation by another.

6. Transfer the tubes to a water bath, if necessary. Incubate until equilibrium has been attained. Remix or shake as necessary to maintain an even suspension. *Note* that the presence of a competing ligand will slow the approach to equilibrium (Section 2.5.1), and that it is possible to go from on-rate limited kinetics (rapid equilibration) to off-rate limited kinetics (slow-equilibration) as the competitor reduces the effective on-rate of the tracer. *Note* that the competitor itself may contribute the rate-limiting step, and that non-competitive agents may produce a very dramatic slowing of the rate of equilibration. It is always advisable to perform experiments to check that equilibrium has been attained, if there is any doubt. As usual, the rule of thumb is to allow 5 times the half-time of the slowest process in the binding reaction. Artefacts caused by failure to achieve equilibrium are described in Section 4.11.

7. Terminate the incubations by processing the reaction mixture, as already described.

8. Measure the free tracer ligand concentration to check the level of depletion. This may only be necessary in the absence of the competing ligand, if depletion is kept to a minimum since tracer depletion is reduced, not increased, by ligand competition. Measure the counter background (4 replicates) and subtract the mean value from the other measurements.

The following method is an extension of that described in *Protocol 15*. Once again, the aim is to minimize the number of manipulations, and hence reduce the pipetting error.

Protocol 17. Exponential dilutions

1. Dilute the receptor preparation, aiming for $R_t = \sim 0.1 \times K_d$ of the tracer ligand. If this is too low, consider using a higher concentration of receptor binding sites, and increasing the tracer concentration to $10 \times R_t$ to combat depletion.

2. Having set aside aliquots of the preparation for measurement of R_t, add the tracer ligand to the bulk receptor preparation to achieve the desired concentration. Keep the receptor preparation uniformly suspended by slow stirring.

3. Set up a series of containers (for example 20 ml scintillation vials), each containing a stirring bar. Place 3.136 ml of receptor preparation (now containing the tracer ligand) into each container, apart from the first. In the first vial, place 3.484 ml of the receptor plus tracer preparation. To this first vial, add the unlabelled competing ligand in a small volume ($< 1\%$ of the total), to obtain the desired initial (highest) concentration. Carry out a series of ten-fold serial dilutions, by pipetting 0.348 ml from the first vial into the second, mixing, pipetting 0.348 ml from the second to the third (do not forget to change the pipette tip between pipettings), etc.

4. Set up a series of tubes, six for each decade of unlabelled ligand concentration to be covered. As described in *Protocol 15*, make serial exponential dilutions to fill in the concentrations within each decade, using the tracer-supplemented bulk receptor preparation. That is, into the first vial containing the receptor, tracer, and the highest concentration of the unlabelled competing ligand, pipette 1 ml of receptor plus tracer. Mix, then pipette out 1 ml into an assay tube. Discard the pipette tip, and repeat the process until the first decade of unlabelled competing ligand concentrations is complete, i.e. until six tubes have been pipetted out. Repeat the process with the second vial, to obtain the second decade of unlabelled ligand dilutions at a constant and uniform tracer concentration. As a check, binding measured on the last tube of decade N should agree with binding measured on the first tube of decade $N + 1$.

5. Set up replicate (minimum four) assays for measurement of total (no added competing ligand; use the tracer-supplemented bulk receptor preparation for this) and non-specific binding (add the standard competitor used for measuring non-specific binding to the assay tubes, followed by the tracer-supplemented receptor preparation).

6. Transfer all the tubes to a water bath and incubate with mixing if necessary to maintain even suspension, until equilibrium has been achieved.

Protocol 17. *Continued*

This will, minimally, require $5 \times t_{1/2}$ of dissociation of the tracer ligand, and may possibly require longer, depending on the kinetics of the competing or modulating ligand.

Because the tracer ligand is added to the receptor before the competitor, this method demands full reversibility of binding. It is advisable to check that equilibrium is actually achieved, by comparisons with the results of incubations in which tracer and competitor are added simultaneously.

7. Set up replicate measurements for determination of R_t, using $100 \times K_d$ of the tracer ligand (with and without blocking inhibitor for determination of non-specific binding).

8. Process and assay the incubations. If performing a centrifugation assay, sample the supernatants for the usual check on depletion of the tracer ligand. If using another assay method, perform supplementary centrifugation assays to accomplish this aim. Measure the counter background and subtract the mean value from the other measurements.

3.6.5 Preliminary analysis and interpretation

i.

In the usual fashion, convert measurements of bound and free tracer ligand to molar concentrations, i.e. for radioligands:

$$RL^* = 10^{-9} \times B/(V \times SA \times 2220) \text{ M} = B/(V \times SA \times 2220) \text{ nM}. \quad (43)$$

ii.

Subtract non-specific from total binding to obtain specific binding. In the usual way calculate means and SEMs (SEM $= \sigma/\sqrt{r}$); $SE_{spec} = (SE_t^2 + SE_{ns}^2)^{1/2}$ for equal numbers of replicates). If tracer ligand depletion was substantial, use a plot of non-specific binding vs free ligand and interpolate it to obtain estimates of non-specific binding for the different free tracer ligand concentrations corresponding to points on the inhibition curve. If tracer ligand depletion exceeded 30%, repeat the experiment with lower R_t, or higher L_t^*. If ligand depletion exceeds 10%, correction or allowance for depletion will be necessary. If ligand depletion exceeds 50%, correction of the data will not be reliable.

iii.

As a check, estimate K for the tracer ligand from the values of R_t, and RL_0^* and L_0^* (the subscript '0' indicates a value measured in the absence of the unlabelled competitor or modulator);

$$K = RL_0^*/((R_t - RL_0^*) \cdot L_0^*). \quad (44)$$

Confirm that the value so determined agrees with your expectations based on more detailed experiments (see Section 3.5).

iv.

Plot RL^*/RL_0^* vs log A (inhibitor/modulator), see *Figures 20* and *21* for examples. RL^* is the specific binding of the tracer in the presence of competitor/modulator concentration A, and RL_0^* is specific binding of the tracer alone. The SEM of this quantity is given by:

$$SE = (RL^*/RL_0^*) \ (SE_1^2/RL^{*2} + SE_2^2/RL_0^{*2})^{\frac{1}{2}}, \tag{45}$$

where SE_1 and SE_2 are the SEMs of RL^* and RL_0^*, respectively, and the numbers of replicates are presumed to be equal. Examine the curve with the aid of the template described in Section 2.4.3. If the points fit, draw a simple Langmuir isotherm through the data. Estimate the IC_{50} value. In the case of a competitive interaction, all of the specific binding of the tracer ligand should be inhibited. In this case, an estimate of the affinity constant of the competing ligand is:

$$KA = 1/K_d A = (1 + K \cdot L^*)/IC_{50}, \tag{46}$$

where K is the affinity constant of the tracer ligand, and L^* is the free concentration of the tracer ligand corresponding to the IC_{50} value. If the interaction is positively or negatively cooperative, no such simple correction is possible.

v.

If tracer ligand depletion was greater than 10%, and if the interaction between tracer and inhibitor/modulator is known to be competitive apply the program DEPCOR (*Protocol 21*, Section 4.10), this will give corrected estimates for occupancy of the receptor site by the competitor ligand as a function of concentration. Depletion correction can only be carried out if R_t is known, or has been measured or estimated.

vi.

Alternative ways to present the data are as a Hill Plot or as a variant of the Scatchard plot.

Define $P = (1 - RL^*/RL_0^*)$. Plot $\log_{10}(P/(1 - P))$ vs $\log_{10} A$ between 10 and 90% occupancy to obtain a Hill plot of the data. If the competition curve follows the simple Langmuir isotherm (*Table 3*), the slope of the curve, n_H, should be 1.0. If the binding curve deviates from the simple Langmuir isotherm, positive or negative cooperativity or binding site heterogeneity are indicated (Section 3.5.4 and Chapter 11) depending on the nature of the deviation ($n_H > 1.0$, positive co-operativity; $n_H < 1.0$, negative co-operativity or multiple sites). Non-unitary n_H values can also arise artefactually, particularly from incorrect estimations of non-specific binding or of the initial level of binding, (RL_0^*), or from ligand depletion. Depletion of the competing ligand will appear to steepen the competition curve if the total concentration, rather than the free concentration, of ligand is used in the analysis of binding. A full account of these artefacts appears in Section 4.

Plot P/A vs P to obtain a normalized Scatchard plot. In the case of a single, non-interacting set of sites, this plot will give a straight line with slope $- KA$, where KA is the affinity constant of the competing ligand, A. In more complex cases, it may be convex upwards (real or apparent positive cooperativity) or concave upwards (real or apparent negative cooperativity, or binding site heterogeneity). Some of the artefacts which can affect Scatchard plots have been mentioned in Section 3.5.4 (*iii.*) and are further considered in Sections 4.1 and 4.8–4.10.

In the case of the competition/modulation experiment, the indirect monitoring of binding by inhibition of tracer binding presents further potential complications. These are at their most serious in their ability to distort multisite competition curves, or mask genuine cases of negative cooperativity, etc. However, in the case of a single site competitor, the effect of tracer ligand depletion is less to distort the slope of the Hill plot or the linearity of the Scatchard plot than to reduce the apparent affinity of the competing ligand. These matters are considered in detail in Section 4.10.

vii.

Non-linear, least-squares fitting of the data is once more the method of choice. Ideally, the model used should be similar to that embodied in COMPDEP (*Protocol 4*), i.e. a model which takes into account both depletion of the tracer ligand and non-specific binding. Models which also allow for depletion of the competing/modulating ligand are also possible, but no amount of sophisticated curve fitting is likely to extract good parameter estimates from a badly-designed experiment.

The analysis of binding experiments in terms of non-competitive binding mechanisms has been introduced in Section 2.2.5 and is further extended in Chapter 11. The use of the ternary complex (floating receptor) model of receptor binding, and effector interaction is also critically evaluated. Other methods of analysis, for example the use of affinity spectra, and the use of non-parametric descriptors are also briefly considered.

If depletion of the tracer and competitor ligands has been avoided a model based on the equations of *Table 3* is suitable for multiple-sites analysis.

Such a model takes the form:

$$RL^* = \sum_i \frac{Ri_t \cdot Ki \cdot L^*}{(1 + Ki \cdot L^* + KiA \cdot A)} + NS \qquad (47)$$

Again, note that this equation should be rewritten in terms of log affinity constants and log ligand concentrations before being used in fitting routines, i.e.

$$RL^* = \sum_i \frac{Ri_t \, 10^{(pi+x)}}{(1 + 10^{(pi+x)} + 10^{(piA+a)})} + NS, \qquad (48)$$

where $pi = \log_{10} (Ki)$, $piA = \log_{10} (KiA)$, $x = \log_{10} (L^*)$ and $a = \log_{10} (A)$.

It has been pointed out by Rothman (39) that the best way to determine the parameters in this and other similar models of binding is to determine inhibition/modulation curves at several considerably different concentrations of L_t^*, spanning the tracer ligand K_d ($0.1 \times K_d - 10 \times K_d$), and to pool and analyse the resulting inhibition curves simultaneously. This provides estimates of the affinity constants of the tracer as well as the competitor ligand. It also provides measures of Ri_t. This approach is efficient, in that it yields lower variances in estimates of the binding parameters KiA than does the expenditure of the same amount of effort on determining an inhibition curve at a single concentration of L_t^*. It provides a stringent test for competitive behaviour. Designing experiments to maximize the information content is discussed in Chapter 11.

viii.
The cold vs hot ligand design is a special case in that two possible routes of calculation can be followed:

(a) The affinity constant of the unlabelled ligand can be calculated from the competition curve as above. Assuming that the unlabelled and labelled ligands bind identically, the value of K so derived can then be used to calculate a series of values of R_t for different points on the competition curve.

(b) The specific activity of the ligand can be adjusted to take account of the dilution of the hot by the cold ligand, i.e.

$$SA' = SA \cdot L_t^*/(L_t + L_t^*), \qquad (49)$$

where L_t^* and L_t are the total concentrations of labelled and unlabelled ligand present. SA' can then be used to calculate the total concentration of bound ligand, as in Section 3.6.5i. The resultant curve of total ligand bound vs total ligand added is then analysed exactly as for a direct tracer saturation curve (Section 3.5).

Further consideration is given to the hot vs cold tracer design in Chapter 11.

4. Artefacts in binding studies, and their avoidance

Artefacts in binding studies fall into several categories:

(a) Artefacts attributable to intrinsic properties of the ligands, for example impurity of the tracer ligand, or competing/modulating ligands; incorrectly estimated specific activity of tracer; instability or losses during storage/handling.

(b) Artefacts intrinsic to the receptor preparation, for example receptor instability (proteolysis, denaturation); occlusion (vesicularization); contamination with endogenous ligands or modulators.

(c) Artefacts caused by interaction of the ligand(s) with the receptor preparation; for example metabolism of the tracer; metabolism of competitors/modulators.

(d) Artefacts due to inadequacies of the procedures, for example incomplete separation of bound from free ligand; loss of bound ligand during separation due to dissociation of ligand from, or physical loss of RL^* complex; non-specific binding to components of the assay system (for example filters, tubes); improper definition of background.

(e) Artefacts due to inadequate experimental design, for example ligand depletion; failure to attain equilibrium.

(f) Artefacts due to inadequate analysis.

4.1 Purity of the tracer ligand

Munson (36, 37) has considered the binding artefacts which may originate from the unwitting use of a tracer ligand which is impure, so that only a proportion of it is capable of undergoing receptor binding (or is 'bindable').
Two situations arise:

(a) The tracer ligand and impurities are labelled with tritium or iodine, etc. in such a way that one or more atoms are completely replaced by the radioisotope. In this case, radiochemical and chemical purity are in effect synonymous. Low purity does not affect the specific activity of the bindable fraction of ligand in this case.

(b) The tracer is only partially labelled, i.e. labelled molecules are in effect diluted with unlabelled ligand: there are now three possibilities:

    i. The tracer is pure, but the unlabelled ligand impure. In this case, the specific activity will be underestimated, but the bindability of the tracer will be 100%.

    ii. The tracer is impure, but the unlabelled ligand pure. In this case, specific activity will be overestimated, and bindability will be less than 100%

    iii. Both tracer and unlabelled ligand are impure. The specific activity may be wrong, and bindability less than 100%.

In the following discussion, we look at artefacts caused by the unsuspected presence of unbindable ligand. We assume the specific activity to be correct.

We consider two types of experiment:

(a) *The direct ligand saturation experiment*, designed to estimate the affinity constant K and the total concentration of binding sites R_t for the tracer ligand alone. In such an experiment, R_t will be correctly estimated, but K will be underestimated (K_d overestimated), provided that other artefacts, for example those due to tracer ligand depletion, do not make the data uninterpretable.

(b) *The cold vs hot tracer design* in which added unlabelled ligand is used to inhibit the binding of tracer ligand, and the results recalculated to provide estimates of bound vs free (see Section 3.6.5). In this case, if the unlabelled ligand is pure, but the tracer is not (the most likely situation), K will be correctly estimated, but R_t underestimated. If tracer and unlabelled ligand are impure, K and R_t will both be wrongly determined. If tracer and unlabelled ligands are both pure, but have different affinities for the receptor there will be artefactual curvature in the calculated Scatchard plot. Both K and R_t may be wrongly estimated in this case.

Under non-depletion conditions, and provided that the concentration of tracer is low relative to K_d, the presence of unbindable ligand may exert minimum effects on the inhibition curves given by unlabelled ligands displacing or inhibiting tracer ligand binding. However, attempts to work at higher L^* concentrations, or under depletion conditions will lead to errors in correcting the binding curves because the true free concentration of the tracer ligand will be overestimated, thus invalidating both shift and depletion corrections. Other errors may arise if impurities in the tracer bind, either to the receptor, with modified affinity, or to non-receptor sites, which show different ligand specificity.

Some examples of insidious artefacts of this nature have been considered by Burgisser and his colleagues (8, 40). These concern the use of racemic radioligands, where one optical isomer binds much more strongly than the other. This can lead to spurious biphasic association and dissociation kinetics, in which the fast component is due to the low affinity isomer and the slow component to the high affinity isomer. Changes in the relative proportions of binding of the isomers occur even after total binding has apparently equilibrated, simulating some of the characteristics of post-binding conformational changes (8). The undetected use of an unresolved racemic radioligand can lead to widely deviant estimates of the apparent affinity constant from direct ligand saturation experiments, and to significant (up to 5-fold) underestimation of the affinities of competitor ligands from inhibition experiments (40).

The artefacts can be circumvented, to some degree, provided that one is aware of the existence of isomers. One way to test this is to measure ligand bindability (36, 37). However, *the optimum solution is to use chemically and radiochemically homogeneous, pure enantiomers, wherever possible.*

4.1.1 Storage and dispensing of ligands

Always follow the manufacturer's instructions for storage (Chapter 1). Do not store dilutions of radioligands unless you know them to be stable. When dispensing from stock solutions of tracer ligands stored at low temperature, allow the bottle to warm to ambient temperature before removing the lid. This minimizes condensation of water vapour in the ligand preparation, which may promote ligand hydrolysis.

4.1.2 Checking ligand purity

Check the tracer ligand purity, by TLC or HPLC, following the manufacturer's instructions (Chapter 1). These methods can also be used on unlabelled ligands, provided that an adequate detection protocol is available. Amersham and NEN both provide data sheets with their products, containing detailed methods.

4.1.3 Measuring bindability

Information on the purity of tracer ligands (particularly radioligands) can be obtained by direct measurement of their 'bindability', i.e. that fraction of the ligand which is capable of being specifically bound by receptor.

In order for this to be possible, it is necessary to find conditions under which a significant proportion of the ligand becomes receptor-bound, i.e. for which $R_t > K_d$. Thus, under normal circumstances, bindability is likely to be measurable only for ligands with $K_d < 10^{-9}$ M, because of the problems associated with obtaining high enough concentrations of binding sites. In addition, the coefficient of non-specific binding should not be greater than ~0.01, nor should the receptor specific activity be much less than 1 pmol/mg protein, so that no more than 1% of tracer ligand is non-specifically bound at a receptor concentration of 10^{-9} M.

Provide that these conditions are fulfilled, bindability may be assessed as follows.

(a) Choose a concentration of tracer ligand approximately equal to K_d.

(b) Measure binding as a function of receptor concentration for $R_t = 0.1 \times K_d$ to $R_t = 10 \times K_d$.

(c) Plot *total binding* vs R_t. If bindability is high, this plot will curve off sharply in the vicinity of $R_t = K_d$ (see *Figure 16*). If bindability is lower, the curvature will be less apparent, owing to non-specific binding of the inactive impurities. The form of the expected curve can be simulated using the program BINDAB (*Protocol 17*). The relationships used in this program can also be employed to fit the experimental data, so deriving an optimized value for the ligand bindability. It is possible to measure the coefficient of non-specific binding separately and incorporate the value so derived in the fitted function as a fixed parameter. In the absence of

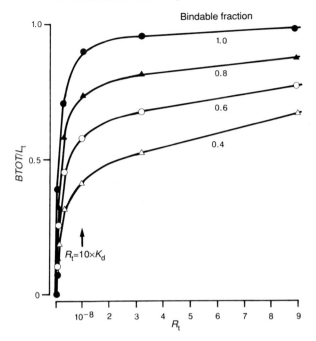

Figure 16. Titration of a radioligand with receptor to determine bindable fraction of ligand. 10^{-9} M of a ligand with affinity 10^9/M was titrated with increasing concentrations of a receptor, specific activity 1 pmol/mg protein. Non-specific binding was 0.01/(mg protein/ml). Total binding was plotted as a fraction of total ligand concentration $BTOT/L_t$ for different bindable fractions as shown. Data simulated using program BINDAB (*Protocol 17*). The arrow marks the concentration of receptor which is 10 times the K_d value.

curve-fitting facilities, the ratio of total binding to total ligand at $R_t = 10 \times K_d$ gives a crude measure of the percentage of the tracer ligand which is capable of receptor binding. A better estimate can be obtained by plotting the reciprocal of the fraction of ligand bound against the reciprocal of the receptor concentration, R_t. For values of $R_t > K_d$, this plot approximates to a straight line, with a *y*-intercept corresponding to the bindable fraction of ligand (41).

Protocol 18. BINDAB program

```
30 REM THIS PROGRAM CALCULATES THE EFFECT ON BINDING OF A LIGAND*
40 REM OF THE EXISTENCE OF A BINDABLE FRACTION ALPHA. NONSPECIFIC BINDING
50 REM OF BOTH BINDABLE AND NONBINDABLE LIGAND IS TAKEN INTO ACCOUNT.
60 REM THE CONCENTRATION OF RECEPTOR RT IS VARIED AT CONSTANT L*T.
65 REM THE CALCULATION IS A VARIANT OF THAT OF PROTOCOL 3
```

Protocol 18. *Continued*

```
 70 INPUT "ASSOCIATION CONSTANT OF LIGAND, K",K
 80 INPUT "TOTAL LIGAND CONCENTRATION L*T",LT
 90 INPUT "TOTAL BINDABLE FRACTION OF LIGAND",ALPHA
100 LTB=ALPHA*LT
110 INPUT "COEFFICIENT OF NONSPECIFIC BINDING,NS", NS
120 INPUT "RECEPTOR SPECIFIC ACTIVITY (PMOL/MG)",S
130 REM CALCULATE PROPORTIONALITY BETWEEN NS BINDING AND RT
140 RO = NS*1E9/S
150 INPUT "INITIAL AND FINAL CONCENTRATIONS OF RT",RTLO,RTHI
160 INPUT "NUMBER OF CONCENTRATIONS PER DECADE",N
170 M=10^(1/N)
180 RT=RTLO
190 PRINT " RT "," L*"," RL*"," BTOT"," BTOT/LT"
200 REPEAT
210 MU=RO*RT
220 C1=LTB+RT+(1+MU)/K
230 C2=RT*LTB
240 ROOT=SQR(C1*C1-4*C2)
250 RL=(C1-ROOT)/2
260 L=(LTB-RL)/(1+MU)
270 NSNB=MU*(1-ALPHA)*LT/(1+MU)
280 BTOT=RL+MU*L+NSNB
285 REM PRINTER CONTROL CHARACTERS FOLLOW. MODIFY AS APPROPRIATE.
290 §% =&0001030A
295 REM NOTE THAT L IS FREE BINDABLE LIGAND. RL SPECIFIC RECEPTOR-LIGAND
296 REM COMPLEX. BTOT IS TOTAL BINDING INCLUDING NS BINDING OF BINDABLE
297 REM AND UNBINDABLE LIGAND.
300 PRINT RT,L,RL,BTOT,BTOT/LT
310 RT=RT*M
320 UNTIL RT>RTHI
330 INPUT "NEW VALUE OF K,Y/N",A$
340 IF A$="Y" THEN GOTO 70 ELSE 350
350 INPUT "NEW VALUES OF L*T Y/N",A$
360 IF A$="Y" THEN GOTO 80 ELSE 370
370 INPUT "NEW BINDABLE FRACTION, Y/N",A$
380 IF A$ = "Y" THEN GOTO 90 ELSE 390
390 INPUT "NEW COEFF OF NON-SPECIFIC BINDING Y/N",A$
400 IF A$="Y" THEN GOTO 110 ELSE 410
410 INPUT "CHANGE INITIAL AND FINAL RECEPTOR CONCNS Y/N",A$
420 IF A$="Y" THEN GOTO 150 ELSE 430
430 END
```

The above procedure has proved valuable in detecting the presence of enantiomers of radioligands. An example is shown in *Figure 17*. Here, a fixed concentration of the muscarinic antagonist [³H]-telenzepine was titrated with increasing concentrations of brain muscarinic receptors (42). The resultant binding curve was well fitted on the assumption that the fraction of the ligand capable of being bound is 0.5. The precise form of the curve is sensitive to the affinity constant of the tracer ligand. However, variations in this parameter can be overcome to a large extent by the introduction of a compensatory scaling factor on the R_t axis, i.e. K and R_t are strongly correlated. The origins of correlations between binding parameters, which are a feature of the analysis of binding experiments, are discussed in Chapter 11 Section 5.2.3. The use of the reciprocal plot to determine bindability is shown in *Figure 17b*.

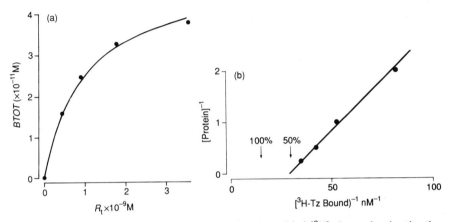

Figure 17. (a) Determination of the bindable fraction of $(\pm)$-[^{3}H]telenzepine by titration with a preparation of muscarinic receptors. Total concentration of ligand was 9×10^{-11} M. Ligand affinity was 1.2×10^9/M. Receptor specific activity was 1 pmol/mg protein, and non-specific binding was 0.02/(mg protein/ml). The curve was fitted on the basis of a bindability of 0.5. (b) Use of reciprocal plot to estimate bindable fraction. The arrows mark the expected intercepts on the *x*-axis if 50% or 100% of the [^{3}H]telenzepine were bindable.

Repeated exposure to the receptor preparation has also been used to deplete an enantiomeric radioligand of the more strongly bound isomer, thus allowing the properties of the more weakly bound enantiomer to be studied independently (40, 42).

4.1.4 Tight-binding radiochemical impurities

Another hazard is the possible accumulation of tight-binding radiochemical impurities (43, 44). If such binding occurs to elements of the assay system, for example filters or assay tubes, pre-exposure of the radioactive ligand to such components may help to eliminate the problem. Thus prefiltration of the ligand solution, or making the dilutions in the microfuge tubes of the type used in the assays, may help to reduce non-specific binding. Repurification of the ligand will often help. The ligand specificity of impurity binding is usually distinguishable from that of genuine receptor-specific binding.

4.2 Specific activity of the tracer ligand

It is crucial to know the specific activity of a radioligand in order to determine both bound and free ligand concentrations accurately. Again Munson has discussed the artefacts which wrongly-determined specific activity may introduce into binding curves (36, 37). Thus, in a direct tracer ligand binding experiment, an incorrectly-determined specific activity will influence the estimation of both R_t and K. If specific activity is overestimated, K will be overestimated (K_d underestimated), and R_t underestimated. The converse will occur if specific activity is underestimated. In the cold vs hot tracer design, in which binding of a low concentration of tracer is inhibited by a series of increasing concentrations of unlabelled tracer ligand, the estimation of K and R_t may be relatively unaffected, provided that the tracer is chemically and radiochemically pure, and 100% bindable, and that the labelled and unlabelled tracer have identical binding properties.

Likewise, estimation of the affinities of unlabelled ligands by competition will not be greatly influenced, provided that depletion conditions are avoided.

Radioactive decay may alter the quoted specific activity of a tracer ligand, in addition to leading to the accumulation of radiolysis products potentially capable of interfering with the assay. If the ligand is not completely labelled, i.e. only a proportion of the molecules contains radioactive atoms, it is theoretically possible to calculate a corrected specific activity using the quoted specific activity on the reference date, and the half-life of the radioisotope.

$$\text{Thus Sp.Act.}(t) = \frac{L^* \cdot e^{-(\log_e(2) \cdot t/t_{1/2})}}{L + L^* \cdot e^{-(\log_e(2) \cdot t/t_{1/2})}} \cdot \text{Sp.Act.}(0), \tag{50}$$

where L^* and L are the initial concentrations of radiolabelled and unlabelled ligands at time zero, and $t_{1/2}$ is the half-life of the radioisotope concerned (see Chapter 1).

There are two basic ways to check the specific activity of a radioligand.

(a) Estimate the K_d by direct binding and by cold vs hot competition under non-depletion conditions ($R_t < 0.1\, K_d$). If the given specific activity is correct, these values should be identical. This assumes that the labelled and unlabelled ligands manifest the same receptor affinity which is usually true for ^{3}H-labelled ligands. In the case of ^{125}I-labelled ligands, it may be necessary to obtain and use the unlabelled, iodinated ligand (see Chapter 1). It is also assumed that the unlabelled ligand is chemically pure. Provided that these conditions are fulfilled, a corrected estimate of the specific activity of the tracer can be obtained by multiplying the quoted value by the expression $(K_d^*/IC_{50}) \cdot (1 + L^*/K_d^*)$, where K_d^* is the apparent K_d measured in the direct binding experiment, calculated using the quoted value of the tracer specific activity, and L^* is the free tracer concentration at the IC_{50} in the cold vs hot experiment, again calculated using the quoted specific activity. This formula applies only if the cold vs hot inhibition study was conducted under non-depletion conditions.

(b) Measure R_t using several different tracer ligands which are known to bind to the same receptor population. Ideally, the values of R_t so obtained should be consistent with one another. Of course, one cannot exclude the possibility that the specific activities of all the ligands so studied may be wrong! Another possibility is that part of the receptor population may be occluded and thus unavailable to highly polar ligands, but not to less polar congeners. To avoid this possibility, use lysed membrane preparations, or better still a solubilized or purified receptor preparation to compare different tracer ligands.

There have been a number of examples of differences in R_t measured on whole cells, or membrane preparations using membrane permeant and impermeant ligands (45, 46). The possibility that such apparent differences are due to incorrect specific activities rather than to genuine differences in accessibility of receptor pools always needs to be eliminated. The determination of ligand specific activity is not a trivial task, and it is not always possible to assume that the manufacturer's quoted specific activity is correct.

4.3 Handling ligands

An obvious problem which may arise is cumulative losses of ligands due adsorption on to tubes or pipette tips. In the case of the tracer ligand, this usually becomes obvious. If in doubt check by performing a series of transfers from tube to tube, measuring the tracer concentration after each one. There is usually only a limited number of high-affinity binding sites for a given ligand on a tube. For this reason, the problem is at its most severe at low ligand concentrations.

The occurrence of handling losses causes more insidious problems when it applies to unlabelled ligands. For instance, severe adsorption of low, but not high, concentrations of a competing ligand on to the tube during the preparation of dilutions prior to performing the competition experiment will lead to overestimation of the true concentrations present in the assays at the low, but not at the high, end of the competition curve. Like other forms of depletion artefact, the effects of this will be to steepen the competition curve. *When a Hill coefficient greater than 1.0 is obtained from a competition study, the occurrence of competitor adsorption should be suspected, and needs to be eliminated as a possibility before the result can be believed.*

A number of stratagems can be adopted to minimize the problems of ligand handling.

(a) In the case of hydrophilic ligands (for example quaternary amines) problems are usually minimal. However, it is advisable to maintain a minimum buffer ionic strength, for example 50 mosmolar.

(b) In the case of hydrophobic ligands, adsorption problems are likely to be more severe. In bad cases, consider diluting the ligand in 50% ethanol.

(c) Adsorption problems are frequently encountered with peptide ligands. It is often possible to ameliorate them by the inclusion of 100 μg/ml bovine serum albumin in the buffer, and by the maintenance of a reasonable ionic strength (for example 100 mM NaCl).

(d) If the handling problems persist, consider siliconizing the tubes used for dilution and assay. This is effective in reducing the adsorption of even quite hydrophobic peptides to Eppendorf tubes. Make up a 5% solution of dichlorodimethylsilane in chloroform (use a fume hood). Fill each tube and stand for 10 minutes. Pour out the siliconizing solution (it can be reused several times) and suck out the last traces. Wash the tubes out with deionized water and then heat them in a 60°C oven for 30 minutes. Store the siliconized tubes in a closed container.

(e) An obvious precaution is to keep ligand concentrations as high as possible during the dilution process. It is usually possible to keep the concentrations 100 times those used in the final assays. It is best to make, and use ligand dilutions fresh, and not to store them for prolonged periods.

(f) Certain compounds, for instance esters such as acetylcholine, hydrolyse quite rapidly in aqueous solution. Keep solutions of such ligands at neutral pH on ice and use them as rapidly as possible. Other ligands, such as catecholamines and 5-HT are readily oxidized, particularly under basic conditions. It is preferable to dilute such ligands under slightly acidic conditions (pH 5). 0.1% Ascorbic acid can be added to provide protection against oxidation. *Note*, however, that because of its reducing properties, ascorbate has occasionally been reported to affect receptor binding properties.

4.4 Receptor instability

The effect of receptor instability is to cause premature cut-off of the binding reaction. This can lead to a serious underestimation of the level of equilibrium binding. The error may vary with the level of ligand occupancy, since high occupancy may actually protect the receptor against degradation. This can lead to apparent positive cooperativity of binding even though none exists. In general, loss of more than 10% of the receptor binding sites within the equilibration period is capable of leading to significant errors in measured binding curves.

The major source of instability of cell surface receptors, in membrane preparations incubated in the usual aqueous buffers at near-neutral pH, is endogenous proteolytic action. Proteolysis also contributes to the instability of detergent-solubilized receptors, and its control is essential for successful purification of receptors. However, the adequacy with which a detergent simulates or preserves the intramembrane receptor–lipid or receptor–protein interactions is usually the dominant factor affecting receptor stability in solution. Detergent solubilization is discussed in detail in reference 7.

Protocol 19. To assess the rate of receptor degradation

1. Dilute the receptor preparation to the desired extent.
2. Split it into two. Preincubate one portion without and one portion with the unlabelled ligand used to measure non-specific binding (use $1000 \times K_d$ concentration)
3. Incubate the preparation at the assay temperature. At intervals, withdraw aliquots and add them to tubes containing a $10 \times K_d$ concentration of tracer ligand. Use separate pipettes for the specific and non-specific binding incubations, to avoid cross-contamination.
4. Incubate the assays for sufficient time to attain equilibrium (Section 3.2.4). Process, and calculate R_t. Plot R_t vs time.

If desired, this experiment can be repeated using receptor prelabelled with tracer. Ligand occupancy often provides some protection against proteolytic degradation of the binding sites. However, although antagonists usually protect, agonists have been reported to accelerate the proteolysis of muscarinic receptors (47), possibly reflecting an induced conformational change. Of course, this may also be used to detect such a conformational change. High affinity agonist binding, which is dependent on the maintenance of receptor–effector coupling (for example receptor–G-protein coupling) is often more sensitive to endogenous proteolysis than antagonist binding, which is usually independent of receptor–effector interactions. However, it should be noted that the preservation of binding activity does not exclude the possibility of proteolytic nicking of parts of the molecule unconcerned with binding (48). A good test for the absence of such partial proteolysis is to label the receptor with an irreversible binding-site-specific ligand after preincubation, and analyse it by gel electrophoresis after a suitable period of incubation (17, 49).

If (proteolytic) instability is detected, there are several possible courses of action.

(a) Try diluting the receptor preparation, to dilute the endogenous proteases.
(b) Try altering the conditions of the assay, for instance, reducing the temperature. Proteolysis and denaturation are much slower at 4°C than at 30°C. However, ligand binding kinetics are also slower. We have found

that the use of an intermediate temperature, for example 15°C represents a good compromise between stability and kinetics when working with solubilized muscarinic receptors.

(c) Try changing the protocol for membrane preparation, for example by the introduction of further washing steps, to remove endogenous proteases.

(d) Try using proteolytic inhibitors.

(e) It may be possible to exploit ligand protection and measure binding curves at relatively high ligand occupancy, thereby protecting the receptor.

A list of proteolysis inhibitors is given in *Table 9*. It deserves a number of comments:

(a) The most serious source of proteolytic instability of receptors in membranes, and particularly after solubilization, seems to be the action of metalloproteases and calcium-activated proteases. When preparing membrane prior to solubilization of receptors, it is always advisable to include a step involving washing with 10 mM EDTA (Chapter 5). 1 mM EDTA should be included in buffers used subsequently. EDTA and/or EGTA-pretreatment can also be useful in promoting stability of receptors in membranes destined for binding studies. However, note that peripheral membrane proteins bound to membranes by Ca^{2+}-mediated interactions may be removed. So may subpopulations of GTP-binding proteins, although an EDTA wash followed by replacement of divalent cations (for example Mg) often enhances GTP effects on binding. Workers in the field are nearly unanimous in the opinion that chelating agents improve receptor stability (cf. 47).

(b) Of the agents listed in *Table 9*, PMSF is probably the second most useful. We use it at a final concentration of 0.5 mM. PMSF should be dissolved in a minimum volume of absolute ethanol or isopropanol. It is very soluble, so add ethanol dropwise with shaking until the PMSF just dissolves. Add the PMSF solution to the ice-cold receptor preparation, or buffer, with constant stirring. Usually, there is momentary precipitation, followed by redissolution. PMSF hydrolyses slowly in aqueous solution, and may need to be renewed after some hours. PMSF hydrolysis yields F^- ions.

(c) In our hands, pepstatin (0.5 μg/ml) has a marginal protective effect on solubilized brain muscarinic receptors.

(d) Bacitracin (100 μg/ml) has been reported to protect cardiac muscarinic receptors in membrane preparations, particularly in combination with EDTA pretreatment (50).

(e) The addition of aprotinin (10 μg/ml) seems a wise precaution in tissues such as pancreas, in which there is a high concentration of kallikrein. There are kallikrein cleavage sites in the C-terminal regions of several of the G-coupled receptors.

Table 9. Protease inhibitors

Inhibitor	Working concn	Protease group
Amastatin	1–10 μg/ml	aminoexopeptidases
Antipain	1–10 μg/ml	trypsin, papain, cathepsin B
Aprotinin	1–10 μg/ml	kallikrein
Bacitracin	100 μg/ml	effective non-specific inhibitor
Benzamidine	up to 10 mM	serine proteases
Benzethonium chloride	0.1 mM	bacteriostatic
Benzylmalic acid	1–10 μg/ml	carboxypeptidases
Bestatin	1 μg/ml	aminopeptidase B, leucine aminopeptidase
Chymostatin	1–10 μg/ml	chymotrypsin, papain
Di-isopropyl fluorophosphate	0.1 mM	serine proteases
Diprotin A+B	10–50 μg/ml	dipeptidyl aminopeptidase
Elastatinal	10 μg/ml	elastases
EDTA	1–10 mM	metalloproteases, Ca-dependent SH proteases
EGTA	1–10 mM	metalloproteases, Ca-dependent SH proteases
Iodoacetate	1–10 mM	SH proteases
Iodoacetamide	1–10 mM	SH proteases
Leupeptin	1–10 μg/ml	serine and SH proteases
Pepstatin A	1–10 μg/ml	aspartic proteases
N-ethyl maleimide	1 mM	SH proteases
Phenylmethyl sulphonyl fluoride	0.1–1 mM	serine proteases
Phosphoramidon	1–10 μg/ml	thermolysin (Zn proteases)
Sodium tetrathionite	5 mM	SH proteases
TPCK	10 mM	chymotrypsin
TLCK	10 mM	chymotrypsin
Trypsin inhibitor	500 μg/ml	trypsin
2-phenanthroline	1–10 mM	metalloproteases, Ca-dependent SH proteases

(f) The use of sulphydryl reagents (iodoacetate, iodoacetamide, N-ethyl maleimide) has been reported to be effective in preserving the morphology of slices of brain tissue, but should be avoided if possible in receptor-binding studies owing to the presence of reactive SH groups in receptors which may become modified.

(g) A high-salt wash may also aid in the removal of peripheral membrane proteins, including proteases. The use of KCl/pyrophosphate treatment in the preparation of a cardiac membrane fraction is described in Chapter 5.

For binding studies, the best tactics are to try in order:

- to use as dilute a membrane preparation as possible
- to use reduced temperatures
- to use EDTA-washed membranes
- to add bacitracin, $100 \mu g/ml$
- to add PMSF, 0.5 mM
- to explore the use of other protease inhibitors

For purification of rat brain muscarinic receptors, we have found that a combination of:

- chymostatin   5 $\mu g/ml$
- leupeptin   5 $\mu g/ml$
- antipain   5 $\mu g/ml$
- pepstatin   0.5 $\mu g/ml$
- EDTA   1 mM
- PMSF   0.5 mM

provides complete protection of the protein. The protease inhibitors (except PMSF), are stored as a 100× concentrated 'solution' at −20°C, and added to the buffers as required. EDTA is included routinely in the buffers. PMSF is made up in ethanol and used fresh. The inclusion of EDTA is essential to prevent degradation of the receptor polypeptide.

4.5 Receptor occlusion

Receptor occlusion is not an artefact, as such, but is better regarded as an intrinsic property of the receptor preparation. In whole cells, occluded (as opposed to cell surface) receptors may be present in vesicles containing newly-synthesized or recycled receptors en-route to the cell surface or in vesicles in transit to the lysosomal compartment. In membrane preparations, however, vesiculation may also be created artefactually by vigorous homo-genization (for example by use of polytron, or gas cavitation, ref. 29).

If there is a significant occluded receptor pool, the consequence may be that extremely hydrophilic tracer ligands (for example quaternary amines) which cannot readily cross hydrophobic barriers detect a smaller total recep-tor concentration than more hydrophobic ligands, which can cross such bar-riers, i.e. R_t measured with a hydrophilic ligand is smaller than R_t measured with a hydrophobic ligand. If such a difference is to be accepted as authentic, it should be shown that the disruption of the vesicle membrane, for example as a result of detergent treatment, or hypotonic lysis, equalizes the apparent values of R_t. If not, one may suspect that the apparent difference is merely due to an inaccurate tracer specific activity.

Detergent permeabilization may be achieved by the addition of a sub-solubilizing concentration of saponin, digitonin, or other detergents. Effective concentrations are around 0.01–0.1% at protein concentrations of 5 mg/ml.

Hypo-osmotic lysis is produced by sudden dilution of vesicles, in an iso-osmotic buffer (for example 0.32 M sucrose) into a low ionic strength, low osmolarity buffer (for example 10 mM Tris–Cl), and incubation on ice for 30 min. This is often supplemented by gentle mechanical shearing, for example Polytron setting 1, Potter–Elvehjem homogenization (0.1 mm clearance), or less commonly by snap-freezing (liquid N_2 or isopropanol/dry ice) and thawing. Examples of the use of hydrophilic ligands to measure the disappearance of receptors from cell surfaces are given in references 45, 46, and 51.

Penetration of ligands into sealed vesicles will also affect the kinetics of ligand binding, and is a possibility that must be taken into account when multiphasic kinetics are encountered. An effect of this kind should be manifested on the association as well as the dissociation rate constant.

It is also theoretically possible that ligand uptake into sealed vesicles might affect equilibrium binding as a result of a local concentration increase. Possibilities of this kind have been discussed by Laduron (16) as a possible explanation of heterogeneity (the appearance of multiple sites) in ligand binding curves. It is not clear that authentic examples of such phenomena are known. However, a more serious risk is that uptake of relatively non-polar ligands either into whole cells, or into vesicles in membrane preparations, may either engender high levels of non-specific binding, or even be capable of mimicking, and thus being confused with, receptor-specific binding.

4.6 Contamination with endogenous ligands

If the concentration of endogenous ligand for a given receptor in the tissues of interest is high, sufficient endogenous ligand may persist in the final tissue preparation to give substantial interference with tracer ligand binding. A test of this is to take some supernatant from the tissue preparation and assay it for inhibition of binding of a low concentration of the tracer (for example 0.1 × K_d). The presence of endogenous ligand may also be suspected if a plot of tracer binding (again with a low concentration of the tracer) vs protein concentration in the assay deviates markedly from linearity for $R_t < K_d$. Another obvious test is to deliberately add an enzyme, such as acetylcholinesterase or adenosine deaminase, which is capable of metabolizing endogenous transmitters, and to look for effects on binding which may result.

One of the best-known examples of interference by endogenous transmitter is provided by the $GABA_A$ receptor, where extreme precautions must be taken to wash brain membrane preparations if the binding of GABA agonists is to be measured. This procedure is given overleaf.

Protocol 20. Washing brain membranes to remove endogenous GABA (adapted from ref. 52)

1. Homogenize brains in ice-cold 0.32 M sucrose at 1:10 weight:volume using a motor-driven Potter–Elvehjem homogenizer.
2. Centrifuge at 100 g for 10 min. Decant and keep supernatant.
3. Centrifuge S/N at 45 000 g for 45 min. Keep pellets.
4. Resuspend pellets in deionized water to their original volume. Homogenize with an Ultra-Turrax or Polytron at setting 5 for 10 sec. Centrifuge at 150 000 g for 30 min.
5. Repeat step 4.
6. Repeat step 4 twice more using 10 mM potassium phosphate buffer, pH 7.4, containing 100 mM KCl.
7. Use the membrane preparation for binding studies immediately.

Interference with muscarinic receptor-binding assays by endogenous acetylcholine is not usually a problem, because of the presence of high levels of acetylcholinesterase in most tissues. However, if AChE inhibitors are added, inhibition by endogenous ACh can become perceptible, particularly when studying high-affinity agonist binding to muscarinic receptors.

4.6.1 'Locked agonist'

A more subtle problem is provided by the phenomenon of agonist 'locking'. This has been described, particularly for β-adrenergic receptors, and may be rather a common phenomenon, applying also to other G-protein coupled neurotransmitter and hormone receptors, and to receptors such as the chemotactic receptor of neutrophils. The basic observation is that under certain conditions, agonist binding is only very slowly reversible ('locked'; ref. 53). This can lead to the artefactual depression of R_t, i.e. to apparent blockade of the receptors by incubation with the agonist. The agonist may be present in the membrane preparation, so that part of the receptor population is isolated in a 'locked' state. In the case of G-protein-coupled receptors, this phenomenon can be reversed by treatment with a low concentration of GTP (10^{-4}–10^{-3} M) which is often subsequently hydrolysed by endogenous GTPase activity. This greatly (~100-fold) accelerates the release of bound agonist from the locked state. The effect is apparently to increase R_t, the total concentration of receptor sites. The occurrence of such phenomena is probably responsible for a number of the reported effects of guanine nucleotides on antagonist binding to G-coupled receptors (54). However, locking cannot account for effects of GTP on K rather than R_t. Locking can also be reversed, although more slowly, by treatment of the membranes with 10 mM EDTA, since the phenomenon appears to require the presence of divalent cations, particularly Mg^{2+}.

Prewashing the membranes with EDTA, therefore, militates against locking artefacts. It is worth noting that the treatment of the β-adrenergic receptor–isoproterenol complex with the sulphydryl reagent NEM induces locking. This provides another reason for avoiding the use of sulphydryl reagents to attempt to control endogenous proteases.

4.7 Instability of the tracer or other added ligands under incubation conditions.

The stability of the tracer ligand should be checked by TLC or HPLC, as described by the manufacturers, at the end of the incubation period. In addition, it is possible to separate the free from the bound ligand (for example by centrifugation if using a membrane preparation) and check its binding to a fresh aliquot of receptor preparation. Further, it is desirable to check that the bound ligand pool is unchanged, by extracting the ligand and assaying its purity as above.

Complete stability of the tracer and added competing or modulating ligands is obviously the ideal. However, successful binding studies can be carried out utilizing metabolizable ligands provided that suitable inhibitors of breakdown can be found. In addition, obvious precautions are to wash crude membrane preparations thoroughly to remove membrane-associated enzymes, to use receptor concentrations which are as low as possible, and to use low rather than higher incubation temperatures.

A selection of additives which have been used to protect ligands in binding studies is given in *Table 10*. The shortness of this list perhaps reflects a tendency for workers to neglect the possibility of tracer metabolism. However, artefacts caused by the processing of the tracer to products which may or may not bind to the receptor preparation with altered specificity are potentially serious, and include the full range of 'bindability' artefacts (Section 4.1.3). In approaching this problem, it is advisable to titrate the enzyme activity to be inhibited to determine the minimum inhibitor concentration which is effective. In this way, interference with ligand binding to the receptor should be minimized.

4.8 Separation artefacts

The main separation artefacts are:

- incomplete recovery of bound ligand
- contamination of bound with free

Munson (36, 37) has discussed the effects of these artefacts on the interpretation of binding studies, again illustrating them with reference to the direct tracer ligand saturation experiment, and the hot vs cold design. In general, incomplete recovery of the receptor–ligand complex will lead to underestimation of R_t without affecting the estimation of K, whilst contamination of the

Table 10. Some agents which protect metabolizable ligands

Ligand	Inhibitor	Target	Concn	Comment	Reference
Acetylcholine	DFP	AChE[a]	200 µM	make up fresh in water	55
	neostigmine	AChE	1 µM	may inhibit mAChR binding	55
	phospholine	AChE	0.5 mM		55
Monoamines (catecholamines 5-HT)	pargyline	MAO[b]	1–100 µM	check to ensure no inhibition of	56
	nialamide	MAO	10 µM	receptor binding; wash out before assay	57
	ascorbic acid		5 mM	antioxidant; has been reported to affect receptor binding	58
	pyrocatechol		1 mM	protecting agent	58
	desipramine	Monoamine uptake	10 µM	can interact strongly with receptors	59
	imipramine		10 µM		
Peptides	protease inhibitors	Endo- and exo-peptidases		see *Table 9*	
	bacitracin		100 µg/ml	see *Table 9*	60
	BSA[c]		100 µg/ml	protecting agent	
Amino acids (GABA/glutamate)	Na-free media	Na-dependent aa[d] uptake		prevents binding to uptake sites	52
Cyclic Nucleotides	IBMX[e]/ theophylline	phosphodiesterases	1 mM		
Histamine	SKF91488	histamine-N-methyltransferase	10 µM		61
GTP	AppNHp[f]	ATPases	100 µM	inhibits NTP hydrolysis	

[a] Acetylcholine esterase; [b] monoamine oxidase; [c] bovine serum albumin; [d] amino acid; [e] isobutylmethylxanthine; [f] adenylyl imidodiphosphate.

bound fraction by free ligand may lead to the detection of an illusory population of apparently low-affinity high-binding capacity sites.

4.8.1 Testing for separation artefacts

(a) Alter the separation conditions, for example check microfuge-based assays using an ultracentrifuge to ensure complete recovery of membrane fragments.

(b) Loss of bound ligand from a filter can either be caused by physical loss of receptor or dissociation of the bound ligand as a result of a half-time for dissociation which is small relative to the washing time (see Chapter 6). This can be checked by comparing filtration assays with centrifugation assays. The latter often permit better recovery of membranes with minimum disturbance of the binding equilibrium at the expense of a poorer specific:non-specific binding ratio.

(c) Dissociation of the bound ligand may be partly compensated for by altering the time needed to do the separation, and then extrapolating back to time zero. If the dissociation rate constant and the separation time are known, a correction of the form $RL_0^* = RL^*/(e^{-k_{21}t})$ should be possible. Rapid cooling by washing the filter with ice-cold buffer usually slows the dissociation time course dramatically. Also, supplementing the washing buffer with a high concentration of multivalent cations may be effective in trapping the RL^* complex. For instance, 10 mM gallamine triethiodide virtually abolishes the dissociation of methylscopolamine from some subtypes of muscarinic receptors (9).

(d) If the definition of non-specific binding is adequate, free ligand contamination of the bound ligand fraction should not be misinterpreted. However, if non-specific binding has been defined by displacement with excess unlabelled tracer, displaceable binding to filters or assay tubes may be confused with receptor binding.

A particular problem may be binding of accumulated radiochemical impurities to tubes or filters. Such impurities may bind with higher affinity than the tracer ligand itself. One way to circumvent this problem is to pre-expose the tracer ligand to the tubes or filters to remove such impurities, i.e. to dilute the tracer ligand in microcentrifuge tubes, if performing centrifugation assays, or to pass the tracer ligand solution through a filter, if performing filtration assays. Displaceable non-specific binding of ligands to components of the assay system, such as filters is a commonly-reported phenomenon. Classic examples of the binding of insulin to glass tubes, and of glucagon to cellulose acetate filters are given by Cuatrecasas and Hollenberg (1). We have noted the binding of relatively high $(10^{-8} M)$ concentrations of muscarinic antagonists to glass fibre filters. Hydrophobic ligands can dry on to filters as a result of the application of vacuum (62). The hydrophobic ligand QNB has been reported to bind with respectable affinity ($K_d = 4 \times 10^{-7}$ M) to mucus (63). In all of these cases, the primary definition of receptor binding by inhibition by the un-labelled tracer is extremely hazardous, since non-receptor as well as receptor-specific binding will be inhibited. In the case of filtration assays, many experimenters try to minimize binding to filters by dilution of the incubation mixture with an excess (e.g. 5-fold) of ice-cold buffer before filtering, in order to reduce the free tracer concentration. Pretreatment of the filters, for example with 0.1–0.3% polyethyleneimine, may also help. This is described in Chapter 6. Another tactic which can be employed is the bipartite centrifugation followed by filtration assay, in which the pellet obtained from a centrifugation assay is resuspended and filtered (64). This procedure is described in Chapter 7.

In all these cases of artefactual 'receptor-specific' binding, close exami-nation of the pharmacology of the supposed binding sites differentiates them from true receptors. From this point of view, it is extremely useful to have close structural analogues of the tracer ligand, such as an inactive enantiomer, which *do not* manifest high affinity for the receptor as well as analogues which, while retaining high affinity, have a distinctly variant chemical structure.

4.9 Improper definition of background

This problem is closely related to that of defining non-specific binding. Over or underestimation of counter background can again distort the tracer ligand binding curve, whether determined by the direct saturation experiment, or by the cold vs hot tracer design. Underestimation of background can lead to the apparent detection of a heterogeneous population of binding sites, whilst overestimation can lead to apparent positive co-operativity (36, 37). For these reasons, it is essential to include an estimation of counter background (using for example four background vials) in every set of data processed. Counter background can be reduced by counting in plastic rather than glass scintil-lation vials. This can be an important consideration when trying to estimate low concentrations of radioligands.

4.10 Ligand depletion

A preliminary review of this problem has been given in the introductory section (2.4.7). At the outset, it should be emphasized that depletion is better avoided than cured. Correction for depletion inevitably reduces the accuracy of parameter estimates. The best way to detect depletion is to measure the free ligand concentration actually present in the assay. It has been empha-sized that this is easily done in the context of centrifugation assays, which terminate the binding reaction without disturbance of the equilibrium, allow-ing the supernatant containing the free ligand to be sampled. Such a measure-ment will detect depletion due both to receptor-specific and non-specific adsorption of the tracer. Filtration assays will normally require supplementation by centrifugation assays to allow the free concentration to be measured. It may be best to use an ultracentrifuge ($100\,000\,g$, 1 hour) to be entirely sure of having completely removed small membrane fragments. Gel-filtration assays may require supplementation by equilibrium dialysis or equilibrium gel-filtration (Chapter 9) measurements to perform the same task. If the ligand is well behaved, it will often be possible to calculate the free ligand concentration as $L_t^* - L_{bound}^*$ provided that the assay method recovers almost 100% of the RL^* complex, and provided that significant dissociation of the bound ligand does not occur during the separation procedure.

Measurement of the free concentration of an unlabelled competitor or modulator is obviously a much more difficult proposition. However, it is not difficult to detect the occurrence of depletion by a form of radioreceptor assay, in which the supernatant from a first binding assay is then used in a second, to see whether its inhibitory potency has been retained.

4.10.1 Effect of depletion on association kinetics

Figure 18 shows two simulated time-courses of binding of a ligand with $K_d = 10^{-10}\,M$, $k_{12} = 10^7/(M\ sec)$; $L_t^* = 3 \times 10^{-10}\,M$; to a receptor ($R_t = 10^{-9}\,M$). The top curve was calculated using program COMPKIN which does not allow for depletion. The second was calculated using program INTRATE, which does. Not only is the final level of binding attained in the realistic, depleted case, 36% of the level which would have been attained without depletion, but also the binding reaches equilibrium in approximately 600 sec in the former, versus 1200 sec in the latter case. Moreover, in the absence of depletion, the time-course is mono-exponential, whereas in its presence, the time-course is more complex. If the lower curve in *Figure 18* is analysed by a simple exponential process of the form:

$$RL^* = RL_0^*(1 - e^{-k \cdot t}), \tag{51}$$

the fit is poor, and the rate constant overestimated.

4.10.2 Effect of depletion on tracer ligand saturation curves

The effect of depletion on the direct tracer ligand binding curve can be compensated for by careful measurement of the free ligand concentration, provided that the tracer ligand is pure and separation of bound from free is complete. The effect of depletion is obviously to reduce the free tracer concentration. If bound tracer is plotted as a function of total rather than free concentration, the effect is to steepen the binding curve.

An example is shown in *Figure 19*. In this rather extreme case, the apparent binding curve, plotted as a function of L_t^* is shifted about one order of magnitude to the right of the true binding curve, at the lower ligand concentrations, at which depletion is around 90%. At the higher ligand concentra-

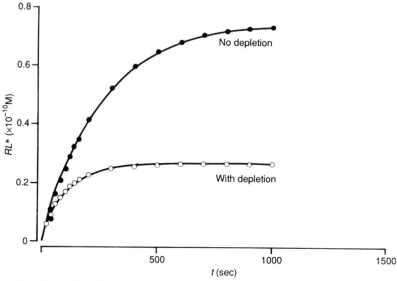

Figure 18. Tracer ligand depletion truncates the association time-course. $R_t = 10^{-9}$ M. $L_t^* = 3 \times 10^{-10}$ M, $k_{12} = 10^7/(\text{M sec})$, $k_{21} = 10^{-3}/\text{sec}$ ($K = 10^{10}/\text{M}$; $K_d = 10^{-10}$ M).

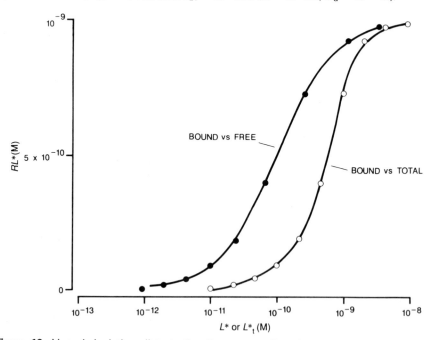

Figure 19. Ligand depletion distorts the direct tracer ligand saturation curve. Plot of bound vs total (○) and bound vs free (●) for binding of a ligand with affinity $10^{10}/\text{M}$ to a receptor present at 10^{-9} M concentration. The binding curve is shifted and steepened unless a correction is made for ligand depletion.

tions the shift is less apparent. At the EC_{50},

$$L_t^* = R_t/2 + L^* \quad \text{(see ref. 1)}, \tag{52}$$

i.e. the true EC_{50} is $\qquad L_t^* - R_t/2. \tag{53}$

Since L_t^* and $R_t/2$ are comparable in size, the true EC_{50}, i.e. the K_d, is poorly determined. Another less obvious effect of depletion is that even if the free ligand concentration is measured accurately, the distribution of the measurements on the concentration axis is distorted. Thus, whilst the intention may have been to obtain measurements at three ligand concentrations per decade, over the range 10^{-11}–10^{-8} M, in practice there are only two points in the critical 10^{-10}–10^{-9} M free concentration range, and four in the relatively non-informative 10^{-12}–10^{-11} range. This significantly decreases the accuracy of the K_d estimate. If the ligand is impure, depletion of the active, bindable ligand at the expense of the inactive impurity will lead to overestimation of the true free concentration, and thus to underestimation of the true affinity.

As has been emphasized in the experimental section, in order to minimize the effects of depletion, it is desirable to keep $R_t < 0.1 \times K_d$: certainly, less than $0.3 \times K_d$. When in doubt, the program EQBIND can be used to simulate the proposed experiment, and thus check its validity. One additional point is that depletion will also affect the level of non-specific binding. The level of non-specific binding appropriate to a given free ligand concentration thus has to be estimated either manually, by reading it off a plot of non-specific binding vs free ligand which has been separately determined, or, preferably, by fitting the binding curve with a function modelling both specific and non-specific binding (Chapter 11).

4.10.3 Effect of depletion on competition curves

The effects of depletion on competition binding curves have been reviewed by Wells et al. (65) and Goldstein and Barrett (66). The effects of tracer ligand depletion are easy to understand in principle. In the absence of the competing ligand, most of the tracer is bound. As the competitor concentration increases, thus displacing bound tracer, the free concentration increases, tending to counteract the inhibitory action of the added competing ligand. The effect is to shift the competition curve to the right on the concentration axis. The effect is additional to the familiar Cheng–Prusoff shift $(1 + L^*/K_d)$, and can be much larger. An example is given in *Figure 20*.

Here the binding of a 10^{-10} M concentration of a high-affinity tracer ligand, $K = 10^{10}$/M is inhibited by a ligand of lower affinity $KA = 10^8$/M. The Cheng–Prusoff correction, based on the total ligand concentration is thus $1 + 10^{-10}/10^{-10} = 2$-fold, i.e. 0.3 log units on the x-axis of *Figure 20*. This is indeed the observed shift for a low concentration of R_t, 10^{-11} M. However, for higher concentrations, 10^{-10}, 2×10^{-10}, 10^{-9} M the inhibition curves shift further to

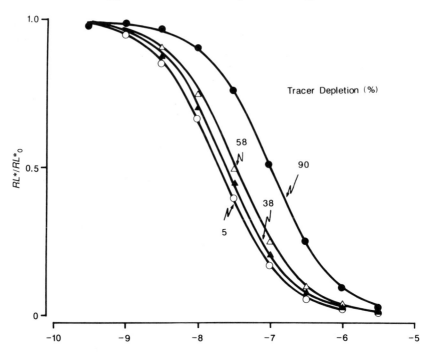

Figure 20. Tracer depletion right-shifts single-site competition curves. Tracer ligand affinity $K = 10^{10}/M$. Competitor affinity $KA = 10^8/M$. Tracer ligand $L_t^* = 10^{-10}\,M$. Total receptor $R_t = 10^{-11}$ (○), 10^{-10} (▲), 2×10^{-10} (△), 10^{-9} (●) M. Percentage depletions at zero concentration of competitor are as marked.

the right, along the x-axis. The corresponding values of depletion are 4.9%, 38%, 58%, and 90%. Whilst the additional shift for ~30% depletion is relatively small, for 90% depletion, the total shift is 10-fold, and is almost totally attributable to depletion. However, the shape of the competition curve is hardly altered. At the most, a very slight flattening is detectable, but the calculated Hill coefficient is still almost equal to 1.0. The simple application of the Cheng–Prusoff correction to this curve would, therefore, lead to a 5-fold underestimation of the affinity of the competitor. In this case, correction is clearly necessary. However, it might be argued that for depletion of up to ~30%, the error introduced is not too large. This is probably true in the case of simple competition for a single set of binding sites. However, in the case of multiple binding sites, serious distortions may still occur.

A useful formula for correcting the IC_{50} to obtain the true K_d or K value of the competitor in such simple cases is due to Goldstein and Barrett (66);

$$KA = (2(L^* - L_0^*)/L_0^* + 1 + K \cdot L^*)/IC_{50}, \qquad (53)$$

where L_0^* is the free tracer concentration in the absence of the competing ligand, and L^* is its concentration at the IC_{50}. The element of this correction factor directly attributable to depletion is $2(L^* - L_0^*)/L_0^*$. If the depletion is all due to receptor binding, it is easy to show that $2(L^* - L_0^*)/L_0^* = D/(1 - D)$, where D, defined as bound/total tracer ligand in the absence of the competitor, is the initial degree of depletion. This formula shows that depletion of 50% ($D = 0.5$) induces a shift of the competition curve to 2-fold higher competitor concentrations (0.3 log units). This effect is additive with the Cheng–Prusoff shift whose value depends on $K \cdot L^*$.

Both of these factors are taken into account in the program DEPCOR (*Protocol 21*) which is designed to correct competition data for tracer depletion. The program requires values of the total concentration of binding sites R_t, and free tracer ligand concentration at each competitor concentration, in addition to the concentration of bound tracer. The assumption is made that the interaction between the tracer and competitor ligands is competitive. Output is provided in the form of corrected values for occupancy of the receptor site by the competitor vs log competitor concentration.

Protocol 21. DEPCOR program

```
 30 REM THIS PROGRAM CORRECTS EXPERIMENTALLY-DETERMINED INHIBITION CURVES
 40 REM FOR THE EFFECTS OF TRACER LIGAND DEPLETION. IT ASSUMES THAT TRACER
 50 REM AND INHIBITOR INTERACT COMPETITIVELY. IF MULTIPLE SITES ARE PRESENT
 60 REM THE TRACER IS ASSUMED TO BE NON-SELECTIVE. YOU NEED TO KNOW THE
 70 REM AFFINITY CONSTANT OF THE TRACER LIGAND, AND TO PROVIDE AN ESTIMATE
 80 REM OF THE CONCENTRATION OF BINDING SITES PRESENT IN THE ASSAY, OR A
 90 REM MEASURE OF BINDING IN THE ABSENCE OF INHIBITOR. YOU ALSO NEED TO
100 REM INPUT THE FREE TRACER LIGAND CONCENTRATION FOR EACH MEASUREMENT.
110 REM THE OUTPUT OF THE PROGRAM CONSISTS OF AN OCCUPANCY FIGURE OF THE
120 REM RECEPTOR BY THE INHIBITOR, AND A 'CORRECTED' INHIBITOR CONCENTRATION
130 REM WHICH IS OBTAINED BY APPLYING A CHENG-PRUSOFF CORRECTION APPROPRIATE
140 REM TO THE FREE LIGAND CONCENTRATION.
150 DIM A(100),B(100),LF(100)
160 INPUT "WHAT IS THE AFFINITY CONSTANT FOR L*", K
170 REM IF THERE IS A KNOWN VALUE FOR RT,READ IT
180 INPUT "DO YOU HAVE A VALUE FOR RT,Y/N"A$
190 IF A$="Y" THEN GOTO 200 ELSE 230
200 INPUT "RT = ",RT
210 GOTO 280
220 REM "OTHERWISE CALCULATE THE VALUE OF RT"
230 INPUT "DO YOU HAVE A VALUE OF BINDING AT A=0",A$
240 IF A$="Y" THEN GOTO 250 ELSE 480
250 INPUT "RL =, LF = ",B,LF
260 RT=B*(1+K*LF)/(K*LF)
270 REM
280 PRINT:PRINT
290 PRINT "ESTIMATE OF RT = ",RT
300 INPUT 'HOW MANY OBSERVATIONS ",N
310 REM INPUT CONCNS OF COMPETITOR,BOUND TRACER,FREE TRACER
```

Protocol 21. *Continued*

```
320 PRINT:PRINT
330 PRINT "TYPE IN CONCNS OF COMPETITOR, BOUND LIGAND, FREE LIGAND"
340 PRINT
350 FOR J= 1 TO N
360 INPUT "COMPETITOR =", A(J), "BOUND =",B(J),"FREE = ",LF(J)
370 NEXT J
380 PRINT:PRINT:PRINT
390 PRINT "LOG ACORR      OCCUPANCY"
400 FOR J= 1 TO N
410 AC=A(J)/(1+K*LF(J))
420 RRL=B(J)*(1+K*LF(J))/(K*LF(J))
430 P=(RT-RRL)/RT
440 TEMP = LOG(AC)
450 PRINT TEMP,P
460 NEXT J
470 GOTO 490
480 PRINT "NO MEANS OF CALCULATING RT"
490 END
```

DEPCOR also allows the correction of competition curves for binding to multiple populations of binding sites, subject to the restriction that the tracer ligand displays the same affinity for all the subpopulations. If this is not the case, the calculation of a corrected curve seems very difficult. In this case, there is no alternative but to fit the data directly to a model of the binding process, such as that embodied in the program COMPDEP (*Protocol 5*). Note that analytical solutions to the range of problems covered by DEPCOR have recently been found by Levitzki and co-workers (67, 68).

The effect of tracer depletion on the form of inhibition curves when multiple sites are involved is rather subtle. An illustration is given in *Figure 21*.

Here, a non-selective high-affinity ligand ($K = 10^{10}$/M) is displaced by a competitor ligand having affinities of 10^7/M and 10^4/M for two equal populations of binding sites. The value of L_t^* is 10^{-10} M.

As R_t increases, thus increasing the extent of maximum radioligand depletion from 5–76%, the effect is to decrease the apparent percentage of high-affinity sites without exerting much effect on the apparent affinity constants of the competing ligand for the two populations. Thus, in contrast to the case of a single population of sites, the outcome is to cause the underestimation of the relative proportion of the high-affinity sites rather than of their affinity for the competitor.

To minimize such depletion artefacts:

(a) If possible, keep $R_t < 0.1 \times K_d$ for the tracer ligand and keep the tracer ligand concentration low. This has the additional value of minimizing the Cheng–Prusoff parallel shift correction.

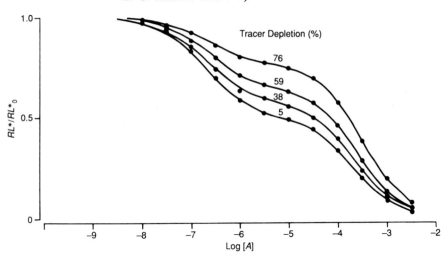

Figure 21. Tracer depletion conceals the high-affinity component of a two-site competition curve. Tracer ligand affinity $K = 10^{10}/M$. Tracer ligand concentration $L_t^* = 10^{-10}$ M. Competitor affinities $KA = 10^7$, $10^4/M$. Percentage of high-affinity sites = 50%. Total receptor $R_t = 10^{-11}$, 10^{-10}, 2×10^{-10}, 4×10^{-10} M. Percentage depletions at zero concentration of competitor A are as marked.

(b) If the above is impossible, increase L_t^* to $10 \times R_t$. Analysis of the data will now involve a significant shift correction. To perform this, it is necessary to understand the mechanism of the interaction between the competitor and the tracer ligand.

(c) If neither tactic is feasible, try to keep depletion to <30% (max. 50%). It will be necessary to analyse the data by non-linear least-squares fitting using a model which explicitly accounts for depletion. *Note* that the LIGAND suite of programs does this (see Appendix 3).

(d) Avoid depletion of the unlabelled ligand. Use the program EQBIND to check whether depletion is liable to be a problem. If it is, reduce R_t, or increase L_t^* to force the binding curve to higher competitor concentrations. Perform a test for depletion of the unlabelled ligand by adding supernatant from one assay to a fresh aliquot of membranes and tracer to see if potency is maintained.

4.11 Incomplete equilibration

The necessity for achieving complete equilibration in both direct tracer ligand saturation, and in cold vs hot ligand experiments has been emphasized. The effects of non-equilibration depend on the type of experiment being performed.

4.11.1 Tracer ligand saturation curve

An example is shown in *Figure 22a*. The values were obtained by running the program INTRATE with $k_{12} = 10^7/(\text{M sec})$, $k_{21} = 10^{-2}/\text{sec}$, $R_t = 10^{-10}\,\text{M}$. The equilibrium binding curve is a Langmuir isotherm with $K_d = 10^{-9}\,\text{M}$. The half-time of dissociation, $\log_e(2)/10^{-2}$ is 70 sec. An incubation time of 350 sec, therefore, ensures attainment of equilibrium at all L^* concentrations.

The effect of premature termination of the binding reaction is to cause an apparent shift of the binding curve to higher tracer concentrations, i.e. an apparent decrease in potency, accompanied by a slight steepening. The potency shift is caused by the failure of the low, but not the high, concentrations of tracer to equilibrate, i.e. there is a shift from off-rate-limited to on-rate-limited kinetics. The steepening only becomes apparent at relatively high occupancies (>75%), and might well be obscured in practice by the larger absolute size of the errors associated with measurement at high L^* concentrations. An attempt to fit the data to a Langmuir isotherm either by Scatchard analysis (*Figure 22b*) or by curve fitting leads not only to overestimation of K_d (underestimation of K) but also to overestimation of R_t. Instability of the receptor may exacerbate this insidious artefact by causing systematic underestimation of the equilibration time (Section 4.4). Such problems are likely to

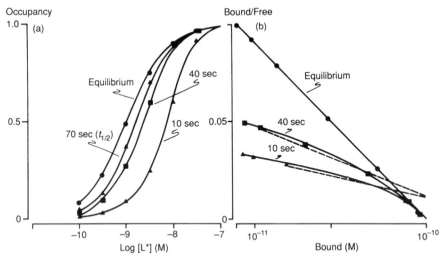

Figure 22. Failure to attain equilibrium causes underestimation of the affinity of the tracer ligand, and overestimation of the total concentration of binding sites. Total receptor concentration $R_t = 10^{-10}\,\text{M}$. Ligand affinity $K = 10^9/\text{M}$. 10, 40, 70, 1000 sec of incubation. (a) Occupancy as a function of free concentration. (b) Scatchard plot of (a). The dashed lines illustrate the overestimation of the concentration of binding sites which is especially significant if only the data at lower concentrations of L^* are examined.

be particularly pronounced with slowly-equilibrating high-affinity ligands ($K > 10^{10}$/M).

4.11.2 Competition experiments

The effects of a competitor on the kinetics of binding of a tracer ligand have been considered by several workers (19, 69–71). Two examples are shown in *Figures 23* and *24*.

The first shows the effect on binding of a tracer ligand (10^{-9}M) with an affinity of 10^{10}/M ($k_{12} = 10^7$/M sec), $k_{21} = 10^{-3}$/sec) of increasing concentrations of a competitor with an affinity of 10^7/M, and a dissociation rate constant k_{31} of 1/sec, i.e. the tracer dissociates SLOWLY while the competitor dissociates RAPIDLY.

The effect is to increase the half-time for association of the tracer from 60 sec in the absence of the competitor to 700 sec at near-receptor-saturating concentrations of the competitor, and the tracer kinetics change from on-rate to off-rate limited. In the absence of competitor, the tracer equilibrates completely within 500 sec. However, if the competition curve is measured at this time point, the effect is a shift to higher apparent potency, accompanied by steepening. The IC_{50} of the competitor ligand is underestimated by a factor of 2 and the Hill coefficient estimated to be 1.16 instead of 1.0. In order to

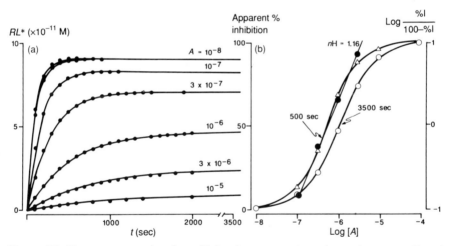

Figure 23. The presence of a low-affinity fast competitor slows the tracer ligand association kinetics (a) leading to overestimation of the competitor affinity and artefactual apparent positive co-operativity (b). For the tracer $k_{12} = 10^7$/M sec, $k_{21} = 10^{-3}$/sec. For the competitor $k_{13} = 10^7$/M sec, $k_{31} = 1$/sec. Total receptor concentration = 10^{-10} M, tracer concentration = 10^{-9} M. The left hand side (a) shows the kinetics of ligand binding in the presence of different concentrations of A. The right hand plot (b) shows the inhibition curve measured at 500 sec (△) and 3500 sec (○). The Hill plot of the 500 sec data (●) is also shown in this figure and illustrates the apparent positive co-operativity (Hill coefficient, 1·16).

obtain error-free measurements, an incubation time of 3500 sec must be used in this case. This once again emphasizes the *primacy of the ligand dissociation rate in the estimation of equilibration time*, and the need to *measure time-courses at high as well as at low competitor concentration*.

A converse example is shown in *Figure 24*. Here, a RAPIDLY-dissociating tracer ($k_{21} = 0.1$/sec) interacts with a very SLOWLY-dissociating high-affinity competitor ($k_{31} = 2 \times 10^{-4}$/sec; $K = 5 \times 10^{10}$/M).

Now, incubation for about 6 h is needed to achieve equilibrium. The binding of the tracer peaks at <100 sec, and then declines with a rate constant determined by the rate of competitor association. Such an experiment provides an indirect method of measuring the association kinetics of the slowly-equilibrating competitor. The effect of premature termination of the incubation is to underestimate the affinity of the competitor, and steepen the competition curve (*Figure 24b*). The errors are more pronounced at the low than at the high competitor concentrations, so the effect is very similar to that seen in the direct tracer binding experiment.

To avoid artefacts of this nature, it is advisable to attempt to measure the dissociation rate constant of the unlabelled ligand, at least roughly (71). A suitable method is the following:

(a) Incubate the receptor preparation with unlabelled competitor at a concentration sufficient to occupy ~90% of the receptor binding sites.

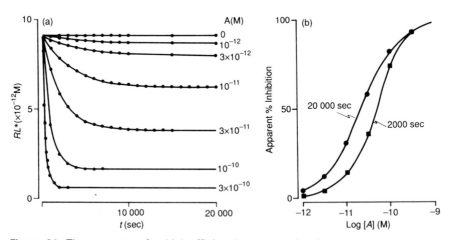

Figure 24. The presence of a high-affinity slow competitor leads to an overshoot in tracer ligand binding (a), giving an underestimation of the competitor affinity, and artefactual apparent positive co-operativity (b). For the tracer, $k_{12} = 10^7$/M sec), $k_{21} = 10^{-1}$/sec. For the competitor $k_{13} = 10^7$/M sec), $k_{31} = 2 \times 10^{-4}$/sec. Total receptor concentration = 10^{-10} M, tracer concentration = 10^{-9} M. In (a), the very rapid rise in tracer ligand binding has been omitted from the plot for the sake of clarity. In (b), the inhibition curves at $t = 2000$ sec (□) and at $t = 20\,000$ sec (○) are plotted, apparent positive co-operativity being observed in the former data.

(b) Separate the bound from the bulk of the free ligand, for example by centrifugation or gel filtration.

(c) Add tracer ligand at $\sim 10 \times K_d$. Measure the association time-course. If the competitor ligand dissociates slowly, its dissociation rate constant will determine the rate of tracer association. Comparison with the rate of tracer association in the absence of competitor will reveal whether or not this is so.

In summary, the ratio of the dissociation rate constants k_{21} of the tracer and competitor ligands is important in determining whether the apparent affinity of a competitor increases or decreases with time of incubation before equilibrium is achieved. If the dissociation rate of the tracer exceeds that of the competitor, the apparent affinity constant of the latter will tend to increase with time. If the dissociation rate of the tracer is less than that of the competitor, the apparent affinity of the latter will tend to decrease with time. If the tracer is 'fast' but the competitor 'slow', kinetic information about the binding of unlabelled ligands can, in principle, be derived from analysis of their effects on tracer ligand time-courses (Chapter 10 and ref. 19).

5. Conclusions

It is easy to be overwhelmed by the complexities and artefacts which can be found in binding studies. However, good planning helps avoid many of the pitfalls. When confronted with a new situation, for instance a new tissue preparation or a novel ligand, it is obviously important to consider the basics: Is the binding in equilibrium? Is the non-specific binding properly defined? Are the receptor and the ligand stable under the assay conditions? Is the separation method working properly? Is the binding saturable? Is the pharmacology appropriate?

In more complex situations, it pays to design experiments to approach the problem from several directions, and to make as many cross-checks as possible. For instance, it is useful to perform both direct ligand saturation binding curves and cold vs hot competition experiments. These will reveal potential problems with the specific radioactivity or purity of the radioligand.

If more than one radioligand is available to label the same receptor, it is often useful to exploit the possibilities offered by this. This is particularly true where there are subtypes of the receptor, and both selective and non-selective ligands are available. For instance, an experiment in which a cold selective ligand is used to inhibit the binding of a hot, non-selective ligand will generate a flat binding curve, with a high- and a low-affinity component. The binding properties of the high-affinity component can then be studied directly using a radiolabelled version of the selective ligand. The two kinds of experiment should yield mutually consistent results.

It is always useful to try to dissect complex binding curves by the use of selective radioligands, and it is always desirable to try to use internally consistent experimental designs. For instance, if a labelled agonist A^* and a labelled antagonist L^* are available for a given receptor, besides the tracer saturation experiments with A^* and L^*, and the self-competition experiments A (unlabelled) vs A^* and L (unlabelled) vs L^* it is useful to perform the cross-competition experiments A vs L^* and L vs A^* at different concentrations of L^* and A^*. Such sets of data make strong demands on the internal coherence of any model of the receptor which one may be trying to develop, and are quick to reveal anomalies which may be important. The authors' own work on the binding of antagonists (refs 9, 12, 24, 31, 42) and agonists (refs 72–74) to muscarinic acetylcholine receptors provide a record of our own attempts to follow this path.

References

1. Cuatrecasas, P. and Hollenberg, M. (1976). *Adv. Protein Chem.*, **30**, 252.
2. Yamamura, H. I., Enna, S. J., and Kuhar, M. J. (eds) (1985). *Neurotransmitter Receptor Binding*, Raven Press, New York.
3. Snyder, S. H. (1986). *Annu. Rev. Physiol.*, **48**, 461.
4. Weiland, G. A. and Molinoff, P. B. (1981). *Life Sci.*, **29**, 313.
5. Molinoff, P. B., Wolfe, B. B., and Weiland, G. A. (1981). *Life Sci.*, **29**, 427.
6. Klotz, I. M. (1989). In *Protein Function, a Practical Approach* (ed. T. E. Creighton), p. 25. IRL Press, Oxford.
7. E. C. Hulme (ed.) (1990). *Receptor Biochemistry, a Practical Approach*. IRL Press, Oxford.
8. Burgisser, E., Lefkowitz, R. J., and DeLean, A. (1981). *Mol. Pharmacol.*, **19**, 509.
9. Stockton, J. M., Birdsall, N. J. M., Burgen, A. S. V., and Hulme, E. C. (1983). *Mol. Pharmacol.*, **23**, 551.
10. Conti-Tronconi, B. M. and Raftery, M. A. (1982). *Annu. Rev. Biochem.*, **51**, 491.
11. Abbott, A. J. and Nelsestuen, G. L. (1988). *FASEB J.*, **2**, 2858.
12. Hulme, E. C., Birdsall, N. J. M., Burgen, A. S. V., and Mehta, P. (1978). *Mol. Pharmacol.*, **14**, 737.
13. Wood, P. L. (1986). In *Neuromethods* (ed. A. A. Boulton and G. B. Baker), Vol. 4, p. 329. Humana Press, Clifton, New Jersey.
14. Strange, P. G. (1987). In *Dopamine Receptors* (ed. I. Creese and C. M. Fraser), p. 29. A. R. Liss, New York.
15. Laduron, P. M. (1987). In *Perspectives on Receptor Classification* (ed. J. W. Black, D. H. Jenkinson, and V. P. Gerskowitch), Vol. 6, p. 71. A.R. Liss Inc, New York.
16. Laduron, P. M. (1984). *Biochem. Pharmacol.*, **33**, 833.
17. Birdsall, N. J. M., Burgen, A. S. V., and Hulme, E. C. (1979). *Br. J. Pharmacol.*, **66**, 337.
18. Klotz, I. M. (1982). *Science*, **217**, 1247.
19. Motulsky, H. J. and Mahan, L. C. (1984). *Mol. Pharmacol.*, **25**, 1.

20. Stephenson, A. (1988). *Biochem. J.*, **249**, 21.
21. Gilman, A. G. (1987). *Annu. Rev. Biochem.*, **56**, 615.
22. Cheng, Y. C. and Prusoff, W. H. (1973). *Biochem. Pharmacol.*, **22**, 3099.
23. Hulme, E. C. (ed.) (1990). *Receptor–Effector Coupling, a Practical Approach.* IRL Press, Oxford.
24. Hammer, R., Berrie, C. P., Birdsall, N. J. M., Burgen, A. S. V., and Hulme, E. C. (1980). *Nature*, **283**, 90.
25. Colquhoun, D. (1987). In *Perspectives on Receptor Classification* (ed. J. W. Black, D. H. Jenkinson, and V. P. Gerskowitch), Vol. 6, p. 103. Alan Liss, New York.
26. Dohlman, H. G., Caron, M. G., and Lefkowitz, R. J. (1987). *Biochemistry*, **26**, 2657.
27. Boast, C. A., Snowhill, E. W., and Altar, A. (eds) (1986). *Quantitative Receptor Autoradiography*. Alan R. Liss Inc., New York.
28. Andreasen, N. C. (1988). *Science*, **239**, 1381.
29. Evans, W. H. (1987). In *Biological Membranes, a Practical Approach* (ed. W. H. Evans and J. B. C. Findlay), p. 1. IRL Press, Oxford.
30. Dawson, R. M. C., Elliott, D. C., Elliott, W. H., and Jones, K. M. (ed.) (1986). *Data for Biochemical Research*. Clarendon Press, Oxford.
31. Birdsall, N. J. M., Chan, S-C., Eveleigh, P., Hulme, E. C., and Miller, K. W. (1989). *Trends Pharmacological Sciences*, **10** (Suppl. Subtypes of Muscarinic Receptors iv), 31.
32. Perrin, D. D. and Dempsey, B. (1974). *Buffers for pH and Metal Ion Control*, p. 156. Chapman and Hall, London.
33. Zivin, J. and Waud, D. R. (1982). *Life Sci.*, **30**, 1407.
34. Munson, P. J. (1983). *Meth. Enzymol.*, **92** (*Immunocytochemical Techniques*, part E), 543.
35. Munson, P. J. and Rodbard, D. (1980). *Anal. Biochem.*, **107**, 220.
36. Munson, P. J. (1984). In *NATO ASI Life Sciences Series* (ed. F. Cattabeni, and S. Nicosia), Vol. 72, p. 1. Plenum, New York
37. Munson, P. J. (1983). *J. Receptor Res.*, **3**, 249.
38. Cornish-Bowden, A. and Koshland, D. E. (1975). *J. Mol. Biol.*, **95**, 201.
39. Rothman, R. B. (1983). *Neuropeptides*, **4**, 41.
40. Burgisser, E., Hancock, A. A., Lefkowitz, R. J., and DeLean, A. (1980). *Mol. Pharmacol.*, **19**, 205.
41. Kermode, J. L. (1988). *Biochem. J.*, **252**, 521.
42. Eveleigh, P., Hulme, E. C., Schudt, C., and Birdsall, N. J. M. (1989). *Mol. Pharmacol.*, **35**, 477.
43. Builder, S. E. and Segal, I. H. (1978). *Anal. Biochem.*, **85**, 413.
44. Honore, B. (1987). *Anal. Biochem.*, **162**, 80.
45. Staehlin, M. and Simons, P. (1982). *EMBO J.*, **1**, 187.
46. Liles, W. C., Hunter, D. D., Meier, K. E., and Nathanson, N. M. (1986). *J. Biol. Chem.*, **261**, 5307.
47. Roskoski, R., Guthrie, R., Roskoski, L. M., and Rossowski, W. (1985). *J. Neurochem.*, **45**, 1096.
48. Rubenstein, R. L., Wong, S. K. F., and Ross, E. M. (1987). *J. Biol. Chem.*, **262**, 16 655.
49. Venter, J. C. (1983). *J. Biol. Chem.*, **258**, 4842.

50. Green, M. A., Vigor, A., and Wells, J. W. (Personal communication.)
51. Strader, C. D. *et al.* (1987). *Cell*, **49**, 855.
52. Yang, S-J. and Olsen, R. W. (1987). *Mol. Pharmacol.*, **32**, 266.
53. Neufeld, G., Steiner, S., Korner, M., and Schramm, M. (1983). *Proc. Natl. Acad. Sci. USA*, **80**, 6441.
54. Severne, Y., Ijzermann, A., Nerme, V., Timmerman, H., and Vauquelin, G. (1987). *Mol. Pharmacol.*, **31**, 69.
55. Gurwitz, D., Kloog, Y., and Sokolovsky, M. (1984). *Proc. Natl. Acad. Sci. USA*, **81**, 3650.
56. Corvera, S., Schwartz, K. R., Graham, R. M., and Garcia-Sainz, J. A. (1986). *J. Biol. Chem.*, **261**, 520.
57. Titeler, M. and Seeman, P. (1980). *Eur. J. Pharmacol.*, **67**, 187.
58. Smith, S. K. and Limbird, L. E. (1982). *J. Biol. Chem.*, **257**, 10471.
59. Niehoff, D. L. and Mudge, A. W. (1985). *EMBO J.*, **4**, 317.
60. Toogood, C. I. A., McFarthing, K. G., Hulme, E. C., and Smyth, D. G. (1984). *Neuropeptides*, **5**, 121.
61. Daum, P. R., Hill, S. J., and Young, J. M. (1982). *Br. J. Pharmacol.*, **77**, 347.
62. Hung, C-R., Hong, J-J., and Bondy, S. (1982). *Life Sciences*, **30**, 1713.
63. Rimele, T. J. and Gaginella, T. S. (1982). *Biochem. Pharmacol.*, **31**, 515.
64. Galper, J. B., Haigh, L. S., Hart, A. C., O'Hara, D. S., and Livingston, D. J. (1987). *Mol. Pharmacol.*, **32**, 230.
65. Wells, J. W., Birdsall, N. J. M., Burgen, A. S. V., and Hulme, E. C. (1980). *Biochem. Biophys. Acta*, **632**, 464.
66. Goldstein, A. and Barrett, R. W. (1987). *Mol. Pharmacol.*, **31**, 603.
67. Horowitz, A. and Levitzki, A. (1987). *Proc. Natl. Acad. Sci. USA*, **84**, 6654.
68. Almagor, H. and Levitzki, A. (1990). *Proc. Natl. Acad. Sci. USA*, **87**, 6482.
69. Sklar, L. A., Sayre, J., McNeil, V. M., and Finney, D. A. (1985). *Mol. Pharmacol.*, **28**, 323.
70. Ehlert, F. J., Roeske, W. R., and Yamamura, H. I. (1986). *Mol. Pharmacol.*, **19**, 367.
71. Aranyi, P. (1980). *Biochem. Biophys. Acta*, **628**, 220.
72. Birdsall, N. J. M., Burgen, A. S. V., and Hulme, E. C. (1978). *Mol. Pharmacol.*, **14**, 723.
73. Berrie, C. P., Birdsall, N. J. M., Burgen, A. S. V., and Hulme, E. C. (1979). *Biochem. Biophys. Res. Commun.*, **87**, 1000.
74. Birdsall, N. J. M., Hulme, E. C., and Burgen, A. S. V. *Proc. R. Soc. Lond. B.*, **207**, 1.
75. Munson, P. J. and Rodbard, D. (1983). *Science*, **220**, 979.

5

Receptor preparations for binding studies

E. C. HULME and N. J. BUCKLEY

1. Introduction

From the point of view of technique in binding studies, receptor preparations fall into two major categories:

First preparations in which the cell surface membrane remains locally or globally intact. These include:

- whole tissues, in particular thin slices
- whole cells
- membrane preparations

Secondly, preparations in which the cell surface membrane is disrupted by the use of a solubilizing agent.

Preparations in the first category lend themselves to direct separation of receptor-bound from free ligand by the application of mechanical techniques such as filtration (Chapter 6) or centrifugation (Chapter 7). Preparations in the second category can be assayed by gel-filtration (Chapter 9), precipitation of the receptor (Chapter 6), or adsorption of the ligand followed by the application of filtration or centrifugation (Chapter 8), or by ion-exchange, in which the receptor, but not the ligand, is selectively adsorbed, for example to a polyethylenimine-coated glass-fibre filter disc (Chapter 6).

1.1 Whole tissues: tissue slices

Tissue slices are usually handled by causing them to adhere to a gelatin-coated or poly-L-lysine-coated glass slide. The separation of bound from free ligand is thus solved by physical removal of the slices from the incubation medium, followed by washing. Because of the existence of diffusion barriers within whole tissues, such slices need to be 5–50 μm thick and must, therefore, be cut with a cryostat. The aim of labelling receptors in slices is normally to study their cellular distribution either by autoradiography (1) or as described by Priestley in a companion book in this series (2), by immunocytochemistry using anti-receptor antibodies, produced as described in Chapter 2. However, it is also possible to carry out binding studies on slices. Normally, the motivation for this is to confirm the pharmacological specificity of sites prior to studies of their distribution by autoradiography. It has also been possible, in several instances, specifically to label receptors in whole tissues, either by injection of a radioligand *in vivo* (1,3) or by perfusion of the tissues *in vitro*. In all of these cases, the normal criteria for receptor-specific labelling need to be rigorously applied (see Chapter 4).

1.2 Isolated whole cells

Receptors on whole cells are typically assayed by filtration or centrifugation techniques. Receptor-bearing cells may either be established lines, or primary cultures. Alternatively, single cell preparations may be obtained by mechanical and/or enzymatic disruption of whole tissues.

Unlike tissue slices, single cell preparations have the advantage of relative physical homogeneity, and lack of diffusion barriers. If they are established clonal cell lines, they are also, in principle, genetically homogeneous. However in practice, many 'clonal' cell lines can be heterogeneous:

(a) because of karyotype variation (i.e. the genotype differs);

(b) because of heterogeneous expression of different differentiated phenotypes (i.e. the phenotype differs);

(c) because the 'same' cell line in different labs can be subject to different selection pressures, probably largely due to differences in media type, or sera and passage regimen. This can lead to a 'construct' phenotype that may be different between laboratories or which can alter as a function of time in the same laboratory.

Obviously, there are aspects of receptor function which can only be studied in intact cells, for example synthesis, post-translational modification, insertion into the cell membrane, and down-regulation. Receptor-response coupling is also often best studied in intact cells. Some responses, such as calcium mobilization (4,5) or channel opening mediated by phosphorylation by C-kinase, or cAMP-dependent kinases (6), or other responses requiring intracellular second messengers or substrates also necessitate an intact, viable cell. Other responses, such as directly receptor-operated ion channels, or GTP binding-protein-mediated effects on enzyme activities or channel conductances (7) can be studied in membrane fragments by biochemical or electrophysiological (patch clamping) techniques (8). Receptor–effector coupling is covered in detail in a companion volume (9).

Now that a number of receptor genes and cDNAs have been cloned, their high-level expression in suitable clonal cell lines, with or without site-directed mutagenesis, is becoming an increasingly popular option for studying function and coupling mechanisms. The high-level expression of cloned receptors is described by Fraser in reference 10.

Large-scale culture may, in time, provide the milligram quantities of receptor protein needed to permit physical and structural studies, for example by crystallography and NMR. The practicalities of animal cell culture are covered in another book in this series (11). The establishment of primary cell cultures from the fetal CNS is described by Saneto and DeVellis in reference 2.

1.3 Membrane preparations

The great majority of routine receptor-binding assays employ one of the many varieties of membrane preparations. These lend themselves well to centrifugation and filtration assays.

The use of broken-cell preparations gets round the problems of non-specific, or carrier-mediated uptake of ligands which are often problematic with assays on intact cells. Thus, the specific:non-specific binding ratio is often improved, so a wider range of ligands can be used. It is also possible to endeavour to wash away endogenous agonists and mediators (for example guanine nucleotides) whose level is unknown, and uncontrolled in whole cells. In addition, soluble or reversibly membrane-associated proteases may be removed, inhibited, or controlled.

Aspects of receptor–effector coupling which depend on interaction with other membrane-associated proteins (for example GTP-binding proteins) are often surprisingly well preserved in membrane fragments. Other aspects of receptor function, depending on the recruitment of small molecules, or enzymes such as C-kinase, or specific receptor kinases (for example β-adrenergic receptor kinase, ref. 12) from the cytoplasm, are obviously lost.

Somewhat more imponderable are the possible effects on receptor function of the absence of the normal transmembrane ionic gradients and potential differences, or of the partial or complete disruption of receptor–cytoskeletal interactions. Fortunately, the dominant non-specific environmental influence on receptor binding seems to come from the lipid bilayer itself, rather than from the intracellular or extracellular milieu. The pharmacological specificity of receptors is, therefore, well preserved in membrane fragments.

A special instance of membrane-binding assays is provided by purified receptors reconstituted in phospholipid vesicles of defined composition, in the absence or presence of specific effector molecules. Examples of reconstitution protocols are provided by Cerione, and by Haga and Haga in *Receptor–Effector Coupling, a Practical Approach* (9).

1.4 Solubilized preparations

Protocols for solubilizing membrane fractions in detergents yielding dispersed preparations of receptors in a form suitable for purification, or physical studies, are described in *Receptor Biochemistry, a Practical Approach* (10). The assay of solubilized receptors is described in Chapters 6, 8, and 9 of the present volume. An excellent review of detergent properties, and some protocols for reconstitution of the nicotinic acetylcholine receptor are given by Jones *et al.* in reference 13.

2. Tissue preparation

2.1 Tissue preparation for autoradiographic or immunohistochemical studies

Successful autoradiographic and/or immunocytochemical localization studies of receptors have been carried out both on prefixed and on unfixed cryostat slices. Prefixation obviously gives better histological preservation. The use of 0.05% glutaraldehyde, but not formaldehyde, (14) has permitted the study of the distribution of muscarinic acetylcholine receptors in rat brain. However, many fixatives are chemical cross-linking agents, and will, therefore, react with cell-surface receptors. This may affect their binding properties, and is certainly expected to modify their coupling properties *vis-à-vis* other membrane-bound proteins. For instance, formaldehyde and glutaraldehyde both react with accessible amino groups. Such reactions may not always destroy the antagonist-binding properties of cell-surface receptors although, for instance, 4% formaldehyde or glutaraldehyde destroys all muscarinic-binding activity. However, no generalizations are possible. Agonist binding properties will normally be affected, through alteration of the thermodynamics of the conformational change induced by binding. Large-scale studies of receptor distribution in tissues, i.e. those not requiring detailed cellular preservation, are best carried out on unfixed slices. The effects of the fixation procedure must be studied at the level of membrane binding before too much effort is invested in distribution studies using a particular procedure.

2.1.1 Tissue fixation

Tissues may be lightly prefixed by perfusion of the whole animal with the fixing agent; 0.05% glutaraldehyde or 0.1% formaldehyde contained in phosphate-buffered saline (PBS) is probably the best initial choice. The technique of whole-body perfusion is well described by Priestley (2). Use of ice-cold fixative is good, especially for brain. It helps to make soft tissues rigid which aids in subsequent dissection and blocking. Small blocks of brain are lightly fixed by immersion in 0.1% formaldehyde in ice-cold PBS for 10 min.

2.1.2 Unfixed tissues

If slices are to be cut from unfixed tissues, morphological preservation is best achieved by coating the tissues with an embedding medium (OCT, Tissue

Tek, Miles Laboratories) before sectioning. In addition, the tissue can be cryoprotected by exposure to 7% sucrose in 0.1 M phosphate, pH 7.4, (2). The brain is best treated by removal from the skull and completely covering with *powdered* dry ice. Use of liquid nitrogen frequently causes fractures in large blocks of tissue.

Protocol 1. Tissue preparation for cutting slices

1. Sacrifice the animal by decapitation or intracardiac perfusion with ice-cold PBS.

2. Remove the tissue of interest. Freeze it by immersion in powdered dry ice or a slurry of isopentane cooled in liquid nitrogen (better for small blocks of tissue, for example peripheral ganglia).

3. Mount the frozen block of tissue on a cryostat chuck with Tissue Tek II OCT mounting medium.

4. Alternatively, freeze the tissue directly to the chuck as described by Priestley (2). Small or fragile pieces of brain tissue may be supported by surrounding them with a concentrated 'brain paste' (1). However, the use of OCT is better; brain paste generally contains receptor-binding sites and even though it is morphologically distinguishable from embedded tissue, this makes photography very difficult.

2.1.3 Cutting slices

The use of microtomes and cryostats for cutting thin slices is well described by Priestley in *Neurochemistry, a Practical Approach* (2):

(a) Section the tissue at −10 to −20°C.

(b) Thaw-mount the slices on gelatin-subbed or poly-L-lysine slides kept at cryostat temperature. Mounted slices may be stored at −80°C until used for labelling and autoradiography.

Protocol 2. Subbing slides

1. 3 × 1″ single frosted slides are most convenient. Use larger slides for larger sections.

2. Mount slides in racks.

3. Sonicate in distilled water plus detergent (for example 5% Decon 75) for 30 min. Wash extensively in running distilled water (at least 2 hours) until the water forms an even film on the slides.

4. Immerse slides in gelatin solution, freshly prepared, for 5 min. Soak gelatin

Protocol 2. *Continued*

(2.5 g) in a small amount of hot distilled water (80°C) until swollen, then add distilled water at 80°C to 500 ml. When the gelatin has dissolved, add 0.25 g of chrome alum. Filter and cool to room temperature before dipping slides. (ALWAYS MAKE FRESH.)

5. Drain and dry in a dust-free atmosphere (laminar flow hood, lab. chimney). Slides can be stored at 4°C. (It is useful to store some subbed slides at −20°C ready for the cryostat.)

6. If slices are frequently lost from slides during subsequent manipulations, a second exposure to subbing solution may help.

2.1.4 Labelling receptors and checking incorporation of label

Initial assessment of the optimal conditions for receptor labelling is best performed on subcellular fractions (Chapter 4). The conditions thus established can be used as a starting point for labelling slices.

Protocol 3. Radiolabelling receptors

1. Choose the ligand. Tritiated ligands have relatively low specific activity (< 100 Ci/mmole; Chapter 1) ^{125}I-iodinated ligands are hotter (> 2000 Ci/mmole). However, iodinated ligands often give higher background, and thus must be used at lower receptor occupancy. In addition their affinities and/or specificities may be different from those of the parent compound. Also, the path-length of the emitted radiation is longer, so that the resolution is lower than that obtainable with tritium.

2. Preincubation may be used to

   • reduce the concentration of endogenous competing ligands

   • block non-specific binding to charged areas of the slide or tissue slice

   Preincubate up to 10 slides in Coplin jars in binding buffer, optionally supplemented with 0.005% polyethylenimine (to reduce non-specific interaction with cationic ligands) for up to 30 min.

3. Incubate slides with unlabelled or labelled ligands in Coplin jars (up to 10 slides per 20 ml medium). Alternatively, apply approximately 50 μl of the solution containing radiolabel to the slice (for example a rat brain slice) and cover with a coverslip cut to size from Parafilm (clean side down!). This is useful in the case of valuable ligands or ligands subject to air oxidation (55). The normal criteria for establishing total, non-specific and specific binding must be applied (Chapter 4). The usual precautions must be taken to protect the integrity of the ligand and receptor during incubation (Chapter 4).

4. Wash the slides in fresh binding buffer at 0°C for 2–30 min depending on ligand dissociation rate. Use several sequential washes (Chapter 4). Finally rinse the slides in ice-cold distilled water.

5. Place the slide on a metal plate cooled in ice, and dry the sections with a hairdryer (cold setting).

6. Count in liquid scintillation counter. Scrape dried slices off the slide with a spatula or razor blade, or remove wet slices by wiping with a Whatman GFC filter. Transfer to a detergent-based aqueous scintillator, and count. Alternatively, dissolve the slice by incubation with Soluene 350 (Packard), 1 ml overnight, then add 20 ml of a non-aqueous scintillator fluid. The use of plastic scintillation vials helps to reduce background.

2.1.5 Detecting labelled receptors by autoradiography

Protocol 4. Autoradiography

1. Tightly appose the slices to tritium-sensitive film, for example LKB Ultrofilm in a darkroom, and expose in X-ray cassettes at room temperature. Allow enough cassettes to permit development after 1, 2, 4, 8 weeks to optimize exposure time. Note that these films lack a protective surface coating. This means that they have to be handled with care to avoid scratching the emulsion. X-ray film is sufficient to resolve gross anatomical features, for example brain nuclei.

2. Develop the films by hand, following the manufacturer's instructions, for example 5 min in Kodak D19 developer at 20°C followed by a 20 sec rinse in a stop bath or distilled water and 5 min in Kodak fixer. Wash for 15 min in running water then hang up to dry.

3. Stain the slices after exposure, for example with cresyl violet (neural tissue, see *Protocol 5*); or eosin or methyl green (general cytoplasmic staining). Avoid stains that leave a granular deposit. Alternatively, expose the slices to nuclear emulsion[a] for better anatomical resolution.

4. Analyse autoradiographs. X-ray films are analysed by densitometry. The images are often colour-coded. This is a specialized topic, well described in reference 1.

---

[a] *Nuclear emulsion.* The preparation of nuclear emulsions is described in several texts on autoradiography (including an Amersham monograph). Emulsions of choice include Ilford K5 and Kodak NTB3. These melt at 45–50°C and are used as a 1:1 dilution with distilled water. Re-use of emulsions should be avoided. The high temperature needed to melt nuclear emulsions usually precludes the direct dipping of labelled slides since receptor–ligand complexes will tend to dissociate under these conditions. This is countered by apposing the labelled sections to coverslips which have been previously dipped and dried. The procedure is described in detail in reference 15. However, if an irreversible label is used, the section may be dipped directly.

Protocol 5. Staining and mounting autoradiographic sections

1. Stain for 2 min in 0.1% cresyl fast violet in distilled water at room temperature.

2. Wash quickly with 2 changes of distilled water.

3. Take slides through 50% alcohol into 70% alcohol. Dehydrate stepwise in 80%, 95%, and absolute alcohol. Clear in xylene.

4. Add a drop of DPX-mounting medium (dibutyl phthalate, 30 ml, Distrene 80, 60 g, xylene, 210 ml) to a clean coverslip. Take the slide out of xylene, wipe its back, apply to the coverslip, and press down the coverslip.

2.1.6 Detecting labelled receptors by immunocytochemistry

The use of immunocytochemical techniques to detect CNS antigens is described by Priestley (2). Immunofluorescence and enzyme-based procedures are both described. These techniques are equally applicable to receptor localization if a suitable antibody is available. The preparation of anti-receptor antibodies is described in Chapter 2.

2.1.7 Use of cultured cells

Cultured cells offer a useful system for studying receptor distribution since the entire cell membrane is accessible to the ligand. However, since cultured cells are usually attached to a hard substrate such as glass or plastic, freezing causes massive loss of morphology. Consequently, cultures can only be usefully employed if the cells can be fixed either prior to or subsequent to labelling. This greatly limits the number of receptor ligands that can be used. Irreversible ligands such as α-bungaratoxin, (used for labelling subsets of nicotonic receptors) or propylbenzilylcholine mustard, (used to label muscarinic receptors) have both been successfully used, and have also been combined with immunocytochemistry to identify labelled cells (16–19). Since these agents are irreversible, extensive washing and chemical fixation can both be used to greatly lower background. Cultures are best manipulated if they can be grown on coverslips that can be placed on the bottom of 24-well dishes or 35 mm plates. At the end of the washing procedures the coverslips can be permanently mounted on 3″ × 1″ slides and processed as usual. Alternatively, cultures can be grown in tissue culture slides which contain pre-formed wells allowing labelling of up to eight different cultures on the same slide. After labelling, the wells and gasket can be carefully removed and the slide treated as a conventional microscope slide. Reversible ligands can also be used as long as the receptor–ligand complex is stable to fixation. This was the case in a study that used [^{125}I]ICYP to label the β-adrenergic receptors on cultured glia (20). In this case, the cultures could not be dipped in nuclear emulsion but had to be apposed to emulsion-coated coverslips; this led to a lower resolution than the use of irreversible ligands as judged by the spread of autoradiograph grains around the outside of the cells. Cultures are rarely improved by histological staining, but are best mounted in a mixture of PBS and glycerol (1:1) and the coverslip sealed with nail polish or better still, molten dental wax.

2.2 Detection of receptor mRNAs by *in situ* hybridization

There have been relatively few studies of receptor mRNA distributions mapped by *in situ* hybridization, although there are no special considerations for the use of *in situ* hybridization to localize any particular mRNA. A recent publication covers the topic of *in situ* hybridization exhaustively (21).

2.2.1 Choice of probe

Basically there are three types of probe:

i. Tailed oligonucleotide probes

These consist of synthetic deoxynucleotide probes that are labelled on their 3′ end with a tail of labelled deoxynucleotides. Such probes have several advantages:

- They can be made simply and cheaply (approximately US$3 per base) and do not need possession of a cDNA clone.
- Labelling is simple (generally less than 2 h).
- Specificity of the probe can be confirmed by generating a series of oligonucleotides corresponding to different regions of the mRNA that is being localized.
- Probes can be generated against domains specific to the mRNA of interest. This is especially important when localizing receptor mRNAs, since many receptors are derived from a family of receptor genes that share a great amount of sequence identity in a number of domains, but which also have other domains that are unique to each receptor.
- Background tends to be very low and non-specific labelling associated with areas of cell density, such as hippocampal cells, is generally not observed.
- As the probes are single stranded, all of the probe is available for hybridizing to the mRNA. With double-stranded probes the antisense strand can hybridize to either the mRNA or the sense strand probe, hence, effectively lowering the concentration of the labelled antisense strand.

Tailing of probes can be done using [^{35}S] α-deoxynucleotides which affords better resolution than ^{32}P-labelled probes since ^{35}S is a weaker β-emitter than ^{32}P.

ii. Nick-translated cDNA probes

These are generated by labelling a full length cDNA probe or a restriction fragment of that probe. Nick translation consists of randomly nicking the cDNA using DNase and simultaneously filling in the nick with $\alpha[^{32}P]dCTP$ using DNA polymerase. The main disadvantages of these probes are:

- Their double-stranded nature (see above).
- Their higher background and non-specific labelling of areas of high density.
- The likelihood that a large cDNA fragment may cross-hybridize to other members of a related gene family (this latter hazard can be avoided by looking for restriction sites that would result in a cDNA fragment from a predetermined region of the gene).

iii. Internally-labelled cRNA probes (riboprobes)
These are generated by inserting a cDNA fragment into a vector containing a promoter for a bacterial RNA polymerase, such as SP6 or T7. A number of vectors exist that contain an SP6 and T7 promoter at alternate ends of the multiple cloning sites into which the cDNA is placed, thus allowing the generation of both sense and antisense probes by simply choosing to use SP6 or T7 polymerase. During the polymerase reaction $[^{32}P]UTP$ is included leading to the synthesis of high specific activity probes. The main disadvantages of riboprobes are: (a) their high background and non-specific labelling, and (b) their synthesis is more involved and can lead to a heterogeneous mixture of probes due to the synthesized strand of RNA being prematurely released from the polymerase.

Whichever probe is used, one rule applies to them all—use fresh radioisotope and use the labelled probes quickly (preferably within 24 h). If an abundant mRNA is being localized some of these considerations can be ignored, but if a rare mRNA is being localized then background considerations rise dramatically in importance, and consequently the quality of the probe can become the limiting factor. Since most arguments favour the use of oligonucleotide probes the following methodology refers to their synthesis and use.

2.2.2 Purification of oligodeoxynucleotide probes

Many workers use the unpurified mix of oligonucleotides as they come off the automated synthesizer. However, it is desirable to work with as clean and homogeneous a probe as possible. A number of different methodologies exist including PAGE, HPLC, and purification through a one-use column (22) commercially available from Applied Biosystems.

PAGE purification is tedious and should only be used if a HPLC set-up is not available, or if the probe is not tritylated and hence not suited to column purification. Convenient probe sizes range from 30–80 mer. A convenient probe length is 48 (the rate of rise of melting temperature (T_m) slows around 40–50 bases), the following methodology refers to the use of 48 mers. If PAGE is to be used then probes should be deprotected by heating to 60°C for 1 h, dried in a vacuum centrifuge, and redissolved in sequencing dye. Probes can then be run on a 20% polyacrylamide gel containing 8 m urea. The position of the probe can be visualized after disassembly of the gel apparatus, wrapping the gel in Saran Wrap, placing on an enhancing screen, and viewing under short wave UV light. The consecutive bands corresponding to 48, 47, 46, 45, 44 mers, etc. can then be seen. The largest and most dominant band is then excised and placed into slots of 1% low melting point agarose gel. The agarose gel is then run sufficiently to let the oligonucleotide exit the polyacrylamide and enter the agarose. The DNA can then be purified by phenolic extraction and ethanolic precipitation. Alternatively, the oligonucleotide can be purified directly from the polyacrylamide by eluting overnight in 0.5 M ammonium acetate, 10 mM Mg acetate, 1 mM EDTA, 0.1% SDS.

Column purification is simple, fast (< 30 min) and cheap (around US$20). The yield is comparable with PAGE purification and demands no specialized apparatus. Complete instructions are provided by the manufacturer. A prerequisite is that the probe must be synthesized with its terminal trityl group left on (trityl-on synthesis).

2.2.3 Terminal deoxynucleotide labelling of oligodeoxynucleotides

Oligos should be resuspended in a small volume of 10 mM Tris/1 mM EDTA, pH 7.6, (TE) so that their final concentration is 50 μg/ml.

Protocol 6. Labelling oligos

1. Mix on ice:
   - 10 μl 5× tailing buffer (500 mM potassium cacodylate; pH 7.2; 10 mM $CaCl_2$, 1 mM DTT)
   - 0.1 μM 48 mer
   - 1 μM [^{35}S]dATP (NEN 1500 Ci/mmol)
   - 100 U terminal deoxynucleotidyl transferase (BRL or Boehringer Mannheim).
   - add H_2O to 50 μl

2. Incubate for 60 min at 37°C (length of time should be determined empirically to determine incubation period necessary to generate a tail length of about 10 bases).

3. Stop with 400 μl TE + 1 μl tRNA (25 mg/ml).

4. Extract with 450 μl phenol/chloroform/isoamyl alcohol (50:49:1); save the aqueous phase.

5. Extract with 420 μl chloroform/isoamyl alcohol (49:1); save the aqueous phase.

6. Precipitate by adding 1/10 volume of 2 M NaCl (40 μl) + 2.5 vol ethyl alcohol (final volume 1 ml).

Protocol 6. *Continued*

7. Precipitate on dry ice for 10 min.

8. Spin in a microfuge for 10 min at 14 000 *g* in the cold room.

9. Wash ppt gently in cold EtOH, remove EtOH and dry pellet in vacuum desiccator.

10 Resuspend pellet in 100 μl TE + 2.5 μl 2.5 M DTT.

11. Count: 1 μl should give 2×10^5 to 5×10^5 d.p.m./μl.

Alternatively, labelled oligonucleotides can be purified by passage through commercially available NenSorb (NEN DuPont) or Quiagen columns by following the manufacturer's instructions.

2.2.4 Hybridization

Sections are prepared as for *in vitro* autoradiography, with no prefixation. Slides should be removed from storage at −80°C and placed on clean aluminium foil for 10 min to allow them to come to room temperature. Solutions must be sterile. Preliminary steps should be conducted in autoclaved plastic Coplin jars (10 slides per Coplin jar). Cultured cells should be grown on glass coverslips and affixed to microscope slides, or grown directly on tissue culture slides. The procedure is described in *Protocol 7*.

Protocol 7. Hybridization with radiolabelled oligonucleotide probe

1. Fix tissue by immersion in 4% formalin/PBS for 10 min.

2. Rinse slides twice in PBS.

3. Acetylate by immersion in 0.25% acetic anhydride in 0.1 M triethanolamine HCl/0.9% NaCl (pH 8.0) for 10 min.

4. Dehydrate by passing through 70% ethanol (1 min), 80% ethanol (1 min), 95% ethanol (1 min), and 100% ethanol (1 min).

5. Defat in chloroform (5 min); defatting should be omitted from cultured cells.

6. Dehydrate in 100% ethanol (1 min) then 95% ethanol (1 min).

7. Dry by standing slides upright in a draining rack.

8. Place 50 μl (sufficient for one rat brain section) of hybridization buffer (*Table 1*) containing $1–2 \times 10^6$ d.p.m. probe/50 μl and 0.1 mM DTT over sections and cover with Parafilm coverslips (see Section 2.1.4). Smaller volumes can be used for small sections.

9. Hybridize overnight in a humidified chamber at 37°C (for example in a tissue culture incubator).

10. Float coverslips off in 1× SSC at room temperature. Transfer slides to

---

Table 1. Hybridization buffer

- *4 × SSC*
- 50% formamide—deionized by passage through a mixed-bed resin (BioRad, AG 501X8, 20–50 mesh)
- 10% dextran sulphate
- 500 μg/ml herring- or salmon-sperm sheared DNA
- 250 μg/ml yeast tRNA
- 1 × Denhardt's solution

1× SSC at 55°C for 15 min. Repeat four times. Transfer slides to 1× SSC at room temperature for 30 min. Repeat. Rinse in distilled water, and blow dry with warm air.

11. Slides are exposed to X-ray film, or dipped in nuclear emulsion, as described in Section 2.1.5.

2.3 Tissue preparation for binding studies

The most important steps are the following:

i. Identify the tissues of interest
Consult an appropriate atlas of anatomy (for example ref. 23) or better still an anatomist familiar with the dissection.

ii. Sacrifice the animal
Stunning by striking the back of the head followed by immediate decapitation with a guillotine is appropriate for rats. Cervical dislocation is acceptable for guinea-pigs and mice. Rabbits are best anaesthetized by exposure to a rising concentration of CO_2 followed by cervical dislocation or decapitation. These procedures are covered by Section 1 of the Animals (Scientific Procedures) Act 1986 (UK) and expert advice and training in these techniques should be sought and any appropriate licence obtained.

One point to note is that to reduce trauma, animals should not be sacrificed in front of their cage mates.

Problems for the operator include bites, and the possible development of allergies. To minimize the latter, it may be advisable to wear disposable surgical gloves and a face mask.

If possible, anaesthesia should be avoided. Anaesthetics are always membrane-perturbing agents. Some exert specific effects on pharmacological receptors. For instance, $GABA_A$ receptors have a specific binding site for barbiturates. However, ether anaesthesia has been used prior to autoradiographic studies (14).

iii. Rapidly dissect and remove the tissue of interest
Place in a Petri dish on ice. If possible, rapidly perfuse with homogenization buffer through a major artery using a hypodermic syringe and needle. Dissect away as much connective tissue and fat as possible. Place the tissue in a weighed beaker containing a known weight of buffer, cooled on ice. Reweigh the beaker at the end of the dissection procedure, to determine the total weight of tissue. If ex-sanguination is critical the whole animal can be perfused with buffer by cardiac perfusion, following the protocol of Priestley (2), before dissection is commenced. This is obviously a sensible precaution if the receptor in question occurs on blood cells as well as in the tissue of interest.

iv. Larger amounts of tissues can be obtained from abattoirs
Fresh tissue is desirable—no more than 30 min post-mortem. If post-mortem stability is a problem, cut the tissue into bite-sized (about 2 cm) chunks before immersing in ice for transport to the laboratory.

v. Human tissues are usually post-mortem or biopsy samples
They are often cooled quite slowly. Successful binding studies can often be performed even on material which has been kept for 24 h at 4 °C. However, there may be considerable autolysis, and possible proteolysis of extra-membraneous regions of the receptor. A useful source of fresh human tissues is the placenta. This has been used fruitfully in a number of receptor studies (see *Table 2*) and may be worth screening.

vi. Freezing whole tissues for binding studies is usually undesirable
However, for some studies, for example those looking at drug effects where multiple tissue samples are accumulated, there may be no alternative. In this case, if possible, snap-freeze the tissue in cryotubes by immersion in liquid nitrogen in a dewar. We have successfully frozen small (about 2 mm) cubes of

---

Table 2. Partial list of receptors present in human placenta

Receptor	Reference
EGF receptor	*J. Biol. Chem.*, **255**, 10737
Calcitonin gene-related peptide R	*Eur. J. Biochem.*, **170**, 373
β_1/β_2 adrenergic	*Receptor Biochemistry: a Practical Approach.* IRL Press, Oxford
Opioid	*Life Sci.*, **44**, 861
PDGF	*FEBS Lett.*, **219**, 331
Insulin	*J. Biol. Chem.*, **262**, 6939
IGF$_1$	*J. Biol. Chem.*, **265**, 17820.

rat brain in glass vials dipped in a solid CO_2/isopropanol freezing mix. Alternatively, small pieces of tissue can be immersed in powdered solid CO_2, or in a slurry of isopentane cooled in liquid N_2. The tissues should be stored at $-70°C$ in the smallest possible container, if necessary containing a small amount of ice to avoid freeze-drying. The tissues can be thawed under the appropriate volume of cold buffer before homogenization. Alternatively, frozen tissues can be disrupted with a polytron without thawing, thus instantly diluting proteolytic enzymes which may be released by freezing damage.

The effect of freezing the tissue on the receptor-binding properties should be checked. Some receptors are quite resistant, others are not. There are species and developmental differences, for example adult rat brain muscarinic acetylcholine receptors are resistant to freezing the tissue, while neonatal rat and frog brain receptors are not. Explore the possibility of cryoprotection, by keeping the tissues in 0.1 M sodium phosphate buffer, pH 7.4, containing 7% sucrose, prior to freezing (2). In general, it is better to freeze receptor preparations as membrane preparations, rather than in the form of whole tissues. Usually membrane preparations can be stored at $-70°C$ for months or even years without deterioration of receptor-binding properties. However, storage at $-20°C$ for more than a few weeks is not recommended.

2.4 Specific protocols

In this section, we give a few specific protocols which we have found to be useful in obtaining preparations of brain, heart, glandular, and smooth muscle receptors.

2.4.1 Removing a rat brain

Protocol 8. Removal of a rat brain

1. Stun and decapitate the rat with guillotine or shears.

2. Lift and snip away the skin between the ears, working forwards towards the nasal region. Remove the strip of skin.

3. Gripping the skull firmly in front of the ears, between the thumb and forefinger of the left hand (if right-handed) and holding the head against an absorbent surface (for example a wad of paper towels), insert the point of a pair of sharp scissors into the foramen magnum. Cut along the midline with the tips of the scissors, extending the cut through the nasal sinuses, so dividing the top of the skull into two parts. Using a pair of scissors with one blunt and one fine point helps to avoid accidents during this procedure. Close the scissors, and insert their points into the cut in the sinus region. Rotate the scissor points through 180° and open, thus splitting the skull down the midline.

Protocol 8. *Continued*

4. With the blunt, angled end of a small spatula, work around the ventral surface of the brain, severing the cranial nerves. Lift the brain out of the skull.

2.4.2 Crude dissection of a rat brain

A crude but rapid blunt dissection of a rat brain into regions can be accomplished in the following manner. You will need a small spatula with a blunt, angled end, a pair of fine-pointed sharp scissors, a pair of fine-pointed curved forceps, and a fine-pointed scalpel. Microdissection of individual rat brain nuclei is described by Palkovitz and Brownstein (24).

Protocol 9. Blunt dissection of a rat brain

1. Place the brain with its ventral side on a piece of filter paper dampened with buffer.

2. Remove the cerebellar hemispheres from the hindbrain with the curved, angled end of the spatula.

3. Section through the pons just behind the occipital cortex to obtain the pons–medulla.

4. Remove the olfactory bulbs.

5. With the round end of the spatula, peel back the cerebral cortices from the midline. With the back of the spatula press them gently on to the filter paper so that they adhere.

6. With the spatula, carefully work round, and then scoop out, the two comma-shaped caudate nuclei, which are readily recognized by their striated appearance.

7. Insert the spatula under the hippocampi and septum and free them from the midbrain, initially working from the midline laterally. Snip off the hippocampi with a pair of fine-pointed scissors, and remove with fine-pointed forceps.

8. With the tip of the spatula, work underneath the midbrain to cut the midbrain, thalamus, and hypothalamus away from the cerebral cortices.

9. From the cortices, scrape off as much of the white matter (corpus callosum) as possible. Lift the cortices away from the filter paper using the spatula. The meninges will remain adherent to the filter paper.

10. Turn the midbrain over. With a fine-pointed scalpel blade, work round the hypothalamus. Steadying it with forceps, separate the hypothalamus from the thalamus by a scissor-cut parallel to the ventral surface of the midbrain.

11. The remaining tissue consists of the midbrain/thalamus. Working in a cold room renders the brain firmer during this dissection procedure.

2.4.3 Removing a rat heart

Protocol 10. Removal of a rat heart

1. Decapitate the rat and lay it on its back on a wad of paper towels.

2. With the aid of a pair of strong toothed-forceps, lift the abdominal skin. Just below the sternum, snip and cut forward up the midline. Peel back the skin flaps.

3. Make a small cut through the diaphragm, and expose the xiphoid cartilage. Grip it firmly with a pair of stout forceps, or artery forceps, and lift. Cut through the ribs on either side of the chest cavity. Fold back and cut away the sternum, and associated ribs, exposing the heart.

4. Gripping the apex of the ventricles with forceps, lift the heart, carefully cutting round it with scissors to free it from the pericardium and chest cavity.

5. Cut through the aorta and vena cava. At this point, buffer can be injected into the vena cava, to help exsanguinate the heart.

6. Snip through the aorta and other vessels. Remove the heart into a cold Petri dish. Inject buffer into the ventricles to remove the rest of the blood. Trim away the remnants of the blood vessels with fine-pointed scissors.

7. The atria can be removed by gripping them with fine-pointed curved forceps, and snipping them away at their junction with the ventricles, with fine-pointed scissors.

2.4.4 Removing rat exorbital lacrimal glands

Protocol 11. Removal of rat exorbital lacrimal glands

1. Decapitate the rat and place head upside down on a wad of paper towels

2. The lacrimals are discrete, dark-brown glands in their own sheaths. The decapitation may expose them. If not, cut the skin up the midline to a point level with the nasal side of the face. Grip the flap, and pull to expose the underlying tissues.

3. The lacrimals are located above and to one side of a prominent pair of muscle blocks, and are usually just visible in the hinge of the skin flap where it remains attached to the head, just below the level of the ear flaps.

Protocol 11. *Continued*

They are likely to be covered in connective tissue and partly surrounded by fat. Their colour distinguishes them from the surrounding tissue.

4. Carefully cut away the connective tissue and fat, and expose the glands. Lift them and cut them free of the head.

2.4.5 Guinea-pig ileum longitudinal smooth muscle

You will need scissors, fine-pointed, curved forceps, and a glass rod or 1 ml glass pipette.

Protocol 12. Preparation of guinea-pig ileum longitudinal muscle

1. Starve the guinea-pigs for 24 hours to clear the intestines.

2. Sacrifice by cervical dislocation. Grip the neck of the guinea-pig between the index and middle finger, with the middle finger across the trachea. Give a sharp, determined flick, as if cracking a whip.

3. Touch the cornea, and ensure the absence of a blink reflex to be certain that the animal is dead before proceeding.

4. Lift the abdominal skin, and make a midline incision through the abdominal wall. Wearing surgical or disposable gloves, with your fingers, grip and rapidly pull the ileum away from the mesentery. Squeeze any faecal material as far down towards the colon as possible. Cut the ileum at either end.

5. Immediately flush the ileum out with isotonic saline contained in a wash bottle. Squeeze out, and put the ileum in a beaker of saline solution.

6. Cut off about a 20 cm section of ileum and work it on to a wetted, round-tipped glass rod. A 1 ml glass pipette is ideal.

7. Hold the top of the ileum section on the glass rod between the index finger and thumb of the left hand (if right-handed). Dampen a piece of paper tissue in saline and gently, with a stroking motion perpendicular to the long axis of the rod, free a circumferential cuff of longitudinal muscle from the end. Grip it with forceps, and gently tear it away by means of longitudinal and lateral tension, working gradually down the section of ileum. This requires a lot of patience, and usually several restarts. The action required is not dissimilar to that involved in removing, intact, a large piece of sunburned skin.

8. Place the longitudinal muscle in a cold, weighed beaker containing buffer.

---

3. Homogenizing tissues

Homogenization of tissues for binding studies is usually straightforward. In many cases, a crude whole-tissue homogenate suffices, although various stages of refinement are possible.

i. The introduction of washing steps

Basically this means centrifugation of the entire membrane fraction followed by resuspension in buffers which may be enriched with additives such as chelating agents and proteolytic inhibitors (see Chapter 4).

ii. Subcellular fractionation procedures

These range in sophistication from subcellular fractionation techniques designed to give more-or-less defined preparations of subcellular organelles (such as synaptosomes and synaptic plasma membranes, post-synaptic densities and endosomes) to affinity-phase partitioning or immunoaffinity techniques designed to provide membrane fractions enriched in a specific receptor or transmitter. Examples of both general and specific fractionation procedures are provided in two companion books in this series. Gordon-Weeks in ref. 2 provides a comprehensive account of the fractionation of neuronal tissues including:

- preparation of synaptosomes
- preparation of synaptic plasma membranes
- preparation of post-synaptic densities

Bailyes, Richardson, and Luzio in ref. 13 give a protocol for the purification of cholinergic synaptosomes using monoclonal antibodies to the Chol-1 antigen, while also discussing the deficits and disadvantages of this approach. Evans in ref. 13 provides a guide to the isolation of subcellular organelles, including endosomes, which are enriched in receptors internalized under the influence of agonists. It is not proposed to duplicate all of these protocols here. However, it is worth commenting that for highly-specific ligands, the effort of purifying brain synaptosomes does not usually justify the relatively modest degree of enrichment of binding characteristically obtained (for example 2-fold for muscarinic receptors). In contrast, the isolation of post-synaptic membranes has been valuable in enabling studies of sodium-independent binding of glutamate analogues to be done (25).

By and large, affinity purification methods have not been systematically applied to membrane preparations used in receptor-binding studies, although there are honourable exceptions (26,27). In general, the yields are insufficient for large-scale binding studies. However, given the increasing availability of high-level expression systems, this situation may change in the future.

3.1 Soft tissues

Tissues such as brain and liver can be homogenized using a Potter–Elvehjem homogenizer with a rotating Teflon pestle. A clearance of 0.2 mm is desirable. Volumes of 50 ml and 10 ml are convenient. Suitable items are provided by most lab. suppliers (for example Jencons, UK). It is useful to have a screw-on handle for the pestle, to permit hand operation for resuspending membrane fractions. A flexible high-torque motor drive is convenient, giving a speed of rotation of about 500 r.p.m. Models with foot pedal-operated switches are available (for example Cole Parmer (USA), Philip Harris Scientific (UK)). We have found it satisfactory to mount the pestle in the chuck of an overhead power drill, rotating at about 300 r.p.m., although the rigidity of the mounting means that care is necessary not to break the homogenizer.

3.1.1 Homogenization of soft tissues

Protocol 13. Potter–Elvehjem homogenization

Perform the following steps at 0–4°C.

1. Weigh the tissue. Take 9 vol (i.e. 9 ml/g wet weight) of ice-cold buffer.
2. Add enough buffer to cover the tissue. With a pair of scissors held vertically, chop the tissues roughly into chunks of about 0.5 cm or less.
3. Add the remaining buffer. Suspend the tissue and pour it into the homogenizer, precooled to 0°C in ice. Do not add more than about 40 ml to a 50 ml homogenizer.
4. Homogenize with 10 or more deliberate down-and-up strokes of the pestle at 300–500 r.p.m. Take care to move the pestle up and down vertically. Hold the homogenizer in a relaxed manner, so that any unevenness in the movement of the pestle can be accommodated. Be careful to avoid jamming the pestle on the downstroke. Do not create evacuated cavities during the upstroke, this may cause receptor denaturation. You are strongly advised to wear protective leather gauntlets, in case the homogenizer breaks, and eye protection during this operation. The end result should be a smooth, creamy homogenate.
5. Pour the homogenate through two layers of cheesecloth drooped over a large (approximately 200 ml) conical filter funnel. Collect the filtrate in a conical flask cooled in ice.

3.2 Tough tissues

Tissues such as heart, salivary glands, and smooth muscle are best homogenized with a Willems Polytron (Brinkman; Kinematica), IKA Ultra-Turrax (Kinematica) or Omni-mixer (Sorvall). These instruments mostly rely on a blender principle, with rotating blades (rotor) either exposed (Omni-mixer) or contained within a cylindrical cavity (stator) through which the homogenate circulates (Polytron and Ultra-Turrax). In the case of the Polytron, the instrument which we use routinely, there is a considerable ultrasonic component in the dispersive process. We find the Polytron PTA10 probe with saw-tooth stator and saw-tooth rotating knives convenient for volumes up to around 200 ml. A range of probe attachments is available which differ in whether they possess rotating knife blades (necessary for tough tissues) and in the degree of aeration which they engender during homogenization (minimal aeration may be desirable). Soft tissues can also be dispersed with these instruments, or with an ordinary household blender. However, the resultant homogenate will not necessarily have the same properties as one produced by Potter–Elvehjem homogenization. For instance, synaptosomes, whose production is dependent on the application of a regular shearing force to nip off the nerve terminals, cannot be prepared from a Polytron homogenate of brain; the sedimentation properties are entirely different. Vigorous Polytron homogenization tends to create numerous small sealed vesicles.

3.2.1 Homogenization of tough tissues

Protocol 14. Polytron homogenization of tough tissues

1. Suspend and chop the tissues as described in Section 3.1. Pour into a beaker or other vessel of suitable size immersed in ice, so that the circulation holes in the Polytron probe are properly covered. The tip should be about 2 cm above the bottom of the vessel.

2. Homogenize with 4×15 sec bursts at setting 7. Leave the homogenate to cool for 2 min between bursts—be patient! If the tissue is lumpy, homogenize at a lower speed setting initially.

3. Inspect the blades, and mechanically remove any entanglements with forceps, to ensure good disruption, if homogenizing a fibrous or stringy tissues, such as ileum longitudinal muscle.

4. Finish the process of homogenization with 10 strokes of a Potter–Elvehjem homogenizer (0.2 mm clearance, 300–500 r.p.m.).

5. The end result should again be a smooth homogenate. Pour through two layers of cheesecloth as in Section 3.1.1.

The above procedures can both be applied to cultured cells. These should be harvested by mechanical removal from culture dishes with the aid of a rubber 'policeman' or Teflon scraper and recovered by a slow-speed spin (10 min, 500 g) before homogenization. The addition of EDTA or EGTA (1 mM) often aids removal of adherent cells. The use of enzymes should, preferably, be avoided. The cells may be swollen by the addition of a hypo-osmotic buffer for 30 min before homogenization using up to 50 strokes of a Potter–Elvehjem (Teflon in glass) or Dounce (glass in glass) homo-genizer.

4. Specific protocols for membrane preparation

In this section we will detail some fairly crude membrane preparations which we have found to be adequate for most membrane-binding studies. All steps are performed at 0–4°C unless otherwise specified. Use Analytical grade reagents, and deionized or double-distilled water.

4.1 Synaptosome preparations from brain

A crude synaptosome (P2) fraction represents a first stage of refinement over the crude homogenate. It is relatively homogeneous, and easy to handle in binding studies.

Protocol 15. Crude synaptosome (P2) fraction from brain

1. Potter–Elvehjem homogenize brain (10 up-and-down strokes) in 9 volumes (w/v) of 0.32 M (iso-osmolar) sucrose containing 20 mM buffer adjusted to pH 7.4 (for example with sodium phosphate, or Hepes). Measure the volume (V).

2. Centrifuge the homogenate in approximately 50 ml aliquots at 1000 g for 5 min. Carefully pour off the supernatant, taking care not to disturb the pellet (designated P1) which is enriched in cell nuclei, unbroken cells, and brain microvessels.

3. Centrifuge the supernatant (S1) in approximately 50 ml aliquots at 10 000 g for 20 min to pellet the crude synaptosome fraction, P2. Pour off the supernatant (S2 fraction).

4. The P2 fraction can be resuspended for binding studies by gentle vortex mixing in approximately 20 ml of sucrose or buffer per pellet. The P2 pellet consists of two layers; a brownish bottom layer enriched in heavier mitochondria, and a light-coloured upper layer enriched in synaptosomes (sometimes designated the buffy coat). With care, resuspension of the upper layer alone can be achieved. The upper layer contains about 50% of the protein content of the original homogenate. Resuspension to half the original homogenate volume thus yields a preparation equivalent to 0.1 g wet weight tissue/ml, corresponding to a protein concentration of about 10 mg/ml. Thus, pour off the resuspended P2 fraction into a measuring cylinder, and resuspend to a volume of V/2 ml. Ensure even resuspension with a few strokes of the Potter–Elvehjem homogenizer.

5. If resuspension in buffered iso-osmolar sucrose is used, the fraction can be used as a starting point for synaptosome preparations using sucrose, Ficoll or Percoll gradients (ref. 2; see *Protocol 16*).

6. Use the P2 fraction immediately in binding studies, or recover it by recentrifugation (100 000 g, 30 min) for freezing and storage. Freeze by immersing the Parafilm-covered tube in a dry ice/isopropanol slurry, and store at $-70°C$; gradual degradation occurs if preparations are stored at $-20°C$.

The P2 fraction obtained is typically around 1.5-fold enriched in synaptic receptors (for example muscarinic acetylcholine receptors) compared to the original homogenate. The total preparation time is approximately 1 hour. Resuspension and recentrifugation of the S1 fraction, which slightly enhances the yield has been omitted in *Protocol 15* in favour of speed.

A further fraction, the crude microsomal fraction, P3, can be obtained by centrifuging the S2 fraction for 60 min at 100 000 g. This produces a further small pellet, which may be somewhat more enriched in receptors (for example 2-fold in mAChRs on a protein basis)

We have found small, but significant, differences in the binding of muscarinic agonists to the three fractions P1, P2, and P3, and in their responsiveness to guanine nucleotides. This may reflect differences in their origin and nature. P1 is enriched in nuclei, cellular detritus, erythrocytes, and brain microvessels, which are known to contain a diverse population of receptors. P2 is enriched in synaptosomes, i.e. sheared-off nerve terminals often with a piece of post-synaptic membrane attached. P3 is enriched in vesicles and membrane fragments, mostly derived from non-synaptic regions of neurones, and from glia.

A more purified synaptosome preparation can be obtained as detailed in *Protocol 16*.

Protocol 16. Preparation of synaptosomes

1. In a 38 ml centrifuge tube (for example Beckman SW27) make a discontinuous sucrose gradient using the following sucrose concentrations, buffered to pH 7.4:

   • 10 ml, 1.2 M

   • 10 ml, 1.0 M

   • 10 ml, 0.85 M

   Layer the less dense on top of the more dense solutions using a Pasteur pipette. Cool the gradients to 4°C.

2. Layer 8 ml of resuspended P2 fraction in 0.32 M sucrose on top of each gradient.

Protocol 16. *Continued*

3. Centrifuge at 100 000 *g* in a swing-out rotor for 2 h at 4 °C.

4. The synaptosome-enriched fraction accumulates at the 1.0/1.2 M interface. Aspirate the overlying material and then carefully remove the interfacial band with a Pasteur pipette. Dilute the synaptosomes 1:1 with buffer, and recover by centrifuging at 100 000 *g* for 30 min.

Numerous variations of this procedure are possible; see reference 2 for more details.

The protocol described above is for synaptosomes, but can be applied to other sealed structures (for example whole cells).

4.2 Lysis of crude synaptosomal fractions

Protocol 17. Hypo-osmotic lysis of synaptosomes

1. Resuspend the fraction, pelleted from iso-osmotic buffer (for example 0.32 M sucrose) in a low ionic strength lysis buffer[a] (for example 5 mM Tris–HCl, pH 8.1) with the aid of 6 strokes of a hand-operated, Teflon-in glass homogenizer of normal (0.2 mm) clearance.

2. Stand the fraction on ice for 30 min to swell. This renders sealed structures fragile, and susceptible to mechanical breakage.

3. Re-homogenize with up to 60 strokes of a tight-fitting, glass-in-glass Dounce homogenizer (0.1 mm clearance).

4. Recover the membranes by centrifugation for 30 min at 100 000 *g*.

[a] The addition of 50 μM CaCl$_2$ to the lysis buffer has been recommended (2). Note, however, that this may activate Ca-dependent proteases.

Cells can be rendered leaky by the addition of low, sub-solubilizing concentrations of agents such as saponin (0.005–0.1%; ref. 28) or digitonin (0.005–0.1%; ref. 29). The effects of such additions on receptor binding needs to be assessed. Cell leakiness can be monitored by measuring the ability of the cell to exclude trypan blue (0.3%). Alternative means of producing cell permeability to small molecules, without producing release of larger molecules, such as cytoplasmic enzymes are to use the fusogenic Sendai virus (100 haemagglutinating units per ml will permeabilize rabbit peritoneal neutrophils; ref. 30) streptolysin O (0.7 International Units per ml; ref. 31) or to use electroporation by high electric fields (32) for which commercially-available equipment is now very widely advertised (for example Braun Biojet M1). A very useful permeabilizing agent is the α-toxin from the *E. Coli* strain Wood 46 (33) which produces efficient permeabilization of eukaryotic cell-surface membrane to small molecules (molecular mass approximately 1000) without impairing intracellular organelles.

Freeze-thawing can also be used to break vesicles, in conjunction with the above techniques. Again, this may release and activate intracellular proteases. Controlled sonication may also prove useful.

The degree of breakage of sealed structures can be monitored directly by phase-contrast microscopy or by assaying the release of cytoplasmic enzymes, such as lactate dehydrogenase. Evans in ref. 13 gives a useful list of cell breakage techniques.

4.3 Washed membrane fractions

In many instances, a relatively crude, washed membrane fraction suffices for binding studies. In our laboratory, we routinely use EDTA-washed membrane fractions as a starting point. EDTA is an effective proteolytic inhibitor. It may also be used in combination with other agents. For instance, a combination of EDTA (1 mM) and bacitracin (200 μg/ml) has been shown to inhibit proteolytic effects on muscarinic receptors in myocardial membrane fractions.

A discussion of proteolysis in binding studies is given in Chapter 4. It is avoided by:

- using fresh tissues
- working at low temperature during membrane preparation
- adding protease inhibitors, particularly EDTA

4.3.1 EDTA-washed membrane preparations

We have used these routinely in studies of muscarinic acetylcholine receptor binding, and as a starting point for receptor purification. The muscarinic receptor appears unproteolysed in such membrane preparations, assessed by affinity labelling and SDS-PAGE.

Protocol 18. EDTA-washed membrane preparations

1. Homogenize the tissue in ice-cold buffer containing 10 mM EDTA. We routinely use 20 mM Na-Hepes, 10 mM EDTA, pH 7.5. The tissues may be brain, heart, or glands. Filter the homogenate through two layers of cheesecloth.

2. Recover the membranes by centrifugation at 70–100 000 g for 30 min at 4°C. This can be done on a large scale in a preparative high-speed rotor (for example Beckman type 19, 6 × 250 ml).

3. Discard the supernatant. Resuspend the tissue to the original homogenate volume in fresh buffer with 10 mM EDTA, and incubate with slow stirring on ice for 30 min.

Protocol 18. *Continued*

4. Re-centrifuge as in step 2.

5. Pour off the supernatant. Resuspend the pellet to the original homogenate volume in a buffer containing a lower, defined concentration of EDTA (for example 1 mM, 0.1 mM).

6. Re-centrifuge as in step 2.

7. Pour off the supernatant. Resuspend the pellet as in step 5. Dispense into 50 ml centrifuge tubes.

8. Centrifuge as in step 2. Pour off the supernatant. Drain the pellets carefully. Snap-freeze the pellets (dry-ice/isopropanol) and store at −70°C; preparations remain stable for several months.

Depending on the aim of the procedure, *Protocol 18* may be cut short. It may be possible to omit the 30 min incubation in 10 mM EDTA (step 3). Multiple washes at lower EDTA concentration are only necessary if a defined EDTA concentration is desired in the final assays, for example if a defined divalent cation concentration is to be used. EDTA-pretreatment solubilizes certain membrane-associated proteins, probably including some proteases. We have found that muscarinic receptor–G-protein interactions are well preserved or even enhanced after such pretreatment, provided that a suitable (0.1–10 mM) concentration of Mg is added back. However, if EDTA pretreatment is deleterious to the effect being studied, a washed membrane fraction can be prepared by the above procedure without EDTA. If the aim is selectively to deplete Ca, EGTA can be substituted for EDTA.

4.3.2 KCl/pyrophosphate-treated membranes from rat heart

The following high-salt washing procedure diminishes the amount of acto-myosin remaining in the preparation and thus dissolves myofibrils. High affinity muscarinic agonist binding to membranes so prepared displays enhanced stability compared with membranes pretreated with EDTA alone (34).

Protocol 19. KCl/pyrophosphate-treated membranes from rat heart

1. Make a 10% homogenate of rat heart in ice-cold 20 mM Na-Hepes, pH 7.5. Filter through cheesecloth as in *Protocol 13* step 5. Measure the volume, designated *V*.

2. To the homogenate, add $0.8 \times V$ of 20 mM Na-Hepes and $0.2 \times V$ of 3 M KCl/250 mM disodium pyrophosphate, cooled on ice and adjusted to pH 7.5, and mix rapidly.

3. Centrifuge at 140 000 *g* for 60 min.

4. Resuspend the pellet to the original homogenate volume, V, in 20 mM Na-Hepes. Supplement with 100 mM EDTA (pre-adjusted to pH 7.5 with NaOH) to give a final concentration of 10 mM EDTA.

5. Continue as in *Protocol 18* step 3.

Examples of preparative-scale enrichment of receptors in membranes containing a low abundance of the receptor in question are rare. An exception is the preparation of an approximately 40-fold muscarinic receptor-enriched membrane fraction from myocardium (35). This large degree of purification was instrumental in enabling the cardiac muscarinic receptor to be purified. A protocol based on this procedure is given by Haga *et al.* in *Receptor Biochemistry, a Practical Approach* (10).

5. Dissociated cells

The establishment of primary culture of neurons and glia isolated from neonatal rat nervous system has been described by Saneto and de Vellis in ref. 2. Both mechanical and enzymatic methods of disruption can be used. The use of clonal cell lines of tumour origin, neuroblastomas, gliomas, and neuroblastoma–glioma hybrids in receptor studies is also well established. Methods for growing and monitoring these lines are well established (see *Animal Cell Culture, a Practical Approach*, ref. 11) and will not be duplicated here. Collections of available cell lines are maintained by the American Type Culture Collection (12301 Parklawn Drive, Rockville, Md 20852) and the ECACC, Porton Down, Salisbury, Wilts., UK. Catalogues of their products are available on subscription. *Table 3* lists some cell lines which express neuro-transmitter and peptide receptors, with appropriate references. In the past few years, such cell lines have been supplemented by a large number of cell lines which have been transfected with cloned receptor genes in expression plasmids (see *Receptor Biochemistry, a Practical Approach*, Chapter 11). Such transfected cell lines are becoming an increasingly important resource for receptor studies. Their use in studying receptor function has recently been reviewed (36).

In this section we give example protocols which have proved useful for the production of relatively pure preparations of dissociated cells from peripheral tissues, in a form suitable for studies of both binding and receptor–response coupling. These protocols require a combination of enzymic digestion (trypsin and/or collagenase) and mechanical disruption. The reader should note that cell-surface receptors are vulnerable to trypic cleavage, and may, therefore, be partly degraded by the enzymic treatment. A period of incubation or culture is, therefore, advisable before experiments commence, to allow the population of cell-surface receptors to recover. The availability of naturally isolated receptor-bearing cells should not be overlooked. Obvious examples

Table 3. Some cell lines expressing pharmacological receptors

Receptor type	Cell line	Reference
Muscarinic AChR	PC12 (phaeochromocytoma) (M4 MAChR)	*Nature*, **297**, 152
	NG108-15 (neuroblastoma/glioma) (M4 MAChR)	*Mol. Pharmacol.*, **32**, 443
	1321N1 astrocytoma (M3 MAChR)	*Mol. Pharmacol.*, **27**, 32 and *BBRC*, **147**, 182
	NIE115 neuroblastoma	*Mol. Pharmacol.*, **27**, 223
	NCB-20 neuroblastoma × brain hybrid	*J. Neurochem.*, **51**, 505
	Flow-9000 pituitary	*FEBS Lett.*, **220**, 155
	SH-SY.5Y neuroblastoma (M3 MAChR)	*Devel. Brain Res.*, **33**, 235
	NB-OK#1 neuroblastoma (M1 MAChR)	*FEBS Lett.*, **226**, 287
	SK-N-SH neuroblastoma (M3 MAChR)	*BBRC*, **154**, 1137
	Fibroblasts (lung)	*Eur. J. Biochem.*, **171**, 401
Nicotinic AChR	BC3H1 muscle	*J. Cell Biol.*, **61**, 398 *J. Physiol. Lond.*, **385**, 325
	PC12	*Nature*, **319**, 368
β-Adrenergic R	N1E115	*Mol. Pharmacol.*, **27**, 223
	S49 lymphoma	*Annu. Rev. Cell. Biol.*, **2**, 391
	DDT1-MF2 smooth muscle	*Proc. Natl. Acad. Sci. USA*, **86**, 4853
	IMR90 fibroblasts (lung)	*J. Cell Physiol.*, **130**, 163
	3T3-L1 pre-adipocytes	*Eur. J. Pharmacol.*, **143**, 35
	A431 (β_2-AR)	*Eur. J. Biochem.*, **167**, 449
α-Adrenergic R	DDT1-MF2 (α_1-AR)	*J. Biol. Chem.*, **262**, 3098
	DDT1-MF2 (α_1-AR)	*J. Pharm. Exp. Therapeut.*, **243**, 527
	HT29 adenocarcinoma (α_2-AR)	*Mol. Pharmacol.*, **32**, 646
	NG108-15 (α_2-AR)	*J. Biol. Chem.*, **262**, 6750 and *FASEB J.*, **2**, 52
5-HT R	N1E115	*Mol. Pharmacol.*, **33**, 303
	NG1085 (5-HT3 R)	*Eur. J. Pharmacol.*, **143**, 291
	CHL fibroblasts	*Nature*, **335**, 254
	NCB20	*J. Neurochem.*, **49**, 183
Histamine	NCB20	*J. Neurochem.*, **51**, 505
Dopamine	NCB20	*J. Neurochem.*, **49**, 183
Opioid	NG108-15 (δ R)	*Mol. Pharmacol.*, **31**, 159, *FASEB J.*, **2**, 52
	NG7315 C (μ R)	*Mol. Pharmacol.*, **33**, 423, *Eur. J. Pharmacol.*, **143**, 127
Vasopressin	LLC-PK1 (VI R)	*Mol. Pharmacol.*, **33**, 432
	Rat mammary tumour (VI R)	*Biochem. J.*, **240**, 189
	A10 smooth muscle (VI R)	*J. Biol. Chem.*, **263**, 2658
	WRK1 (VIa R)	*J. Cardiovasc. Pharm.*, **8** (suppl), 7

Table 3. Some cell lines expressing pharmacological receptors

Receptor type	Cell line	Reference
Bradykinin	DDT1-MF2	*J. Biol. Chem.*, **262**, 3098
	NCB20	*J. Neurochem.*, **51**, 505
	NG115 401L	*EMBO J.*, **6**, 49
Substance P	PC12	*J. Neurochem.*, **49**, 253
TRH	GH3 pituitary	*Mol. Pharmacol.*, **33**, 592
Thrombin	CHL fibroblasts	*Nature*, **335**, 254 and refs therein
	HEL/Dami	*Cell*, **64**, 1057
EGF	A431	*Receptor Biochemistry: a Practical Approach*. IRL, Oxford
NGF	Human neuroblastoma LANI	*J. Neurochem.*, **49**, 475
	PC12	*EMBO J.*, **6**, 1197
Insulin	L6 myocytes	*Exp. Cell Res.*, **134**, 297
	Fibroblasts	*J. Neurol. Sci.*, **80**, 229
IGF1	Neural cell lines (SK-N-SH, SK-N-MC)	*Endocrinology*, **122**, 145
	WI38	*J. Cell. Physiol.*, **133**, 135
	L6 myoblasts	*J. Biol. Chem.*, **262**, 12745
IGF2	L6 myoblasts	*J. Biol. Chem.*, **262**, 12745
	Rat hepatoma	*BBRC*, **151**, 815
CSF 1	Transfected NIH3T3*	*Mol. Cell. Biol.*, **7**, 2378
TGF-β	Retinoblastoma	*Science*, **240**, 196
PTH	Osteosarcoma	*J. Biol. Chem.*, **263**, 3864
Atrial natriuretic peptide	Human phaeochromocytoma	*BBRC*, **148**, 286

The table gives some cell lines expressing receptors involved in transmembrane signalling, with some recent references. It is not intended to be comprehensive, but instead, to act as an entrée to the literature.
* Widely used in expression studies.

are erythrocytes (β-adrenergic receptors, 37; P2 purinergic receptors, 38), neutrophils (chemotactic peptide, 39); platelets (α_2-receptor, 41; 5-HT$_2$ receptor, 42; Paf, thromboxane α_2, ADP, thrombin, 40,43), mast cells (histamine receptor, 44), lymphocytes (45), macrophages and sperms (46).

Less obvious, and relatively unexplored, is the possible presence of pharmacological receptors in free-living micro-organisms, such as protozoa, slime moulds, and yeasts, which are potential vehicles for the expression of cloned receptor genes, and which certainly express G-proteins homologous to those of the higher eukaryotes (47).

5.1 Isolation of rat lacrimocytes

Berrie's method is used for obtaining a dispersed acinar cell preparation from the lacrimal gland, and is modified from that of Parod *et al.* (48), and similar to that introduced by Kanagasuntheram and Randle (49) for the preparation of dispersed parotid acinar cells.

Protocol 20. Isolation of rat lacrimocytes

1. Remove the exorbital lacrimal glands from 6 rats (*Protocol 11*). Place several layers of filter paper dampened with saline on the Teflon disc of a McIlwain tissue chopper, and cut the tissue into approximately 1 mm cubes.

2. Place the tissue fragments in 20 ml of Krebs–Ringer–bicarbonate solution:

   - NaCl, 123 mM
   - KCl, 5.0 mM
   - $MgCl_2$, 1.3 mM
   - $CaCl_2$, 0.8 mM
   - KH_2PO_4, 1.4 mM
   - $NaHCO_3$, 26.0 mM
   - glucose 10 mM

   Gas with 5% CO_2/95% O_2 before use, and periodically during incubations. All manipulations are carried out at 37°C in a vigorously shaking water bath (approx 1 r.p.s.) in an atmosphere of 95% O_2/5% CO_2.

3. Centrifuge at 1500 g for 1 min, and discard the supernatant.

4. Resuspend the fragments in 10 ml medium containing 0.5% BSA and 0.4 mg/ml trypsin (Sigma catalogue number T0134, Type IX, 15 550 units/mg protein).

5. Incubate for 15 min.

6. Centrifuge at 1500 g for 1 min and discard the supernatant.

7. Resuspend in 10 ml medium with $MgCl_2$ and $CaCl_2$ replaced by NaCl and containing 0.5% BSA, 2 mM EGTA and 0.3 mg/ml soybean trypsin inhibitor (Sigma, catalogue number T9003, Type I-S).

8. Incubate for 5 min.

9. Centrifuge at 1500 g for 4 min and discard the supernatant.

10. Repeat steps 7, 8, and 9.

11. Resuspend in 16 ml of KRB containing 0.5% BSA.

12. Centrifuge at 1500 g for 1 min and discard the supernatant.

13. Repeat steps 11 and 12.

14. Resuspend in 16 ml of KRB containing 0.5% BSA and 0.19–0.27% collagenase (Sigma, catalogue number C0773, Type VII) depending on the activity of the enzyme (600–800 units/mg protein).

15. Incubate for 60 min. In order to facilitate tissue fragmentation, the suspension is sucked into and expelled from a 5 ml Finnpipette tip every 15 min.

16. Filter suspension through nylon mesh (300 μm, Nitex) into 10 ml medium containing 4% BSA.

17. Centrifuge at 50 g for 5 min and discard the supernatant.

18. Resuspend in 10 ml medium, without BSA.

19. Centrifuge at 50 g for 5 min and discard the supernatant.

20. Repeat steps 18 and 19.

21. Resuspend in 18 ml medium.

22. Equilibrate cell suspension with an initial 60 min incubation.

Microscopic examination shows that the cells are present either singly, or as small clusters of up to 8 cells/cluster. A sample of the cells should be examined for ability to exclude 0.3% trypan blue; viable cells exclude the dye. They remain viable for several hours.

5.2 Isolation of cardiac myocytes

The following procedure is based on the protocol of Farmer *et al.* (50), and is designed to produce a preparation which is calcium-tolerant, i.e. which remains viable in the presence of physiological concentrations of extracellular calcium. It is said to preserve the glycocalyx, promote healing of gap-junctions, and avoid sodium loading, all factors important for avoiding the 'calcium catastrophe'. Myocytes from fetal hearts can be cultured, and have been used successfully in binding studies (51).

Protocol 21. Preparation of rat cardiac ventricular myocytes

All steps are performed at 37°C. All glassware is siliconized before use.

1. Heparinize (500 units i.p.) adult rats (250–300 g). After about 30 min, remove hearts.

2. Place the hearts on a retrograde perfusion apparatus described in reference 52 and perfuse with Krebs–Henseleit solution at 60 mmHg for 15 min:

- NaCl, 118 mM
- KCl, 4.74 mM

Protocol 21. *Continued*

- KH_2PO_4, 0.93 mM
- $MgSO_4$, 1.2 mM
- $NaHCO_3$ 25 mM
- glucose, 10 mM
- $CaCl_2$, 2.5 mM
- Gas with 95% O_2/5% CO_2 to pH 7.4

3. Perfuse with Ca-free Krebs–Henseleit solution at 8 ml/min for 4.5–5 min. This is necessary to loosen intercellular cement at the base of the intercalated discs.

4. Perfuse with the same Ca-free medium with 0.1% collagenase (type II Worthington or type I Sigma), 0.1% BSA (type V, Sigma, dialysed against Krebs solution plus 50 μM $CaCl_2$), 0.1% hyaluronidase (type I-S, Sigma) plus 50 μM $CaCl_2$ at 6–8 ml/min.

5. After 5 min remove hearts from perfusion apparatus by cutting at the atrial/ventricular junction. Slash twice vertically towards but not through the apex, and place in 25 ml conical flasks containing 33 ml of Ca-free Krebs buffer with 0.1% collagenase, 0.1% hyaluronidase, 2% BSA, and 50 μM $CaCl_2$. Shake gently for 5 min.

6. Separate dissociated myocytes by sieving through 300 μm nylon mesh (Nitex) into centrifuge tubes containing 2 volumes of Krebs plus 2% BSA plus 50 μM $CaCl_2$.

7. Repeat digestion as in step 5, each time filtering off the myocytes.

8. Collect the myocytes from the second and third digestions by centrifugation at 50 g for 1 min. The first digestion provides few viable cells.

9. Wash the sedimented myocytes twice by resuspension in Krebs plus 2% BSA plus 50 μM $CaCl_2$.

10. Resuspend myocytes in Krebs plus 1.8 mM $CaCl_2$, or in Modified Eagles Medium with 1.8 mM $CaCl_2$ and 10% fetal calf serum (Gibco).

Myocytes are considered to be viable if they exclude 0.3% trypan blue. The yield is up to 60% of tissue wet weight.

5.3 Isolation of smooth muscle cells

The procedure is generically very similar to the two examples given above. Details are given by Benham and Bolton (53)

6. Protein assays

The Lowry technique (in ref. 54) usually suffices for the measurement of membrane protein concentrations. In the case of solubilized preparations containing detergents, the reaction can be performed in the presence of 1% SDS to avoid the formation of precipitates. Alternatively, the protein can be precipitated using trichloroacetic acid before assay. The practical detection range of this assay is 10–100 µg, so it is not very sensitive. Note that Tris buffers interfere with the assay procedure by forming a precipitate, and Hepes buffers give a high blank value. Reference 54 should be consulted for further information.

Protocol 22. Lowry protein assay

1. Make up reagents:

- A. 2% Na_2CO_3 (anhydrous) + 0.1 M NaOH
- B. 0.5% $CuSO_4 \cdot 5H_2O$ in 1% NaK tartrate, adjusted to pH 7.0 with NaOH
- C. Folin–Ciocalteau reagent (Fisons)

2. Set up a series of tubes containing 0, 10, 20 . . . 100 µg of bovine serum albumin in 10 µl of buffer equivalent to that used for unknowns.

3. Set up unknowns containing protein in same range.

4. Mix 1 ml A with 50 ml B. Add 1 ml of the combined reagent per tube of standard or unknown.

5. Add 0.1 ml of 11% SDS to each tube if there is a problem with precipitation of detergent at this stage. An alternative is to precipitate the protein before adding Lowry A + B. To do this, precipitate the protein by addition of 30 µl of 1% sodium deoxycholate followed by 1 ml of ice-cold 12% trichloroacetic acid to each tube. Incubate for 10 min and recover the precipitate by spinning at 1000 g for 20 min, then add Lowry A + B as above. This procedure is useful in the case of SDS-insoluble detergents, for example alkyl trimethylammonium salts, or if interfering substances (for example Tris, Hepes), are present.

6. Mix well and stand for 10 min. Membranes usually dissolve completely at step 4, if not, SDS (1%) can be added.

7. Add 0.1 ml of C. Mix. Stand for 30 min to 2 h.

8. Read at 750 nm in a 1 ml cuvette. Alternatively, dilute by addition of deionized water and use a larger cuvette.

9. Store at 4°C to retard the development of colour if you want to keep the tubes for longer periods before reading them.

A variant of this assay, which is more sensitive and convenient although more expensive, is the micro-BCA assay from Pierce which uses bichinchonic acid as the colourimetric reagent.

A sensitive, although more tedious assay, relies on base-hydrolysis of the protein samples, followed by reaction of fluorescamine (Fluram, Roche; also supplied by Sigma) with the primary amino groups of the amino acids. This assay can detect down to 1 μg of protein.

Protocol 23. Fluorescamine assay for proteins

1. Place the protein samples in 0.2 ml of deionized water, or a non-interfering buffer (for example 20 μM Na-Hepes, pH 7.5) in borosilicate glass tube of about 5 ml capacity, or use Pierce screw-cap septum vials. If necessary, the protein sample can be gel-filtered into the buffer to remove interfering substances, using a small column of G50F (see Chapter 9). We have found that the presence of 0.1% digitonin does not interfere with this assay.

2. Add 0.2 ml of 1 M NaOH.

3. Cap, or cover, the tubes individually with aluminium foil and autoclave at 120°C (15 p.s.i) for 20 min taking care to avoid condensation in the tubes.

4. Add 0.4 ml of 0.5 M HCl followed by 2 ml of 0.5 M sodium borate buffer, pH 8.5. Mix vigorously.

5. Add fluorescamine solution (20 mg/100 ml dry acetone; Sigma), 0.3 ml. Mix for several seconds.

6. Immediately read fluorescence at 475 nm with excitation at 390 nm.

For a less sensitive assay, the pH of the borate buffer should be increased to 9.5, and the NaOH hydrolysis and HCl neutralization steps omitted.

The measurement of low concentrations of protein in detergent solutions may seem a trivial problem. In practice, it can be quite difficult. An approach which we have found most satisfactory is to separate the protein from contaminants by FPLC gel-filtration on a Superose 12 column (Pharmacia) in 50 mM Tris-Cl/0.1% SDS with monitoring of the resultant profile at 214 nm; at this wavelength the main absorbance is due to peptide bonds. The protein content of the sample can be measured after integration of the elution profile, and comparison made with the appropriate standards.

This method is very sensitive, allowing accurate measurement down to 0.1 μg of protein, and also providing some insight into the polypeptide composition of the sample. Because the procedure incorporates a molecular weight fractionation, low molecular weight contaminants do not interfere. Dilute samples can be concentrated before injection on to the column. A disadvantage of the method is that the throughput of samples is low. Because the entire elution profile can be examined, the results have much more credibility than colourimetric or fluorimetric measurements in which the contributions of protein and contaminants cannot be disentangled. For this reason, it is our method of choice for use with valuable samples. Other examples of sensitive protein assays are given in references 10 and 54: the latter reference is particularly informative.

References

1. Boast, C. A., Snowhill, E. W., and Altar, C. A. (eds) (1986). *Quantitative Receptor Autoradiography*. Alan R. Liss Inc., New York.
2. Turner, A. J. and Bachelard, H. S. (eds) (1987). *Neurochemistry: a Practical Approach*. IRL Press, Oxford and Washington.
3. Frey, K. A., Hichwa, R. D., Ehrenkaufer, R. L. E., and Agranoff, B. (1985). *Proc. Natl. Acad. Sci. USA*, **82**, 6711.
4. Berridge, M. J. (1987). *Annu. Rev. Biochem.*, **56**, 159.
5. Conn, P. M. and Means, A. R. (eds) (1987). *Methods in Enzymology.*, Vol. 141. Academic Press, London and New York.
6. Levitan, I. B. (1988). *Annu. Rev. Neurosci.*, **11**, 119.
7. Gilman, A. G. (1987). *Annu. Rev. Biochem.*, **56**, 615.
8. Sakmann, B. and Neher, E. (1984). *Annu. Rev. Physiol.*, **46**, 455.
9. Hulme, E. C. (ed.) (1990). *Receptor–Effector Coupling: a Practical Approach*. IRL Press, Oxford and Washington.
10. Hulme, E. C. (ed.) (1990). *Receptor Biochemistry: a Practical Approach*. IRL Press, Oxford and Washington.
11. Freshney, R. I. (ed.) (1986). *Animal Cell Culture: a Practical Approach*. IRL Press, Oxford and Washington.
12. Benovic, J., Strasser, R. H., Caron, M. G., and Lefkowitz, R. J. (1986). *Proc. Natl. Acad. Sci. USA*, **83**, 2797.
13. Findlay, J. B. C. and Evans, W. H. (eds) (1987). *Biological Membranes: a Practical Approach*. IRL Press, Oxford and Washington.
14. Rotter, A., Birdsall, N. J. M., Burgen, A. S. V., Field, P. M., Hulme, E. C., and Raisman, G. (1979). *Brain Res. Rev.*, **1**, 141.
15. Scott Young III, W. and Kuhar, M. J. (1979). *Brain Res.*, **179**, 255.
16. Buckley, N. J. and Burnstock, G. (1986). *J. Neurosci.*, **6**, 531.
17. Hassall, C. J. S., Buckley, N. J., and Burnstock, G. (1987). *Neurosci. Lett.* **74**, 145.
18. Greene, L. A., Sytkowskiy, A. J., Vogel, Z., and Nirenberg, M. W. (1973). *Nature*, **243**, 163.
19. Messing, A. and Gonatas, N. K. (1983). *Brain Res.*, **269**, 172.
20. McCarthy, K. D. (1983). *J. Pharmacol. Exp. Ther.*, **226**, 282.
21. Conn, P. M. (ed.) (1989). *Meth. Neurosci.*, Vol. 1. Academic Press, New York.
22. McBride, L. J., McCollum, C. J., Davidson, S., Efkavitz, J. W., Andrus, A., and Lombardi, S. J. (1988). *Biotechniques*, **6**, 362.
23. Paxinos, G. and Watson, C. (1986). *The Rat Brain in Stereotaxic Coordinates*. Academic Press, London and New York.
24. Palkowitz, M. and Brownstein, M. J. (1988). *Maps and Guide to Microdissection of the Rat Brain*. Elsevier, New York.
25. Butcher, S. P., Collins, J. F., and Roberts, P. J. (1983). *Br. J. Pharmacol.*, **80**, 355.
26. Olde, B. and Johannson, G. (1985). *Neuroscience*, **15**, 1247.
27. Flanagan, S. D., Johannson, G., Yost, B., Ito, Y., and Sutherland, I. A. (1984). *J. Ligand Chromatography*, **7**, 385.
28. Mick, G. I., Bonn, T., Steinberg, J., and McCormick, K. (1988). *J. Biol. Chem.*, **263**, 10667.

29. Weigel, P. H., Ray, D. A., and Oka, J. A. (1983). *Anal. Biochem.*, **133**, 437.
30. Barrowman, M. M., Cockroft, S., and Gomperts, B. D. (1986). *Nature*, **319**, 504.
31. Howell, T. W. and Gomperts, B. D. (1987). *Biochim. Biophys. Acta*, **927**, 177.
32. Baker, P. F. and Knight, D. E. (1984). *Trends Neurosci.*, **7**(4), 120.
33. Lind, I., Ahnert-Hilger, G., Fuchs, G., and Gratzel, M. (1987). *Anal. Biochem.*, **164**, 84.
34. Berrie, C. P., Birdsall, N. J. M., Hulme, E. C., Keen, M., and Stockton, J. M. (1984). *Br. J. Pharmacol.*, **82**, 853.
35. Peterson, G. and Schimerlik, M. (1984). *Prep. Biochem.*, **14**, 33.
36. Buckley, N. J., Hulme, E. C., and Birdsall, N. J. M. (1990). *Biochim. Biophys. Acta*, **1055**, 43.
37. Yarden, Y., Rodriguez, H., Wong, S. K. F., Brandt, D. R., May, D. C., Burnier, J., Harkins, R. W., Chen, E. Y., Ramachandran, J., Ullrich, A., and Ross, E. M. (1986). *Proc. Natl. Acad. Sci, USA*, **83**, 6795.
38. Harden, K., Hawkins, P. T., Stephens, L., Boyer, J. L., and Downes, C. P. (1988). *Biochem. J.*, **252**, 583.
39. Korchak, H. M., Wilkenfeld, C., Rich, A. M., Radin, A. R., Vienne, K., and Rutherford, L. E. (1984). *J. Biol. Chem.*, **259**, 7439.
40. Zschauer, A., van Breemen, C., Buhler, F. R., and Nelson, M. T. (1988). *Nature*, **334**, 703.
41. Kobilka, B. K., Matsui, H., Kobilka, T. S., Yang-Feng, T. L., Francke, U., Caron, M. G., Lefkowitz, R. J., and Regan, J. W. (1987). *Science*, **238**, 650.
42. Geaney, D. P., Schacter, M., Elliot, J. M., and Grahame-Smith, D. G. (1984). *Eur. J. Pharmacol.*, **97**, 87.
43. Casals-Stenzel, J., Muacevic, G., and Weber, K-H. (1987). *J. Pharm. Exp. Ther.*, **241**, 974.
44. Arrang, J-M., Garbarg, M., Lancelot, J-C., Lecomte, J-M., Pollard, H., Robba, M., Schunack, W., and Schwartz, J-C. (1987). *Nature*, **327**, 117.
45. Klaus, J. (ed.) (1987). In *Lymphocytes: a Practical Approach*. IRL Press, Oxford and Washington.
46. Singh, S., Lowe, D. G., Thorpe, D. S., Rodriguez, H., Kuang, W-J., Dangott, L. J., Chinkers, M., Goeddel, D. V., and Garbers, D. L. (1988). *Nature*, **334**, 708.
47. Lochrie, M. A. and Simon, M. I. (1988). *Biochemistry*, **27**, 4957.
48. Parod, R. J., Leslie, B. A., and Putney, J. W. (1980). *Am. J. Physiol.*, **239**, G99.
49. Kanagasuntheram, P. and Randle, P. J. (1976). *Biochem, J.*, **160**, 547.
50. Farmer, B. B., Mancina, M., Williams, E. S., and Watanabe, A. M. (1983). *Life Sci.*, **33**, 1.
51. Nathanson, N. M. (1983). *J. Neurochem.*, **41**, 1545.
52. Powell, T., Terrar, D. A., and Twist, V. W. (1980). *J. Physiol. Lond.*, **302**, 131.
53. Benham, C. D. and Bolton, T. B. (1983). *J. Physiol.*, **340**, 469.
54. Peterson, G. L. (1983). In *Methods in Enzymology* (ed. C. H. Hirs and S. N. Timasheff), Vol. 91, p. 95. Academic Press, New York.
55. Mantyh, P. W., Hunt, S. P., and Maggio, J. E. (1984). *Brain Res.*, **307**, 147.

6

The use of the filtration technique in *in vitro* radioligand binding assays for membrane-bound and solubilized receptors

JIAN-XIN WANG, HENRY I. YAMAMURA, WAN WANG, and WILLIAM R. ROESKE*

1. Introduction

It has been nearly a century since the receptor hypothesis was first proposed for the mechanism of the effects of drugs and toxins on biosystems (1,2). In 1926, A. J. Clark established his receptor occupation theory (3) describing a quantitative interaction between drugs and their receptors. For a period of almost 40 years putative drug receptors were detected by measuring the functional response elicited by drugs. This method must make assumptions that relate the drug–receptor binding properties to the function measured. The first attempt to directly measure the receptors with radiolabelled ligands can be traced back to the 1960s when [³H]atropine (4) and [³H]propranolol (5) were used to label the muscarinic uptake sites in the intestinal smooth muscle and the adrenergic uptake site in atria, respectively. In 1970, radioisotope labelled adrenocorticotropic hormone (ACTH) (6), angiotensin (7) and α-bungarotoxin (8) were used to measure ACTH, angiotensin, and nicotinic cholinergic receptors, respectively. These pioneering experiments demonstrated that radioisotopically labelled hormones or toxins could be used to study directly the interaction of hormones or toxins with their specific membrane sites. However, some of these studies were limited by technical difficulties such as a low signal-to-noise ratio.

In order to directly measure drug receptors, it was necessary to develop a new technique that had a sensitivity high enough to detect the receptors which existed as a very small portion of the total cellular tissue (sometimes less than 1/10 000 of total cellular tissues). In addition, this technique had to be highly efficient in the separation of the bound from the free ligand. Furthermore,

* To whom correspondence should be sent.

the assay should have a low non-specific background resulting from the radioligand binding to cellular components other than the receptors (9). Several techniques have been developed to separate free from bound ligand, such as centrifugation, equilibrium dialysis, gel filtration, and precipitation of the ligand–receptor complex or the free ligand. Some of these methods have poor efficiency, are laborious, and possess high background. In 1969, a filtration technique was introduced for insulin binding to liver membrane receptors (10). After being adapted and improved by many investigators (11), the rapid filtration method soon became a major technique frequently used in receptor-binding studies, due to its rapidity, high efficiency, low non-specific background, and ease of operation. Filtration techniques have been used extensively in the study of membrane receptors and other membrane associated proteins, such as ion channels and enzymes. These techniques have also been used, with slight modification, to measure the binding of radioligands to soluble receptor proteins (*Table 1*).

In this chapter we will discuss the methodology of the rapid filtration technique, mostly based on our own experience and knowledge from the study of cholinergic receptors (*Table 2*). Some experimental factors which may influence the accuracy and precision of the assay will also be discussed. Finally, a comparison of the filtration technique with other separation techniques is provided to demonstrate the advantages and shortcomings of this method.

2. Methodology

2.1 General principles

After incubation of a radioligand with membrane particles, a fraction of the ligand will bind to its specific receptor to form a ligand–receptor complex. Meanwhile, most of the radioligand is still free in the aqueous phase. The distribution of total ligand between the free and the bound form is determined by the law of mass action. In general, less than 10% of the radioligand should be bound to the membrane receptor. In order to measure the small

Table 1. Assays utilizing the rapid filtration technique

Preparations	Example
Membrane-associated proteins	membrane-bound receptors
	ion channel proteins
	enzymes
	uptake and transporter receptor
	ion uptake
Soluble proteins	solubilized membrane receptors
	cytoplasmic soluble receptors

Table 2. Radioligand binding studies of cholinergic receptors[a]

Receptor type		Radioligand	K_d (nM)[b]	Reference[c]
Muscarinic receptor	Antagonist non-selective	[³H][(−)]Quinuclidinyl benzilate	0.02–0.06	Yamamura and Snyder (1974) (12)
		[³H][(−)-N-methylscopolamine	0.15–0.30	Birdsall and Hulme (1976) (13)
		[³H][(+)]Benzetimide	0.4	Laduron et al. (1979) (14)
		[³H]Propylbenzilylcholine mustard	10–20[d]	Burgen et al. (1974) (15)
	M₁-selective	[³H]Pirenzepine	4–10	Watson et al. (1982) (16)
	M₂-selective	[³H]AF–DX 116	20	Wang et al. (1987) (17)
	Agonist	[³H][(+)-cis-Methyldioxolane	1–2	Ehlert et al. (1980) (18)
		[³H]Oxotremorine-M	1–2	Harden et al. (1983) (19)
Nicotinic receptor	Antagonist	[¹²⁵I]α-Bungarotoxin	5	Changeux et al. (1970) (8)
		[³H]Tubocurarine	2	DeRobertis et al. (1967) (20)
	Agonist	[³H][(−)-Nicotine	2–15	Romano and Goldstein (1980) (21)
		[³H]Methylcarbamylcholine	1	Abood and Grassi (1986) (22)
		[³H]Acetylcholine (in the presence of muscarinic blockers)	15–30	Schwartz et al. (1982) (23)

[a] Table 2 is modified from Vickroy et al. (1986) (24). Only the frequently used radioligands are listed.
[b] K_d values were obtained from the literature, which vary between laboratories depending on the species of animals, tissues, and the assay conditions used. If more than one binding site was found for a radioligand, only the K_d values for the high affinity sites are given in this table.
[c] References are to the papers that have described the pharmacological characteristics of the radioligand with feasible methods.
[d] Reversible binding followed by irreversible alkylation of the receptors.

fraction of the radioligand that is associated with the receptors, the free radioligand is immediately separated from the bound ligand after the assay incubation. This is achieved by rapid filtration of the entire incubation mixture through a filter with a certain pore size. A number of filters are commercially available. The choice is an empirical decision based on the ratio of specific to non-specific binding. The separation process is accelerated by operating under reduced pressure. To complete the separation of the free radioligand from the bound receptors, membrane particles are rinsed with a sufficient volume of ligand-free buffer to reduce the free ligand trapped by the membrane particles and the filters during the filtration process. The rinsing process should not affect ligand bound with high affinity to the membrane receptors, however it may be wise to do a rinse time study to determine the proper volumes and frequency of rinsings. Separation of the free and bound ligand with this technique can be accomplished within seconds with both high efficiency and with a high specific/non-specific binding ratio.

2.2 Filtration machines

Several types of filtration machines are commercially available. These have different capacities, screen sizes, and degrees of automation. *Figure 1* shows a sample of one such filtration machine, the Cell Harvester manufactured by

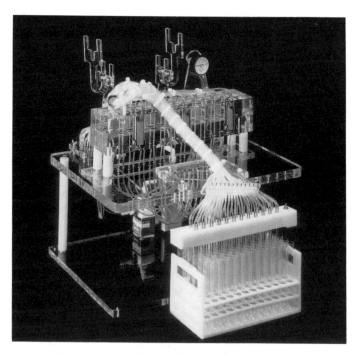

Figure 1. The Cell Harvester, manufactured by Brandel Inc.

Brandel Inc. A schematic diagram of the connection of a filtration machine with a vacuum pump and a buffer pump is presented in *Figure 2*. The machine contains a two-part central filtration unit which has 24 ⅝″ diameter metal screens in one part and 24 rubber O rings in another part. A filter paper is placed on the surface above the screens. The machine can then be sealed by closing the two parts and clamping the adjustable latches. Each screen and its corresponding O rings, once sealed, form an isolated chamber for filtration. The central filtration unit is connected on one side to metal probes through plastic tubing. These metal probes are arranged and fixed in a plastic board. For each screen, there are two parallel metal probes. One probe aspirates the contents of the incubation tubes to the filter. The second metal probe is used to refill the incubation tubes with fresh buffer from a buffer reservoir by switching on a buffer pump (the starter is at the top of the board). Below the central filtration unit is a collection manifold, connected to a vacuum pump through two consecutive pressure-resistant bottles. One bottle is used to collect the reaction mixture and the rinsing buffer. The second bottle is used to protect the pump from being flooded by any radioactive solution which could enter the pump if the first bottle becomes full. A vacuum gauge is provided on one of the two metal-tube-board holders facing the operator. The vacuum pump often comes with its own vacuum gauge.

2.3 Filtration procedure

Protocol 1. Filtration using the cell Harvester

1. Turn on the vacuum pump and aspirate the contents of the 24 incubation tubes simultaneously on to each corresponding filtration chamber. (The membrane particles will be retained on the surface of the filters whereas most of the free ligand will pass through the filters and leave the filtration chambers.)

2. Refill the culture tubes with ice-cold fresh buffer. (The buffer is used to wash the tubes and consequently rinse the filters by being reaspirated into the filtration apparatus.)

3. Repeat the rinsing step several times depending on the characteristics of the radioactive ligand.

4. Turn the vacuum pump off after the filtration is accomplished.

5. Open the clamps and remove the now radioactive filter paper using forceps. (The filter paper clearly shows 24 round indentations resulting from the pressure of the rubber rings on the filter paper.)

6. Punch these areas (easily and precisely accomplished) into scintillation vials. (The radioactivity on the punched filter paper can be extracted and counted in a liquid scintillation spectrophotometer.)

Protocol 1. *Continued*

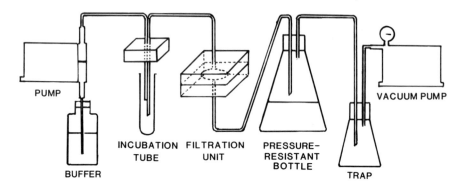

PUMP INCUBATION TUBE FILTRATION UNIT PRESSURE-RESISTANT BOTTLE VACUUM PUMP

BUFFER TRAP

Figure 2. Schematic diagram of the organization of the filtration apparatus.

7. Add 10 ml of an aqueous scintillation cocktail to the punched-out filter disc, and mix thoroughly.

8. Stand the vials for several hours, mixing occasionally until the filters become transparent, and a homogeneous supernatant is obtained.

9. Allow any air bubbles to dissipate before counting the filters.

10. Thoroughly wash the filtration machine before using it for the next filtration protocol.

Some workers prefer to dry the filters in a microwave oven before adding scintillator. However, this may run the risk of baking the radioligand on to the filters, thus reducing counting efficiency.

A recent addition to filter-counting methodology is a development of solid-phase scintillation technology, namely the use of melt-on scintillator strips ('MeltiLex TM'; Wallac(Pharmacia)). The basis of the technique is that scintillator is brought into intimate contact with radioactivity trapped on the dry filter. In principle, this enables the whole filter to be counted in a suitable β-plate counter, allowing parallel acquisition of data from many filter positions. A limiting factor in this technique may be the efficiency of detection, and the absence of quench corrections for individual samples.

The precision and accuracy for the determination of specific binding of the radioligand can be affected by conditions used with the rapid filtration technique. Some of these experimental factors will be discussed below.

2.4 Factors affecting the separation of free from bound radioligand

2.4.1 Dissociation of the radioligand–receptor complex

When a large amount of the ligand-free buffer is used to rinse the filter paper, dissociation of the bound ligand will occur due to the resultant dilution of the free ligand concentration. Assuming that no reassociation of the radioligand occurs during the filtration process, dissociation of the bound ligand is simply a function of the dissociation rate constant of the ligand–receptor complex and the filtration time. This relationship is described by the following equation:

$$B_t = B_0 e^{-K_{21}t}, \tag{1}$$

here, B_0 is the specific binding of the radioligand before the dissociation starts; B_t is the specific binding measured at time t after the dissociation; and k_{21} is the dissociation rate constant for the ligand–receptor complex.

Although the filtration process takes only a few seconds, dissociation of the bound radioligand from the receptor sites is still unavoidable. Many investigators find that dissociation of the receptor–ligand complex of less than 5–10% during the filtration process is acceptable. If dissociation of the receptor–ligand complex occurs at a 5 or 10% level (i.e. $B_t = 0.95 B_0$, or $B_t = 0.9 B_0$), the filtration time allowed can be estimated from the following calculations:

$$t = -\log_e (0.90)/k_{21}, \text{ at a 10\% level,} \tag{2}$$

$$t = -\log_e (0.95)/k_{21}, \text{ at a 5\% level.} \tag{3}$$

The value of the dissociation rate constant for different radioligands reported in the literature varies from 0.01/min to 0.99/min, or about 2×10^{-4}/sec to 2×10^{-2}/sec. The theoretical relationship between the dissociation rate constant within this range and the filtration time which causes a 5 or 10% loss of the specific binding of the radioligand is plotted in *Figure 3*.

The two theoretical curves separate the plane into three areas: the area below the 5% curve (area A), the area above the 10% curve (area C), and the area between the two curves (area B). In a given assay system, if the dissociation rate constant of the radioligand and the filtration time is located in area A, the possibility for a loss of specific binding of the radioligand would be expected to be less than 5%. Therefore, *area A can be considered a safe area.* In contrast, under combinations of conditions (either too fast a dissociation of the radioligand or too long a filtration time, or both) located in area C, a significant dissociation of the receptor–ligand complex (>10%) will occur during the filtration process. Therefore, *area C is unacceptable for using the rapid filtration technique.* Area B represents a marginal condition. The possible loss of specific binding of the radioligand varies between 5–10%. Careful selection of the filtration conditions is necessary to improve the quality of the assay.

Alternatively, since most of the filtration process can be accomplished in 5–15 sec, *Figure 3* can be arbitrarily divided into three columns: column I ($t < 5$ sec), column II (5 sec $< t < 15$ sec), and column III ($t > 15$ sec). The border between column I and column II crosses the 5% curve, at a point

The use of the filtration technique

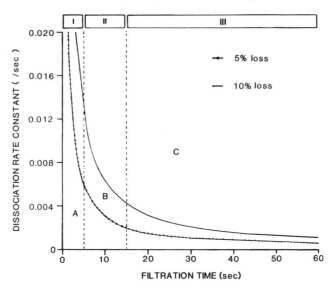

Figure 3. Relationship of dissociation rate-constant and the filtration time which causes 5 or 10% loss of receptor–ligand complex. Curves are computer generated using the equations (2) or (3) for the loss of receptor–ligand complex at a 5% or 10% level, assuming that the dissociation process only occurs during the filtration procedure.

corresponding to a k_{21} value of about 0.6×10^{-2}/sec, suggesting that *any ligand with a dissociation rate constant larger than 0.6×10^{-2}/sec will dissociate from the receptor to an extent of at least 5%. These ligands are not suited to the filtration technique.* On the other hand, the border between column II and column III crosses the 10% curve at a point of 0.4×10^{-2}/sec. A maximum of about 10% of the specific binding could be lost during the filtration if a ligand displays a dissociation rate constant smaller than 0.4×10^{-2}/sec. *These ligands with k_{21} values less than 0.4×10^{-2}/sec are candidates for using the rapid filtration technique.* Ligands with a k_{21} value between 0.4–0.6×10^{-2}/sec are again in a marginal situation. Therefore, *Figure 3* can be very useful for predicting whether or not a radioligand with a given dissociation rate constant is suitable for using the rapid filtration technique.

In some cases, the dissociation rate constant of a ligand is unknown. The overall dissociation constant (K_d) of the ligand can be used to estimate the suitability of the ligand. Many radioligands have an association rate constant (k_{12}) of about 10^{+6}/sec/M or about $5 \times 10^{+7}$/min/M. Hence, a k_{21} value of 0.6×10^{-2}/sec corresponds to a K_d value of 10–20 nM. Consequently, the specific binding of a ligand with a dissociation constant of more than 20 nM cannot usually be determined precisely with the rapid filtration technique. However, it should be emphasized that the overall K_d value is not always a reliable criterion. For example, [³H]AF–DX 116, the radioligand for M_2

220

muscarinic cholinergic receptors, has a dissociation rate constant of about 0.8/min at 25°C in membranes from rat cerebral cortex and heart, which limits the use of rapid filtration technique at this temperature (17). However, the dissociation rate is dramatically decreased at 0–4°C to 0.02–0.04/min or 3×10^{-4}–6×10^{-4}/sec. The change of the k_{21} value from high temperature to low temperature is about 30–40-fold. This means that if the filtration technique is used at 0–4°C and the separation accomplished within 10 sec, less than 1% of the bound ligand should dissociate. Therefore, the filtration technique can be satisfactorily used for the assay of [³H]AF–DX 116 binding at 0–4°C. However, if one looks at the change of the overall K_d value, there is only a slight decrease of 30% (from 30 nM to 20 nM). A significant loss of the radioligand bound by using the filtration technique would still be assumed based on the K_d values at 0–4°C. Another classic example showing the inconsistency of the change of the K_d value and the k_{21} value are the binding parameters of [³H] (−)QNB. [³H](−)QNB displays a higher affinity at a higher incubation temperature (25), whereas this ligand dissociates more rapidly at higher temperatures compared with its very slow dissociation rate at 0–4°C.

To decrease the temperature, the filters and the rinsing buffer should be kept at 0–4°C before and during the filtration. The filters can be kept in ice-cold buffer or rinsed with ice-cold buffer before the filtration starts. The tissue can be either incubated at 0–4°C or rapidly immersed in an ice-cold bath after incubation at a higher temperature. Because of the possibility of the binding kinetics being altered by suddenly changing the incubation temperature in the latter case, the incubation mixture, after incubation at a higher temperature, can be directly loaded on to the ice-cold filter followed by immediate rinses with ice-cold buffer. Since less than 5 mg tissue are usually applied to the filter, the membranes are rapidly cooled by the large volume of ice-cold buffer. The tightly-sealed filtration wells provide good heat-insulation which retains the low temperature of the membranes and the filters during the filtration process.

In general, many radioligands dissociate more slowly at lower temperatures than at higher temperatures. Using an ice-cold buffer for rinsing is generally recommended.

2.4.2 Rinsing volume and number of rinses

Rinsing the filters after filtration is a critical step. Usually the volume and the number of rinsings should be determined carefully. A typical study for the proper rinsing time is shown in *Figure 4*.

Incubations are usually performed in plastic or glass culture tubes with a total incubation volume of 0.25–2.0 ml. Filters can be rinsed with 1–5 ml ice-cold buffer each time. The rinsing process is often repeated several times. There is no precise way to estimate the amount of the free ligand trapped in the membranes. The termination of the rinsing process is at the point where a profound decrease of non-specific binding of the ligand is achieved at no significant sacrifice of the specific binding of the radioligand.

In this laboratory, we prefer loading the tissue on to the filters directly, followed by rinsing with buffer. Some investigators prefer diluting the incubation mixtures before loading on to the filters. This method is considered as a step for terminating the incubation and facilitating the subsequent separation of the free from the bound ligand. However, the dilution of the incubation mixture may take a longer, and more variable, time than directly loading and immediately separating the free and the bound ligands. Appreciable dissociation of the receptor–ligand complex may occur. This is especially true when a relatively low affinity ligand is used to label the receptors. Therefore, dilution with buffer before the filtration process does not seem to have any advantage and should be discouraged.

Rinsing the filters ideally depletes only the free ligands which are trapped by the membrane particles and the filters during the separation. However, in most cases it seems that rinsing the filters also causes dissociation of the ligand molecules which are non-specifically bound to other membrane components for which the ligand has a low affinity. Therefore, the precision of determination of specific binding of the ligand is improved by the increased specific/

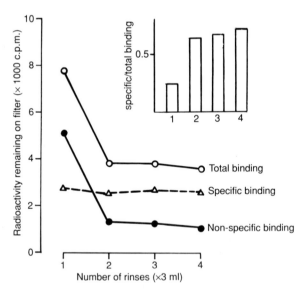

Figure 4. Effect of different numbers of rinses of the filter on the binding of [³H]AF–DX 116 to rat cerebral cortical membranes. The binding of [³H]AF–DX 116 (5 nM) to rat cerebral cortical membranes was performed at 25°C and filtered at 0–4°C through Whatman GF/B filters. The filters were rinsed 1–4 times with 3 ml of ice-cold buffer each time. Inset: the ratio of specific binding/non-specific binding of [³H]AF–DX 116 after rinsing for different times with ice-cold buffer.

non-specific binding ratio. However, the 'free concentration of ligand', calculated from the difference between the total ligand used and the total binding determined experimentally, will then be higher than the free ligand concentration. However, if the 'total binding' measured *without* rinsing the filters (i.e. including the 'trapped' free radioligand) is less than 5–10% of the total radioligand added, we are reasonably certain that this difference should not cause a significant bias in the determination of the free concentration of the radioligand. Therefore, the estimated dissociation constant of the ligand (K_d value) should be reasonably accurate. If the binding assay is not performed under optimal conditions, because of difficulties in the experimental design, the error in the determination of the free concentration of the ligand may lead to an underestimation of the affinity of the ligand (i.e. a higher K_d value could be obtained). Since the low affinity binding of the radioligand to the non-specific membrane components is also tissue concentration-dependent, errors in the determination of the free concentration of the ligand may contribute to apparent tissue concentration-dependence of K_d values, a phenomenon which has been observed by several investigators (26, 27).

2.4.3 Rinsing buffers

The ideal buffer for rinsing the filters is the same as that used for the incubation. This choice introduces no other factors which could cause dissociation of the ligand–receptor complex in addition to the dilution of the free ligand concentration. In some cases, saline may be used as an economical substitute for the buffer. However, careful control experiments should be performed before substituting saline for the incubation buffer for rinsing. The binding of the ligand should not be sensitive to the changes of the ionic strength or species.

2.4.4 Vacuum pressure

It is obvious that rapid filtration cannot be achieved under normal atmospheric pressure. To accelerate the filtration process, reduced pressure is often applied to the filtration apparatus. In general, the lower the vacuum pressure, the faster the filtration process. This method helps to reduce the possibility of the dissociation of the ligand from the specific binding site. This is even more important when the ligand has rapid dissociation kinetics. The Cell Harvester apparatus usually comes with a vacuum pump for maximal reduction of 760 mm (or 30 inches) of mercury vacuum pressure. A vacuum pressure of 500–600 mm (or 20–25 inches) of mercury vacuum pressure can often be reached for most of the binding studies with membrane preparations. However, it is possible that lower vacuum pressure could increase the chance of the loss of small membrane particles during the filtration. This could be an important factor affecting the binding assay when soluble receptor preparations are used (see Section 2.5 of this chapter). In this situation preliminary experiments should be performed to determine a suitable pressure, which should be the minimum vacuum pressure to allow an efficient separation.

The Cell Harvester filtration machine ideally applies uniform vacuum pressure to each filter screen. This uniformity of the vacuum pressure reduces experimental variation between the samples. Nevertheless, significant differences in the filtration time between screens can be observed when the Cell Harvester has become clogged by tissue or other debris. This effect sometimes results in a systematic error in the determination of the binding of the radioligand, and is usually caused by the blockage of the screen with membrane tissue. Routine cleaning of the screens with 2% detergent or 0.2 M NaOH helps to eliminate the contamination of the screens with tissue. If slower filtration is observed with a few screens, the obstructed screens can be changed. However, if a general reduction of the filtration efficiency (increase of the aspiration time) is observed, we recommend changing the entire filtration unit. Sometimes, the connecting tubing develops slow air leaks. Changing the Tygon tubing may restore the full power of the vacuum pump and increase the efficiency of the filtration.

2.4.5 Tissue concentration

The tissue concentration used in the radioligand binding assay can be a critical factor affecting the filtration speed. A significant increase in the filtration time may be observed when higher concentrations of tissue homogenates are loaded on to the filters. The limitation of the amount of the tissue which can be used without a significant reduction of the filtration speed may vary between different tissues and different preparations. Based on our experience, more than 10 mg of tissue/tube *cannot* be used in the assay. For instance, we have noticed that increasing the amount of rat heart homogenate from 2 mg tissue/tube to 10 mg tissue/tube prolonged the filtration time from 10 sec to 30–60 sec. Therefore, using the lowest possible concentration of the tissue for the ligand binding assay will help to achieve the maximum efficiency of separation with the rapid filtration technique.

2.5 Use of the rapid filtration technique in the assay of soluble receptors

The most popular methods used for the assay of radioligand binding to soluble receptors include gel filtration (Chapter 9), equilibrium dialysis (28), polyethylene glycol (PEG) precipitation, or ammonium sulphate precipitation followed by centrifugation, protein absorption, or free ligand absorption (29 and Chapter 8). Direct filtration of the incubation mixture is generally inappropriate, inasmuch as most of the soluble proteins can easily pass through the filters during the filtration (30). However, the traditional methods are usually time-consuming and less reproducible than the filtration technique used for the assay of membrane associated proteins. Two modified filtration methods have been suggested for the study of soluble receptors.

2.5.1 Precipitation with PEG followed by filtration

It has been reported that antibody-bound and unbound peptide hormones can be separated by distribution into an aqueous polymer two-phase system (31). This observation led to the development of a convenient method for separating free from antibody-bound hormones in the radioimmunoassay (32). This method has also been successfully used for the assay of radioligand binding to soluble receptors (33, 34) or solubilized membrane receptors (35, 36).

Protocol 2. Precipitation with PEG followed by filtration

1. Incubate soluble receptors, or detergent solubilized receptors, with radioligand in an ionic buffer.

2. Add a small volume of ice-cold γ-globulin, at the end of the incubation, to the incubation tubes to give a final γ-globulin concentration of 0.8–1.0 mg/ml.

3. Place the culture tubes on ice.

4. Add a PEG solution to give a final concentration of 10–15%. (γ-globulin can be suspended in the PEG solution and added to the incubation tubes simultaneously.)

5. Thoroughly mix the contents of the culture tubes and allow to stand for 10–20 min.

6. Filter the contents of the tubes under reduced pressure on cellulose acetate (EH) Millipore filters.

7. Wash the filters with 3–5 ml of 8% PEG.

8. Dry and count using a liquid scintillation method.

The concentration of PEG used to separate the free and the bound ligands is critical. Desbuquois and Aurbach (32) reported that a minimum of 10% PEG is required to assure complete precipitation of antibody-bound insulin and parathyroid hormones. 100% precipitation of bound hormones could be obtained with 10–12.5% PEG. Slightly different concentrations of PEG have been used by several investigators for different radioligands under different experimental conditions. However, as the PEG concentration was increased so the solubility of free ligands in PEG also increased. The solubility of free ligands in PEG varies between the different ligands. For example, free insulin had a lower solubility than parathyroid hormone in PEG. Therefore, increasing the concentration of PEG precipitates free parathyroid hormone to a greater extent than insulin (31). Co-precipitation of free ligand with bound ligand can be a serious problem with the PEG precipitation method. Cuatrecasas (33) reported a significant precipitation (> 5%) of free [^{125}I] insulin at a concentration of PEG higher than 8%. It has also been found that the PEG precipitation method is not of use in assay of glucagon binding, since a significant fraction of the free glucagon is precipitated (37). Therefore, the proper concentration of PEG required to precipitate the ligand–receptor complex should be determined for each radioligand under the investigator's own conditions.

The molecular weight of polymers also affects the separation. The higher the molecular weight of the polymer, the lower the concentration required for phase separation (31). PEG with molecular weights of 6000–8000 are often used (32, 35, 36).

In this method, γ-globulin is used as a carrier for the precipitation reaction, since it is known to be precipitated by PEG (31). γ-Globulin is more suitable than is serum albumin, which does not precipitate at pH 7 or higher (31, 32). A suitable concentration of γ-globulin is 0.8–1.0 mg/ml (0.08–0.1%, w/v). Higher concentrations of γ-globulin also increase the precipitation of free hormone, due to physical trapping of the labelled hormone by the precipitate (32). In certain cases, protamine sulphate (1%) has been used instead of γ-globulin (38).

The influence of pH on the precipitation of the receptor–ligand complex is not clear. While no change of precipitation of antibody-bound hormones was observed in a pH range from 9 to 7 (32), it was noticed by Cautrecasas (33) that the ligand–receptor complex is less effectively precipitated if the pH of the buffer containing γ-globulin is above 8 or below 7. Fortunately, most radioligand binding assays are performed between pH 7 and pH 8; pH may also affect the solubility of the free ligand in PEG. It was found that the solubility of growth hormone and parathyroid hormone, but not insulin and arginine vasopressin, was decreased as the pH was lowered from 9 to 7 (33).

Small molecules, such as sucrose and ions, are usually the essential ingredients of an incubation buffer. The phase distribution of PEG was not affected by the presence of sucrose and sodium (31). The solubility of protein in PEG was also found to be independent of salt concentrations (39). However, ionic strength and species may be important for the solubility of free ligands in PEG. For instance, the solubility of parathyroid hormone was increased by the presence of sodium and potassium. The effect of monovalent ions is influenced by different halide counter ions (31). Therefore, selecting a proper buffer system may increase the efficiency of the separation of the free from the bound ligand. In cases where detergent-solubilized membrane receptors have been used, the presence of detergent has not interfered with PEG precipitation of the hormone–receptor complex.

The recovery of the receptor–ligand complex with PEG precipitation methods was found to be higher than that obtained by the gel filtration method (33). The difference between the two methods was attributed to the dissociation of the ligand–receptor complex during chromatography. Baron and Abood (40) compared the binding parameters of [^{3}H](−)QNB to solubilized muscarinic receptors using equilibrium dialysis, charcoal absorption of free ligand, and PEG precipitation followed by filtration. The greatest recovery of the bound ligand was obtained using equilibrium dialysis. Slightly less binding activity was detected using the PEG method, whereas charcoal absorption gave a significantly lower apparent yield. The three methods gave similar values for the dissociation constant and for the Hill coefficient. The PEG method was used for further studies due to its rapidity and high yield.

The precipitation reaction usually takes 10–20 min to complete. A new equilibrium between the free and bound ligands may be established during the precipitation. The filtration process also lasts a few minutes due to the high viscosity of the PEG solution. However, the dissociation of the ligand–receptor complex could also be less serious than in the usual filtration process, since the diffusion of the ligand in the PEG solution may be slower than in an aqueous solution.

In summary, compared with the other methods for the assay of ligand binding to soluble receptors, the method of PEG precipitation is more rapid, inexpensive, and reproducible.

Besides using PEG, the ligand–receptor complex can also be precipitated with saturated solutions of ammonium sulphate. Some investigators have successfully used these methods to measure the specific binding of radio-ligands to solubilized muscarinic receptors (41,42). The non-specific binding can be reduced by rinsing the filters with saturated ammonium sulphate. This method compared very favourably with the Sepharose column assay technique (43). However, this method may not be suitable for most of the binding assay systems for peptide hormones because of the possibility of precipitating both free and bound hormones. Moreover, it takes a longer time to perform than the PEG precipitation method (44).

2.5.2 Filtration with PEI-treated filters

Bruns *et al.* (45) have suggested a modification of the route filtration technique, which allows retention of both solubilized membrane–bound receptors or naturally soluble receptors on filters. The glass fibre filters are pretreated with a 0.3% solution of polyethylenimine (PEI) for several hours (usually 2–3 hours is enough since longer times may cause decomposition of the glass fibre filters). The filters are placed into the filtration machine without rinsing. The contents of each incubation tube containing solubilized receptors are then loaded on to the filters under vacuum and washed with ice-cold buffer. With this method, the authors have determined ligand binding for several solubilized membrane-bound receptors such as the muscarinic, adenosine A_1, α_2-adrenergic, β-adrenergic, dopaminergic D_2, opiate, bradykinin, and benzodiazepine receptors, as well as naturally soluble oestradiol receptors. For muscarinic, adenosine, α_2-adrenergic, dopaminergic, and oestradiol receptors, specific binding measured with the PEI-treated filters was 84–110% of specific binding measured with gel filtration, demonstrating that the technique gave recovery of the bound ligand equal to the gel filtration technique.

The mechanism of retention of soluble receptors on PEI-treated glass-filters is thought to be ionic. Most of the integral membrane proteins tend to be acidic. For instance, the β-adrenergic receptors have a pI of 5.8 (46), the muscarinic receptors have a pI of 4.3–4.5 (47), and the nicotinic receptors have pI values of 5.2 and 6.0 (48). PEI binds strongly to silica and probably in a similar way to glass fibres, which are also negatively charged. The PEI-treated glass fibres can then strongly bind polyanions. The binding of soluble receptors to PEI-treated filters was found to be resistant to washing and insensitive to changes in the ionic strength. However, in order to be retained on PEI-treated filters, a receptor should be acidic, or at least not basic. Bruns *et al.* (45) also suggested the omission of the PEI treatment in the case of basic proteins, since the glass filters have negative charges and are known to bind basic proteins strongly.

In a preliminary study, we have used both methods (PEG-precipitation and PEI-treatment) to determine the specific binding $[^3H](-)QNB$ and $[^3H]$pirenzepine to digitonin-solubilized muscarinic cholinergic receptors. The results obtained are similar with both methods. However, it is easier and faster to use the PEI-treatment method.

3. Evaluation of filtration techniques

3.1 Comparison of the rapid filtration technique with other separation techniques

The rapid filtration technique compares favourably with other methods for receptor binding assays because of its simplicity, rapidity, low non-specific binding, and high reproducibility. We will briefly discuss the advantages of the rapid filtration technique over the other three widely used separation techniques, namely, centrifugation, equilibrium dialysis, and gel filtration. A comparison of the major characteristics of the filtration technique and other methods is presented in *Table 3*.

Table 3. Comparison of the rapid filtration technique with other methods for separation of the free and the bound ligand

Method	Complete separation	Simplicity of operation	Time of separation	Specific to non-specific ratio	Reproducibility of results
Rapid filtration	good	good	sec	high	good
Centrifugation	fair	fair	min	fair	fair
Dialysis	poor	poor	day	low	poor
Gel filtration	fair	fair	min	high	fair
Precipitation	fair	fair	min	fair	fair
Absorption	fair	fair	min	fair	poor

3.1.1 Centrifugation

Both the filtration and the centrifugation techniques have been listed as methods for the rapid separation of the free and bound ligand. Centrifugation can be accomplished in minutes using the desktop microcentrifuge. However, compared with the filtration technique, the major compromise of the centrifugation procedure involves the less-efficient surface washing of the pellets. The background of the binding is consequently much higher than when using the filtration technique due to the incomplete separation of the free ligand trapped in the pellets. Nevertheless, the centrifugation technique is a good alternative to the filtration technique for those ligand binding studies with a rapid dissociation of the ligand–receptor complex which limits the use of the rapid filtration technique in the assay.

3.1.2 Equilibrium dialysis

Equilibrium dialysis is considered to be the method which gives least disturbance of the ligand–receptor complex during the separation of the free and bound ligand, and can be used for both membrane particles and soluble proteins. With this method, the free ligand is equally distributed on both sides of the filter membrane rather than being washed out, as in the filtration method.

The specific binding of the radioligand usually amounts to less than 10% of the total radioligand used. Therefore, the surplus of the radioactivity in the receptor side of the dialysis membranes, compared to the buffer side, rarely exceeds 20%. This affects the precise determination of specific binding. This method is now used only for the determination of binding of radioligands with very low affinity, when both centrifugation or filtration techniques are unsuitable. As sometimes encountered in the filtration procedure, binding of the radioligand to the membranes is a major artefact in the experiments using equilibrium dialysis. The extent of this non-specific binding needs to be determined for each radioligand. Moreover, the equilibrium dialysis procedure can take hours to days to complete.

3.1.3 Gel filtration chromatography

Gel filtration chromatography is often used for the binding assay of soluble receptors. Dissociation of the bound ligand during the column elution is a major concern with this technique. In some cases the results show less recovery of the ligand–receptor complex than using the PEG precipitation method. Preparation of gel columns before the experiments and regeneration of the columns after experiments are quite laborious. However, if the binding assay of a soluble receptor cannot be performed successfully with a PEG precipitation, or with the PEI filter treatment, then gel filtration is an alternative worth trying.

3.2 Disadvantages of the rapid filtration technique

Although rapid filtration has many advantages over other methods, it has its own shortcomings which can limit the use of the technique. Errors can be introduced in the estimation of the binding parameters of the radioligands to specific binding sites.

3.2.1 Error in the estimation of the free concentration of radioligands

As mentioned above, rinsing the filter greatly increases the specific/non-specific binding ratio and, therefore, improves the precision of the determination of the specific binding of the ligand. However, in most cases, washing the filter causes loss of low affinity or non-specific binding of the ligand to other membrane components. Since the free concentration of the radioligand is usually calculated from the difference between the total ligand used and the total binding of the ligand measured, the filtration technique tends to exaggerate the free concentration of the ligand and, therefore, overestimates the K_d value, in other words, underestimates the affinity of the ligand. One method for testing this problem is to directly measure the free concentration of the reaction mixture in some culture tubes. Alternatively, free concentrations can be measured in parallel experiments using the centrifugation method.

3.2.2 Error in the estimation of the receptor concentration

The use of the filtration technique is limited to ligands with relatively high affinity. In the case of a rapidly dissociating ligand ($k_{21} > 0.6$/min or a $K_d > 20$ nM), significant loss of the ligand–receptor complex can occur. The receptor concentration could, therefore, be underestimated. If the dissociation of the bound ligand during the filtration is a serious problem, then the receptor concentration should be carefully checked using different separation techniques.

Loss of small size membrane particles during filtration is another reason for the possible underestimation of the receptor concentration. Most of the particles will be captured by either Whatman GF/B (pore size 1 μm) and GF/C (pore size 1.2 μm) glass fibre filters. However, small particles can pass through the filter during the filtration under reduced pressure. This can be controlled by measuring the protein content in the tubes compared with that retained on the filters. If soluble receptor preparations are involved, a modification of the filtration method should be adopted (see Section 2.5). The use of filters with smaller pore size, such as Whatman GF/F (pore size 0.7 μm) is recommended. In this case, however, the results from this method should be compared with those obtained using other methods.

3.2.3 Binding of the free radioligand to filters

Another major compromise related to the use of filtration techniques is the non-specific binding of the ligand to the filters during the filtration. Some ligands tend to bind to filters (49–51). In some experiments binding of the ligand to the filters can be monitored by simultaneously passing the ligand, with no tissue present, through the filters followed by the same rinsing process. The radioactivity found on the filter can be considered as the binding of ligand to the filters during the filtration. However, this method does not always provide a suitable result. In some cases, the non-specific binding of the ligand to the filters, determined by this method, can exceed the total non-specific binding found in the incubation tubes, where the radioligand is incubated with tissue in the presence of a high concentration of inhibitor. This occurs because the free concentration of the ligand in the control tube is usually higher than the free concentration of the ligand in the incubation tube. On the other hand, there are less non-specific binding sites for the ligand on the filters that are covered by the membrane particles than on empty filters. This difference can be reduced by including bovine serum albumin or denatured tissue in the control tubes.

In many cases, it is not always necessary to have a tissue-free control in each experiment. If the total binding of radioligand in the incubation tubes is less than 10% of the total ligand added, the free concentration of the radioligand in the total binding tube can be considered the same as that in the non-specific binding tube. Since the membrane concentration in both tubes is the same, it is possible to assume that the chance for the free radioligand to bind to the filters will be the same in both tubes. Thus, counts from the filter binding of the radioligand can be subtracted from the total binding counts to estimate the specific binding of the radioligand. However, in some cases, it has been found that non-specific binding of radioligand to filters can be 'specifically' displaced by unlabelled inhibitors (49, 51). This causes lower filter binding in the presence than in the absence of inhibitor. Therefore, proper controls may have to be performed simultaneously in order to accurately estimate the specific binding of the radioligand.

The existence of non-specific binding of the ligand to the filter increases the non-specific background, therefore affecting the precision of estimates of specific binding of the ligand. Several methods can be used to reduce the filter binding of the ligand. First, the filter can be preincubated with a high concentration of a structurally similar compound to saturate the binding sites on the filters. Because of the competition of the radioligand with this compound on the filter surface, the opportunity for the radioligand to bind to the filter will be reduced. Similarly, if the filter binding of the radioligand is due to an acidic radical or a basic radical in the ligand molecule, preincubation of the filters in a basic solution or an acidic buffer will decrease the binding of the ligand to the filter. Another frequently used method is to immerse the filters in 0.1%

polyethylenimine solution (PEI) for a short time before filtration. PEI associates with the glass fibres and blocks the binding of the ligand to the filter. This method has proven very effective in many cases. Alternatively, the filters can be presoaked with 1% albumin. Albumin will prevent the binding of the ligand to the filters. Any free ligand which is trapped by albumin will be eventually washed away together with the albumin.

3.2.4 Contamination between filter strips

Repeated filtration experiments using the same apparatus can cause a cumulative radioactive contamination of the screens. The contaminant can be transferred from one filter strip to another. Higher counts can be obtained in later compared with earlier samples. The variation of the determination can become more serious when high concentrations of the radioligand are used, or many filtration experiments are done on the same day. Contamination can be reduced by thoroughly washing the filtration machine between consecutive filtrations. In this laboratory, the filtration machine is routinely washed with 3–5 litres of distilled water between two consecutive filtrations. We have observed less than 10% variation of the counts between the first and the last filtration. Since the contamination of the screens is related to the concentration of the radioligand used in each tube, it is expected that more contamination occurs in the screens where the higher concentrations of the radioligand have been used than in the screens that received lower concentrations of the radioligand. A random arrangement of the incubation tubes in each experiment will avoid systematic errors caused by contamination of the screens.

Summary

The rapid filtration technique is a major advance in the development of the receptor assay. The positive features of this technique, namely rapidity, complete separation, simplicity, and reproducibility permit the study of membrane receptors as well as other proteins, including ion channels, enzymes, and soluble receptors. However, the rapid filtration technique has its limitations which can lead to errors in the interpretation of the results. Careful examination of the conditions used in the receptor-binding assay will help to improve the accuracy and the precision of determinations using the rapid filtration technique.

Acknowledgements

This study was supported by a Grant-In-Aid from the American Heart Association, Arizona Affiliate, and in part by USPHS grants DK-36289 and HL 20984. We thank Sue Waite for reviewing this manuscript, and sharing with us her personal experience of *in vitro* receptor-binding studies. We also thank Pam Abrams for her excellent secretarial support.

Jian-Xin Wang, Henry I. Yamamura, Wan Wang, and William R. Roeske

References

1. Langley, J. N. (1906). *Proc. R. Soc. London, Ser. B,* **78,** 170.
2. Ehrlich, P. (1913). *Lancet,* **ii,** 445.
3. Clark, A. J. (1926). *J. Physiol. (Lond.),* **61,** 530.
4. Paton, W. D. and Rang, H. P. (1965). *Proc. R. Soc. London, Ser. B,* **163,** 1.
5. Potter, L. T. (1967). *J. Pharmacol. Exp. Ther.,* **155,** 91.
6. Lefkowitz, R. J., Roth, J., Pricer, W., and Pastan, I. (1970). *Proc. Natl. Acad. Sci. USA,* **65,** 745.
7. Lin, S. Y. and Goodfriend, T. L. (1970). *Am. J. Physiol.,* **218,** 1319.
8. Changeux, J. P., Kasai, M., and Lee, C. Y. (1970). *Proc. Natl. Acad. Sci.,* **67,** 1241.
9. Bennett, J. P., Jr. and Yamamura, H. I. (1985). In *Neurotransmitter Receptor Binding* (ed. H. I. Yamamura, S. J. Enna, and M. Kuhar), p. 61. Raven Press, NY.
10. Cuatrecasas, P. (1969). *Proc. Natl. Acad. Sci. USA,* **63,** 450.
11. Yamamura, H. I., Enna, S. J., and Kuhar, M. J. (ed.) (1978, 1st edn; 1985, 2nd edn), *Neurotransmitter Receptor Binding.* Raven Press, NY.
12. Yamamura, H. I. and Snyder, S. H. (1974). *Mol. Pharmacol.,* **10,** 861.
13. Birdsall, N. J. M. and Hulme, E. C. (1976). *J. Neurochem.* **27,** 7.
14. Laduron, P. M., Verwimp, M., and Leysen, J. E. (1979). *J. Neurochem.,* **32,** 421.
15. Burgen, A. S. V., Hiley, C. R., and Young, J. M. (1974). *Brit. J. Pharmacol.,* **50,** 145.
16. Watson, M., Roeske, W. R., and Yamamura, H. I. (1982). *Life Sci.,* **31,** 2019.
17. Wang, J.-X., Roeske, W. R., and Yamamura, H. I. (1987). *Life Sci.,* **41,** 1751.
18. Ehlert, F. J., Dumont, Y., Roeske, W. R., and Yamamura, H. I. (1980). *Life Sci.,* **26,** 961.
19. Harden, T. K., Meeker, R. B., and Martin, M. W. (1983). *J. Pharmacol. Exp. Ther.,* **227,** 570.
20. DeRobertis, E., Azcurra, J., and Fiszer, S. (1967). *Brain Res.,* **5,** 45.
21. Romano, C. and Goldstein, A. (1980). *Science,* **210,** 647.
22. Abood, L. G. and Grassi, S. (1986). *Biochem. Pharmacol.,* **35,** 4199.
23. Schwartz, R. D., McGee Jr. R., and Kellar, K. J. (1982). *Mol. Pharmacol.,* **22,** 56.
24. Vickroy, T. W., Watson, M., Yamamura, H. I., and Roeske, W. R. (1986). In *Receptor Binding in Drug Research* (ed. R. A. O'Brien), pp. 297–319. Marcel Dekker Inc., NY and Basel.
25. Lin, M., Wang, J.-X., Roeske, W. R., and Yamamura, H. I. (1987). *J. Pharmacol. Exp. Ther.,* **242,** 991.
26. Chang, K.-J., Jacobs, S., and Cuatrecasas, P. (1975). *Biochem. Biophys. Acta,* **406,** 249.
27. Fields, J. Z., Roeske, W. R., Morkin, E., and Yamamura, H. I. (1978). *J. Biol. Chem.,* **253,** 3251.
28. Klotz, I. M. (1981). *Protein Function, a Practical Approach* (ed. E. Creighton), p. 52. IRL Press, Oxford and Washington.
29. El-Refai, M. F. (1984). In *Receptor Biochemistry and Methodology,* Vol. 1 (ed. J. C. Venter and L. C. Harrison) p. 99. Alan R. Liss, Inc., NY.
30. Laduron, P. M. and Iien, B. (1982). *Biochem. Pharmacol.,* **31,** 2145.

31. Albertsson, P. (1960). *Partition of Cell Particles and Macromolecules*. Almquist and Wiksells, Stockholm.
32. Desbuquois, B. and Aurbach, G. D. (1971). *J. Clin. Endocrinol.*, **33**, 732.
33. Cuatrecasas, P. (1972). *Proc. Natl. Acad. Sci. USA.*, **69**, 318.
34. Strauss, W. L., Ghai, G., Fraser, C. M., and Venter, J. C. (1979). *Arch. Biochem. Biophys.*, **196**, 566.
35. Haga, T., Haga, K., and Gilman, A. G. (1977). *J. Biol. Chem.*, **252**, 5776.
36. Shreeve, S. M., Roeske, W. R., and Venter, J. C. (1984). *J. Biol. Chem.*, **259**, 12398.
37. Goldstein, S. and Blecher, M. (1976). In *Methods in Receptor Research* (ed. M. Blecher), p. 119. Marcel Dekker, Inc., NY and Basel.
38. Wenger, D. A., Parthasarthy, N., and Aronstam, R. S. (1985). *Neurosci. Lett.*, **54**, 65.
39. Fried, M. and Chun, P. W. (1971). In *Methods in Enzymology* (ed. W. B. Jakoby), Vol. 22, p. 238. Academic Press, NY.
40. Baron, B. and Abood, L. G. (1984). *Life Sci.*, **35**, 2407.
41. Hurko, O. (1978). *Arch. Biochem. Biophys.*, **190**, 434.
42. Baumgold, J., Merril, C., and Gershon, E. S. (1987). *Mol. Brain Res.*, **2**, 7.
43. Haga, K. and Haga, T. (1983). *J. Biol. Chem.*, **258**, 13575.
44. Cremo, C. R., Herron, G. S., and Schimerlik, M. I. (1981). *Anal. Biochem.*, **115**, 331.
45. Bruns, R. F., Lawson-Wendling, K., and Pugsley, T. A. (1983). *Anal. Biochem.*, **132**, 74.
46. Shorr, R. G. L., Lefkowitz., R. J., and Caron, M. G. (1981). *J. Biol. Chem.*, **256**, 5820.
47. Hulme, E. C., Berrie, C. P., Haga, T., Birdsall, N. J. M., Burgen, A. S. V., and Stockton, J. (1983). *J. Receptor Res.*, **3**, 301.
48. Alpert, A. and Regnier, F. E. (1979). *J. Chromatogr.*, **185**, 375.
49. Cuatrecasas, P., Hollenberg, M. D., Chang, K.-J., and Bennett, V. (1975). *Recent Progress Hormone Res.*, **31**, 37.
50. Watson, M., Yamamura, H. I., and Roeske, W. R. (1983). *Life Sci.*, **32**, 3001.
51. Vickroy, T. W., Roeske, W. R., and Yamamura, H. I. (1984). *Life Sci.*, **35**, 2335.

7

Centrifugation binding assays

E. C. HULME

1. Introduction

Centrifugation assays are best applied to assaying the labelling of receptors in membrane preparations using radioligands. The principle is elementary. At the end of the assay period, the membranes bearing the labelled receptors are sedimented, and thus physically removed from the supernatant, containing the bulk of the unbound radioligand. After pouring off the supernatant, and subjecting the pellet to washing procedures of greater or less rigour, the amount of bound ligand is assayed, typically by liquid scintillation spectroscopy, or γ-counting.

In the simplest case, centrifugation assays can be regarded as equilibrium assays. This is because a small volume of the supernatant, proportional to the tissue concentration in the assay and typically around 1% for a suspension of 1 mg membrane protein/ml, is entrapped in the membrane pellet after sedimentation. This ensures that the binding equilibrium, once attained, is maintained throughout the separation procedure, i.e. that providing that washing is confined to superficial rinsing of the surface of the pellet and the tube, the equilibrium is minimally disturbed within the pellet itself. This is one important aspect in which centrifugation assays differ from filtration assays. In the latter, partial ligand dissociation is an inevitable result of washing the filters to reduce non-specific binding. However, there are two variants of the centrifugation assay in which the binding equilibrium is deliberately disturbed:

(a) In the first the pellet is resuspended, diluted and then recovered by filtration (1). The aim is to reduce the amount of entrapped ligand, and thus to improve the ratio of specific–non-specific binding. During the filtration step dissociation of the ligand will clearly tend to occur just as in a standard filtration assay.

(b) In the second, labelled membranes are centrifuged through a layer of oil (2). The labelled ligand remains in the aqueous supernatant, whilst the membranes are pelleted. The extent to which equilibrium is disturbed depends on the oil–water partition coefficient of the ligand, and on the sedimentation time of the membranes.

These variants on the simple centrifugation assay suffer from the main drawback of filtration assays, namely that rapidly-dissociating components of binding may be lost and remain undetected. These problems are averted in the 'unwashed pellet' version of the assay. However, the price to be paid is a worse apparent ratio of specific:non-specific binding. Typically, 1–10% of the added ligand may be entrapped, and/or non-specifically bound. The implications of this for the working range of ligand affinities have been discussed in Chapter 4. Except under exceptional circumstances, these constraints mean that the lower limit of the ligand affinity constant compatible with a microcentrifugation assay is approximately $3 \times 10^7/M$.

Another drawback of the centrifugation assay is the rather long and relatively indeterminate separation time. Centrifugation times usually range from 1 to 5 min. Even under favourable conditions, the effective sedimentation time is never likely to be less than about 20 sec. Thus centrifugation assays cannot be regarded as suitable for defining kinetic time courses with half-times of less than 1 minute, even under the most favourable circumstances. It may, however, be possible to estimate the sedimentation time, and apply a correction.

Given the above deficiencies, what are the advantages of centrifugation assays?

They are:

(a) Simplicity of equipment. Microfuges are cheap, and widely available.

(b) Cheapness and convenience. Binding, separation, and scintillation counting can all be performed inside Eppendorf-type microfuge tubes. A high throughput is possible. Utilization of expensive scintillation cocktails is minimized.

(c) In the simplest, superficially-washed pellet version of the assay, the binding equilibrium is not disturbed. Even rapidly-dissociating components of binding are measurable.

(d) The supernatant is readily available for sampling and assay, allowing a routine check for non-specific and specific ligand depletion.

(e) Recovery of labelled membranes is usually very good (>90%). When we have compared the methods, centrifugation has given a better recovery of sites than filtration assays, partly owing to the third point, but also to mechanical loss of membranes through filters.

2. Equipment

2.1 Choice of microcentrifuges

A number of features need to be considered:

i. g Force

To be capable of sedimenting membrane fractions effectively in less than

5 min, the microfuge needs to develop at least 10 000 *g*, and preferably 15 000 *g*. If still higher *g* forces are required, it is possible to use an ultracentrifuge, although this is cumbersome for routine assay. An alternative is the Beckman Airfuge. This provides very high centrifugal forces, but has a rather low capacity, with small tubes.

ii. Centrifuge capacity

To ensure a reasonable throughput of assays, the rotor should take at least 10 tubes. For practicality, it is necessary to have the capability to centrifuge 50 tubes more or less simultaneously. This either means having several small machines, or a machine which will accept a high-capacity rotor.

iii. Rotor type

Both angle and drum-type rotors are available. The drum rotors take removable racks or blocks in which the tubes are placed. They accept more tubes but usually develop less centrifugal force than the smaller fixed-angle rotors. The blocks can also be used for tube handling/incubation which is convenient. They are also available for a range of tube sizes. Angle rotors tend to produce more compact, better-shaped, more adherent pellets, particularly in microfuge tubes which are conical rather than spherical near the tip, i.e. which have a high radius of curvature. These are preferable to the blunter, rounded tubes which are sometimes encountered.

iv. Temperature control

Temperature control is an expensive option for microfuges. However, it is worth considering for a relatively expensive, high-capacity machine. The extra weight and bulk will decrease portability. For routine assays at temperatures in the range 25–37°C, running microfuges at room temperature seems adequate. For low-temperature work, microfuges can be placed in a cold room. There is obviously some tendency for heat to be generated during long spins. This is minimized if there is adequate air circulation through the rotor compartment, although this also produces the possibility of aerosol formation if a tube breaks. In practice, sedimentation tends to be completed before a rise in temperature becomes a problem.

v. Timing of the run

Digital control is preferable for precision.

vi. Fast braking

This needs to be checked carefully, to make sure that pellets are not disturbed. It is useful if the microfuge has a slow-braking option. The same effect can be produced by turning off the power supply at the end of the run, and allowing the rotor to coast to a halt.

2.2 Microcentrifuges and suppliers

In *Table 1* we summarize some major suppliers of microfuges suitable for membrane-binding assays. A large number of machines are available. Custom-made microfuges with fixed angle rotors taking 1.5 ml Eppendorf tubes cost between £600 and £1000. Alternative rotors taking different tube sizes are usually available. These centrifuges produce up to 15 000 *g*. Drum rotors taking up to 60 tubes represent a more expensive option, but are more economical on a cost-per-tube basis. They have some disadvantages, enumerated in Section 2.1, but also have the virtue of flexibility. Some of the insert racks take angled tubes, thus allowing pellets to be centrifuged as if in a fixed angle rotor. These may represent a good option. Given the flexibility of being able to use a range of rotors, a relatively large capacity, refrigerated bench-top centrifuge generating around 12 000 *g* with a drum rotor may be a good alternative to dedicated microfuges with fixed angle rotors, even though the latter generate a slightly higher *g* force.

3. Performing and processing microfuge assays

In this section, we detail the considerations which need to be specifically taken into account for the successful performance of a microcentrifugation binding assay. General protocols for setting up and incubating binding assays have already been given (Chapter 4, Section 3) and should be followed.

3.1 Minimum ligand affinity

Ligand affinity needs to be at least 3×10^7/M for routine assays in which the pellet is superficially washed (Chapter 4). This requirement may be relaxed somewhat for assays in which the receptor preparation is centrifuged through an oil layer (2) or where the pellet is resuspended, washed, and then recentrifuged or filtered (1). These procedures may be worth trying if the ligand off-rate is slow enough. For guidelines, see Chapter 4, Section 2.3.7.

3.2 Minimum receptor concentration

The constraints on minimum receptor concentration have been discussed in Chapter 4, Section 3. For accurate determination of a full ligand saturation curve using a ^{3}H-ligand, the minimum acceptable receptor concentration is approximately 5×10^{-11} M. Ideally, the concentration of binding sites should not exceed $0.1 \times$ the K_d of the tracer ligand, in order to avoid significant radioligand depletion.

3.3 Minimum protein concentration

The minimum concentration of membrane protein in a centrifugation binding assay is normally approximately 100 µg/ml. If it is less than this, a properly compact pellet will not form. Although it has proved possible to assay recombinant muscarinic receptors expressed in Chinese hamster ovary (CHO) cells by centrifugation at membrane protein concentrations as little as 10 μg/ml, the failure to form a properly compact pellet means that loss of bound ligand is inevitable during the washing procedure unless a very slowly-dissociating ligand is used. The maximum practical membrane protein concentration is about 2.5 mg/ml. If this is exceeded too large a proportion of the added ligands will become entrapped in the pellet. Furthermore, parts of the pellet will tend to break away during washing. Note that purified receptors reconstituted into phospholipid vesicles using the PEG precipitation procedure (3) can be assayed readily using microcentrifugation, and tend to form compact, well-conditioned pellets.

3.4 Choice of tubes

It is convenient to perform the assays in polypropylene Eppendorf-type microfuge tubes (see Chapter 4). Different sizes are available, for example 1.5 ml, 0.75 ml, 0.5 ml, 0.25 ml, and different microfuge rotors can be obtained to fit them. For easy rinsing of pellets, and general ease of manipulation, coupled with maximization of the recovery of the receptor–ligand complex in a binding assay, we have found the 1.5 ml size extremely convenient. To encourage good pellet formation, it is necessary to have tubes with a markedly conical tip. It is also important that they should have well-made tight-fitting caps and robust hinges, since several capping and uncapping operations are necessary during the course of the assay. We have found that some hinges become friable, especially when operations are carried out at 4°C.

3.5 Choice of buffer and centrifugation time

Membranes pellet best from buffers containing near-physiological concentrations of cations (*Table 2*). Thus, good, compact pellets are formed at around 14 000 g within 1 minute from Krebs–Henseleit solution (Chapter 5). Buffers containing 100 mM of an alkali metal salt, or 10 mM of a salt of a divalent cation (Mg^{2+}/Ca^{2+}) tend to behave similarly. However, pelleting from dilute buffers not enriched with inorganic cations is relatively poor, requiring up to 5 min of centrifugation at 14 000 g. Even then, the pellets formed are relatively voluminous, and sloppy. The mechanical properties of the pellets are particularly poor if the buffers contain EDTA, without added divalent cations. However, they are substantially improved by the addition of a millimolar excess of cations such as Mg^{2+} or Ca^{2+}. Pelleting from concentrated salt solutions (1 M NaCl) or iso-osmolar sucrose is also poor. Here the use of an ultracentrifuge and higher g-forces will be necessary. The presence of 10 mM Mg^{2+} sometimes seems to be more important than a high monovalent ion concentration. In CHO cell membranes, we have found that only 65–75%

Table 1. Microcentrifuges suitable for centrifugation binding assays

Make	No.[a] tubes	Rotor	g	Adjustable speed	Temp. control	Timing	Brake	Supplier	Cost per tube[b]
Beckman									
Microfuge E	12	angle	15 850	–	–	analog	+	Beckman	58
Microfuge 11	12	angle	12 400	+	–	analog	+	Beckman	146
Microfuge 12	60	drum (tilted)	12 200	+	–	analog	+	Beckman	49
Burkard Fugette/									
Koolspin μP	12	angle	32 000	+	–/+[c]	digital	adj.	Biotech Instruments	156/263[c]
Damon/IEC									
Micro-MB	12	angle	12 700	–	–	incremental	+	Damon/IEC	61
Centra M	24	angle	17 000	–	–	analog	+	Damon/IEC	33
Denley BM402	16	angle	11 500	+	–	digital	+	Denley Instruments	47
Eppendorf									
5415	18	angle	14 000	+	–	analog	+	Anderman	49
5413	40	drum	11 500	–	–	analog	+	Anderman	28
Eagle ES20	48 6 × 12	angle/ drum	15 000	+	–	analog	adj.	Eagle Scientific	28
Heraeus									
Biofuge A	24	angle	14 900	+	–	analog	+	Heraeus	32
Biofuge B	8 × 10	drum	11 600	+	–	analog	+	Heraeus	

Hermle								
Z-229	12/24	angle	15 600/13 200	—	—	analog	+	Anderman	27
Z-230	24	angle	13 000	+	—	analog	+	Anderman	37
Z-365	8 × 11	drum (tilted)	10 030	+	—	analog	adj.	Anderman	30
ZK-365				+					48
Hettich								
Mikrotiter	12/24	angle	11 300	+/−	—	analog	+	Arnold Horwell	30
Mikrorapid/	12/24	angle	10 300	+	−/+	analog	+	Arnold	
Mikrorapid K	6 × 10	drum	12 000	+	−/+	analog	+	Horwell	
Jouan								
M14.11/	28	angle	12 560	+	−/+	analog	+	Jouan	49
Mr14.11	6 × 10	drum	12 070	+					
MSE								
Microcentaur	12	angle	13 400	low/high	—	analog	+	MSE Scientific	54
Sorvall								
Microspin	12/24	angle	12 300	−/+	−	analog	+	Dupont	27/37

This summary of characteristics is only a guide, and is not comprehensive. The manufacturers should be consulted for further information. Likewise, the costings are also approximate. The higher cost per tube of some machines is justified by additional features not mentioned in this table, but also not directly relevant to centrifugation binding assays.

[a] based on 1.5 ml microfuge tubes.
[b] approximate cost per tube, including rotor (£ sterling).
[c] −/+ indicates that the feature is optional. Prices without and with the option are given in some cases.
[d] This table was complete at the time of writing, but some further developments have occurred within the past six months.

of the receptors are sedimented after 5 min at 15 000 *g* if less than 3 mM Mg is present in the incubation medium. Interestingly, comparably low recoveries are observed if separation is carried out by the filtration technique. These findings are important in that an increasing use is being made of membranes containing isolated recombinant receptor subtypes for binding experiments.

When in doubt, recovery of the receptor-rich membranes can be checked by comparing the results of microfuge assays with those of the same assays replicated using an ultracentrifuge at 100 000 *g*. Alternatively, the use of a Beckman airfuge may be considered if pelleting at lower *g* is not obtained. The disadvantage of the airfuge is its limited capacity.

Table 2. Centrifugation times for various buffers

Buffer	Minimum centrifugation time (14 000 g)
Krebs–Henseleit	1 min
20 mM Tris/Hepes/phosphate + 100 mM NaCl/KCl, etc.	2 min
20 mM Tris/Hepes/phosphate	5 min
20 mM Tris/Hepes/phosphate + EDTA	5 min (sloppy pellet, consider use of ultracentrifuge or airfuge)
20 mM Tris/Hepes/phosphate + Mg/Ca (1 mM)	5 min

3.6 Processing the assays after incubation

3.6.1 Superficial washing of the membrane pellet

Protocol 1. Standard microcentrifugation assay

1. Centrifuge the assay tubes (see *Table 2* for a guide to centrifugation times).

2. Remove the tubes from the rotor and place them in racks.

3. Open the tubes carefully. Sample the supernatant (e.g. 0.1 ml from 1.0 ml) for determination of the free ligand concentration. Check the temperature of the supernatant to make sure that it has not risen excessively.

4. Carefully pour off the supernatant from a pair of tubes, holding the tubes by the caps, and tipping them with the fingertips of the other hand. (*Figure 1*). Observe precautions for the disposal of radioactive waste.

5. Carefully dip the tubes, including the caps, under the surface of a litre beaker of water or buffer. Hold the tubes in a near-horizontal position, letting the buffer run gently down the sides to avoid disturbing the pellets, then fully immerse the caps also. Withdraw the tubes, and pour out the wash solution, shaking the tips of the tubes gently to get the last drops out.

6. Repeat step 5 with two fresh beakers of buffer. Renew the contents of the beakers periodically, for example every 100 tubes.

7. Place the tubes in an inverted position in a test-tube rack. Leave for about 20 min to drain.

8. Shake out, or blot off, any buffer still adhering to the rim of the tubes and caps. Return the tubes to a vertical position.

9. If the ligand is a γ-emitter, cut off the tips of the tubes with a heated scalpel blade in a fume hood, and transfer them to γ-counting tubes. Some γ-counters are able to count Eppendorf tubes directly.

10. If the ligand is a β-emitter, the pellet will need to be solubilized. To each tube, add 100 μl of a tissue solubilizing agent, such as Soluene-350 (Packard), using a repetitive pipette. Cap the tubes and leave overnight at room temperature. Soluene-350 has a higher water-tolerance than Soluene-100.

11. When the pellets have dissolved, usually after overnight incubation at room temperature, although the time varies, add 1.2 ml of a non-aqueous scintillation cocktail (for example Beckman NA) to each tube, and vortex-mix. Check that a homogeneous phase is formed.

12. To count the Eppendorf tubes, insert them into wide-necked (for example Beckman) 20 ml plastic scintillation vials. These usually do not need to be capped. Scintillation count in the normal fashion. The scintillation vials can be re-used. Check the validity of d.p.m. calculations based on external standard ratios or H-numbers by using internal radioactive standards.

13. It is possible to control for the internal volume of the pellet, and for the amount of entrapped incubation medium, by including in the incubation media a radioactive ion such as ^{86}Rb (100 000 d.p.m./ml) which becomes entrapped in the pellet, and which can be subjected to double-label counting along with the receptor-bound ligand. Using a ^{3}H-ligand, with ^{86}Rb to control for the internal volume of the pellet, about 10% overlap from the entrapped ^{86}Rb into the ^{3}H channel can be obtained with appropriate scintillation counter window settings. Expression of the results as a ratio of ^{3}H:^{86}Rb d.p.m. significantly reduces the error due to variations in pelleting of membrane fractions.

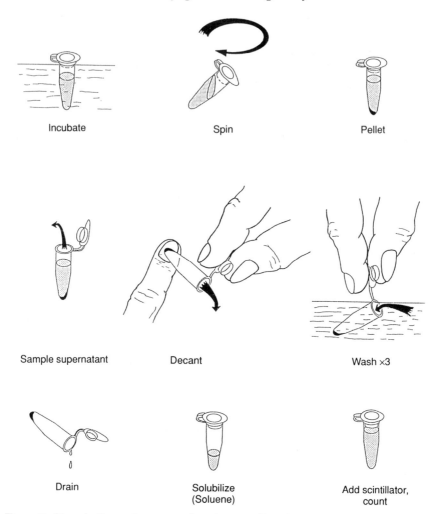

Figure 1. Steps in the performance of a microcentrifugation assay.

3.6.2 Resuspension and filtration of the pellet (modified from Galper *et al*, ref. 1)

***

Protocol 2. Mechanical resuspension of the membrane pellet

1. Add about 100 μl of glass beads to a 1.5 ml microcentrifuge tube (approximately 80 mesh; suitable beads are those supplied for gas chromatography by BDH Chemicals Ltd, Poole, UK)

2. Add 250 μl of buffer containing the radioligand, plus other additions.

3. Add 250 μl of membrane preparation at around 5 mg/ml.

4. Vortex-mix and incubate for an appropriate time at an appropriate temperature (see Chapter 4).

5. Microfuge at 14 000 g for 5 min.

6. Remove the supernatant by suction.

7. Vortex-mix pellets for 10 sec to resuspend; at this stage, binding equilibrium should continue to be maintained.

8. Wash the contents of the tube on to a Whatman GFC filter, or similar (see Chapter 6), under vacuum with three successive aliquots of 2 ml of ice-cold buffer, using a repetitive pipette. Wash the filter well with three further aliquots of ice-cold buffer, trying to wash the glass beads away from the sides of the filter well, and towards the centre of the filter.

9. Count the filter (Chapter 6).

The glass beads help to resuspend the pellet in the residual, entrapped, ligand-containing buffer. Equilibrium is, therefore maintained at this point. Dissociation can occur during the filtration step, but the extent of this is minimized by the use of ice-cold wash buffer. A possible drawback might be the binding of the ligands to the glass beads. This needs to be specifically checked.

3.6.3 Centrifugation through an oil layer

Here the bulk of the aqueous phase, containing the free ligand, is stripped away by the centrifugation of the receptor-bearing membranes through an oil layer. The density of the oil needs to be intermediate between that of the membranes and water.

Protocol 3. Centrifugation through an oil layer (see reference 2)

1. Mix 50 ml of dibutyl phthalate with 50 ml of dioctyl phthalate (Aldrich) to yield a mixture with a final density of 1.012

2. Pipette 0.5 ml of oil into a series of 1.5 ml microfuge tubes.

3. Carry out the binding incubation in a separate set of tubes, using a final volume of 0.5 ml (see Chapter 4).

4. Carefully layer the incubation mixture over the oil layer.

5. Spin for 2.5 min at 14 000 g.

6. Aspirate off the supernatant and oil. Superficially wash the pellets and sides of the tubes. Drain.

7. Count the pellets, as before.

Because of the comprehensive washing of the membranes resulting from their centrifugation through the oil layer, it has been suggested that the above protocol may be suitable for assaying relatively low affinity receptor–ligand interactions ($K < 10^7/M$). However, note that this is still a relatively little used technique. Many high affinity ligands are hydrophobic, and have a high oil:water partition coefficient, and might well be partially lost into the oil phase during centrifugation. Varying the separation time by altering the amount of oil or the centrifugal force may control for this.

3.7 Estimation of the effective sedimentation time

If a ligand is available whose dissociation rate-constant is comparable to the centrifugation time it may be possible to use the dissociation process as an internal clock, to estimate the effective sedimentation time of the assay. Thus, if the dissociation process is initiated from equilibrium at time 0, for example by addition of an excess of an unlabelled competing ligand, and samples are taken and sedimented at different times, t, the dissociation process will be described (in the simplest case) by:

$$\log_e (RL^*/RL_0^*) = -k_{21} \cdot (t + t_s);$$

where t_s is the effective sedimentation time, defined by the elapsed time between turning on the microfuge, and the formation of a pellet in which low-molecular weight ligands are no longer exchangeable with the bulk solution, RL^* is the measured concentration of receptor–ligand complex measured after incubation for time t, and RL_0^* is the equilibrium value of RL^*, before addition of the unlabelled competing ligand (see Chapter 4, Section 3); t_s may be estimated from a log–linear plot, or by curve-fitting.

References

1. Galper, J., Haigh, L. S., Hart, A. C., O'Hara, D. S., and Livingston, D. J. (1987). *Mol. Pharmacol.,* **32**, 230.
2. Sivaprasadarao, A. and Findlay, J. B. C. (1987). In *Biological Membranes, a Practical Approach* (ed. J. B. C. Findlay and W. H. Evans), p. 287. IRL Press, Oxford.
3. Cerione, R. (1990). In *Receptor–Effector Coupling, a Practical Approach* (ed. E. C. Hulme), p. 59. IRL Press, Oxford.

8

Charcoal adsorption for separating bound and free radioligand in radioligand binding assays

PHILIP G. STRANGE

1. General description of the technique

As discussed extensively in this book, there is a requirement for accurate, reproducible, rapid, and cheap methods for separating bound and free radioligand in radioligand receptor binding assays. In other chapters (Chapters 6, 7) methods have been described for the assay of membrane bound receptors based on collection of the bound radioligand by filtration or centrifugation of the membranes. When studying a soluble receptor (either a naturally soluble cytosolic receptor, such as a steroid hormone receptor, or a detergent solubilized receptor, formerly membrane bound) these methods are not applicable. A popular technique in such studies is the charcoal adsorption assay which relies on the ability of charcoal to selectively remove the free radioligand from the mixture of bound and free radioligand.

The adsorption of small organic molecules by charcoal has been exploited in a number of contexts. Organic chemists use charcoal to decolorize preparations during organic synthesis, and charcoal is widely used, industrially, in purification processes. Charcoal has been used for adsorption of nucleotides during the preparation of radiolabelled species (1), and has also been used to remove fatty acids from albumin (2) and steroids from serum (3). More recently, charcoal has been used for separating bound and free ligand in steroid receptor assays (4) and in radioimmunoassays (5). Our use of charcoal adsorption in a radioimmunoassay for cyclic GMP (6) prompted the use of the same separation technique in assays for solubilized D_2 dopamine receptors (7). As will be discussed below charcoal adsorption provides an excellent separation method in radioligand binding assays of solubilized D_2 dopamine receptors, and it has also been used for solubilized muscarinic acetylcholine receptors (8) and serotonin (5-HT$_2$) receptors (9).

The charcoal used for these assays should, strictly speaking, be termed 'activated carbon', but the term charcoal will be retained here as it is used in most papers on the topic. Activated carbon is made by removing the volatile components of carbonaceous material, for example peat or wood in the absence of air, the resulting carbon (charcoal) is activated using high temperature steam. The activation process creates pores (mainly micropores (< 1 nm) and mesopores (1–25 nm)) of molecular size in the carbon, and vastly increases its surface area. The large surface area then acts as an efficient adsorbent for small molecules via physical adsorption based on Van der Waals forces, and chemisorption dependent on specific chemical interaction. The adsorption of molecules depends on their size, chemical nature (non-polar molecules are adsorbed more readily than polar ones), and the pH (information from Norit (UK) Ltd, Glasgow).

Therefore, if charcoal is added to the mixture of free radioligand and receptor (with bound radioligand) in a typical binding assay the free radioligand is selectively adsorbed to the charcoal which can then be removed from suspension by centrifugation. Sampling of the supernatant for radioactivity should largely reflect radioligand bound to receptors. The selective adsorption of free radioligand may reflect the size difference between receptor and free radioligand; the latter being adsorbed readily into the micropores, the former being absorbed less readily into mesopores, especially if these are saturated with albumin (see below).

Typical methods for the assay of solubilized brain D_2 dopamine receptors are given in *Protocols 1* and *2*.

Protocol 1. Preparation of working solutions

1. Cool solutions to 4°C for use.

2. Solubilize D_2 dopamine receptors from bovine caudate nucleus as in refs 13 and 14 and keep at 4°C.

3. Make up the assay buffer: Hepes (20 mM), EDTA (1 mM), EGTA (1 mM), pH 7.4.

4. Make up the [³H]spiperone working solution: This contains [³H]spiperone (Amersham or NEN) at a concentration of about 10 nM in assay buffer (step 3). The exact concentration is determined on the day of the assay by scintillation counting of aliquots. For convenience the [³H]spiperone obtained from the supplier is diluted into ethanol to a concentration of 5 µM, then divided into aliquots of 0.5 ml and held at −20°C. The 5 µM solution can then be further diluted on the day of the assay.

5. Make up the Butaclamol solution: (+) or (−) Butaclamol (Research Biochemicals Inc., Natick, Mass. USA, or SEMAT, St Albans, UK) (10 µM) in assay buffer (step 3). For convenience these compounds are held as a frozen (−20°C) stock at 100 µM in buffer A containing 0.25%

acetic acid (used to dissolve the drug powder), and diluted on the day of the assay.

6. Make up the charcoal suspension: Mix together one part (by weight) of charcoal (Norit GSX, from BDH), one part (by volume) of albumin solution (see below), and nine parts (by volume) of assay buffer (step 3). Stir for 30 min at 4°C, using a small magnetic stir-bar, it should be vigorous enough to keep the charcoal in suspension but not so vigorous as to cause frothing.

7. Make up the albumin stock solution: bovine serum albumin (RIA grade from Sigma) (22%), NaCl (0.85%), NaN$_3$ (0.1%) (all w/v), and keep at 4°C for convenience.

Protocol 2. Radioligand binding assay and charcoal adsorption

1. Perform all manipulations at 4°C.

2. Set up triplicate tubes (LP3, Hughes and Hughes) containing [^{3}H]spiperone working solution (10 nM) (50 μl) and 50 μl of (+)- or (−)-butaclamol (10 μM). Start assay by adding 400 μl of soluble D$_2$ dopamine receptor preparation to each tube, vortex mix and leave at 4°C for 4–16 h.

3. Take a maximum size of 12 tubes per group and add 100 μl of charcoal suspension to each tube. This can be done conveniently with a Flow Laboratories Syringe Phaser equipped with a 2 ml plastic disposable syringe and a wide gauge needle. It is particularly important to add the same amount of charcoal to each tube and this can prove a problem as the charcoal tends to fall out of suspension. Therefore, the charcoal working suspension should be stirred constantly, rapidly drawn up into the syringe, and rapidly dispensed. With batches of 12 tubes this can be achieved successfully. Following the addition of charcoal, vortex-mix tubes, and immediately centrifuge (12000 g, 2.5 min, Damon IEC microcentrifuge).

4. Pipette 300 μl of supernatant from each tube directly into a scintillation counting vial (5 ml volume, Hughes and Hughes) and determine radioactivity after mixing with 3 ml of Optiphase (Fisons).

The difference in radioactivity between tubes containing (+)- and (−)-butaclamol represents specific receptor binding as discussed in refs 13 and 15. The radioactivity in the presence of (+)-butaclamol represents non-specific binding and should ideally be a small proportion (<25%) of the total binding. Non-specific binding will consist of non-specific binding of radioligand to protein present in the solution plus residual radioligand not removed by the charcoal adsorption (see Experimental point 2.(b) p. 250. For other receptors a suitable choice of ligand and concentration thereof will have to be made for defining specific radioligand binding, and it is to be emphasized that the choice must be carefully made on the basis of experimental data obtained under the conditions of the assay.

2. Experimental points

(a) We have routinely used Norit GSX charcoal, whereas other workers have used Norit A charcoal successfully (8, 10). Norit GSX is a steam-activated, acid-washed carbon preparation recommended for pharmaceutical applications, whereas Norit A is also a steam-activated carbon but has been micronized to produce very fine particles (information from Norit Ltd).

(b) Odell (11) has emphasized that in the use of charcoal adsorption for radioimmunoassay a balance is struck between adsorption of free ligand (should be high) and adsorption of protein containing the bound ligand (should be low). This is a function of the physicochemical properties of the charcoal as outlined above. Use of a high charcoal concentration might increase protein adsorption and this could be a particular problem when working with solutions of antibody or receptor of low concentration. In the assays described here (*Protocols 1* and *2*) the charcoal is coated with albumin; this should obviate uptake of receptor by the charcoal. It is likely that the albumin occupies the mesopores in the charcoal and so prevents uptake of receptor and bound ligand into these pores whilst leaving the micropores available for adsorption of free ligand. In the assay outlined here which uses approximately 1.6% charcoal, more than 90% of the free radioligand is adsorbed. In a comparative study (8, 10) using 5.5% charcoal the bound radioligand was equivalent to that measured with other assays (see *Table 1*), so that there is unlikely to be significant loss of receptor-bound radioligand whereas adsorption of free radioligand is very efficient.

(c) Charcoal coated with dextran (Sephadex) has been used by some workers and claimed to render the charcoal more sticky and thus easier to remove by centrifugation into a pellet (11). In a comparative study for assays of solubilized muscarinic acetylcholine receptors, dextran-coated charcoal performed no better than albumin-coated charcoal (8). Albumin coating may indeed be essential when working with very low receptor protein concentrations to avoid loss of receptor on the charcoal.

(d) It is important that the charcoal pellet is obtained quickly and cleanly by centrifugation, partly to avoid dissociation of receptor-bound radioligand (see below). This is also important in obtaining an efficient removal of free radioligand. For example, if particles of charcoal (with adsorbed free radioligand) remain in suspension after centrifugation these will be sampled along with the bound radioligand and increase the apparent non-specific radioligand binding. This might be a problem if a particularly viscous assay medium is used, for example fractions from a sucrose density gradient, or in studies on the reconstitution of receptors into vesicles where sucrose was added to binding assays to prevent sedi-

250

mentation of vesicles during the centrifugation to remove the charcoal (12).

(e) The addition of charcoal to the mixture of bound and free radioligand removes the free radioligand from the equilibrium so disturbing the equilibrium. Other assays for solubilized receptors (except equilibrium dialysis) also disturb the equilibrium. Therefore, in the present assay dissociation of bound radioligand will, in principle, proceed until the charcoal has been removed from suspension and a pellet formed. Dissociation will still proceed after this point, but radioligand that has dissociated from receptors will be sampled as if it were bound. Thus it is important to minimize dissociation of bound radioligand while the charcoal is in suspension. Hence manipulations are performed at 4°C with the minimal centrifugation time required to produce a pellet, and assay tubes are processed in small batches in order to minimize the time difference between adding charcoal to the first and last tubes. Dissociation does not seem to be a major problem with most brain receptors where high-affinity radioligands are available which have slow dissociation rates at 4°C but it could be a problem with lower-affinity ligands. We have in fact been able to perform assays at 25°C for solubilized brain D_2 dopamine receptors using [^{3}H]spiperone binding and the charcoal technique. The higher incubation temperature enables a 30 min assay incubation time to be used. The charcoal separation was performed at room temperature with ice-cold charcoal suspension to attempt to minimize dissociation. The number of receptor sites assayed was reduced in these assays compared with assays at 4°C, presumably partly due to dissociation and partly due to the instability of soluble receptors at the higher temperature.

(f) Because of the possible inconsistencies noted above it is important to maintain a set of experimental conditions when once established. If the system is then altered in any way the conditions should be revalidated.

(g) A study has been published comparing the use of different assay techniques (charcoal adsorption, gel filtration, ammonium sulphate precipitation, adsorption on DEAE-cellulose filters, and equilibrium dialysis) for solubilized muscarinic acetylcholine (8) and D_2 dopamine receptors (10). Data taken from these studies are summarized in *Table 1*, and show that the charcoal technique performs well both in terms of the amount of specific binding assayed and the degree of non-specific binding. It is also cheap and reproducible with few drawbacks so that for routine use in assays of solubilized receptors it can be considered a useful assay.

Acknowledgements

I would like to thank Sue Davies for typing the manuscript and Mr Epherson of Norit (UK) for helpful information.

Table 1. Comparison of different assay techniques for solubilized receptors

Assay	Muscarinic acetylcholine receptor radioligand binding		D_2 dopamine receptors radioligand binding	
	Specific binding (% of value given by gel filtration)	Non-specific binding (% of total binding)	Specific binding (% of value given by gel filtration)	Non-specific binding (% of total binding)
Gel filtration	100	1	100	24
Charcoal adsorption	91	3	141	17
Ammonium sulphate precipitation	59	3	65	20
Equilibrium dialysis	—	—	57	91
DEAE-cellulose filter trapping	60	21	—	—

The data are taken from (8, 10) for muscarinic acetylcholine receptors and D_2 dopamine receptors solubilized from dog striatum by digitonin. Gel filtration was performed on Sephadex G-50, charcoal adsorption used 5.5% charcoal with 1.1% albumin, ammonium sulphate was used at 50% saturation and 2.5 cm DEAE-cellulose discs were used. For the muscarinic receptor assay [^{3}H]dexetimide was used, non-specific binding was almost entirely due to residual radioligand not removed by the assay method. For D_2 dopamine receptors [^{3}H]spiperone was used, non-specific binding for gel filtration and charcoal adsorption was due largely to the protein added, only about 10% being due to residual radioligand not removed by the assay.

References

1. Zimmerman, S. (1963). *Meth. Enzymol.*, **6**, 258.
2. Chen, R. F. (1967). *J. Biol. Chem.*, **242**, 173.
3. Westphal, W., Burton, R. M., and Harding, G. B. (1975). *Meth. Enzymol.*, **36**, 91.
4. Stott, C. A. (1975). *Meth. Enzymol.*, **36**, 34.
5. Niswender, G. D., Akbar, A. M., and Nett, T. M. (1975). *Meth. Enzymol.*, **36**, 91.
6. Dobson, P. R. M. and Strange, P. G. (1984). *Meth. Enzymol.*, **109**, 827.
7. Withy, R. M., Wheatley, M., Frankham, P. A., and Strange, P. G. (1981). *Biochem. Soc. Trans.*, **9**, 416.
8. Gorissen, H., Aerts, G., Ilien, B., and Laduron, P. (1981). *Anal. Biochem.*, **111**, 33.
9. Ilien, B., Schotte, A., and Laduron, P. M. (1982). *FEBS Lett.*, **138**, 311.
10. Gorissen, H., Ilien, B., Aerts, G., and Laduron, P. (1980). *FEBS Lett.*, **121**, 133.
11. Odell, W. (1980). *Meth. Enzymol.*, **70**, 274.
12. Wheatley, M. and Strange, P. G. (1984). *FEBS Lett.*, **166**, 389.
13. Hall, J. M., Frankham, P. A., and Strange, P. G. (1983). *J. Neurochem.*, **41**, 1526.
14. Strange, P. G., and Williamson, R. A. (1990). In *Receptor Biochemistry, a Practical Approach* (ed. E. C. Hulme), pp. 79–97. IRL Press, Oxford.
15. Strange, P. G. (1983). *TIPS*, **4**, 188.

9

Gel-filtration assays for solubilized receptors

E. C. HULME

1. Introduction

There are two basic strategies which can be applied to the assay of soluble receptor–ligand complexes. The first is to render the complex insoluble, and then to apply the filtration or centrifugation techniques which are described elsewhere in this book to separate bound from free ligand (Chapters 6, 7); a variant on this approach is to absorb the free ligand, and then to assay the residual binding (Chapter 8). The second is to use solution technology, namely equilibrium dialysis, or gel-filtration, to assay the levels of bound and free ligand.

As a technique, equilibrium dialysis suffers from several disadvantages. The most serious is that it relies on depletion of the free ligand to achieve a measurable signal. Because the concentrations of receptor sites in binding assays are typically nanomolar, or less, the attainment of depletion conditions requires a ligand affinity of 10^9/M or more, in most cases. However, these are precisely the conditions which allow non-equilibrium methods, such as precipitation or gel-filtration, to be used (Chapter 4). For this reason a cumbersome technique such as equilibrium dialysis, which requires a large number of individual dialysis cells to be set up, is usually rejected in favour of simpler, less elaborate methods.

Other disadvantages of equilibrium dialysis are its relative slowness, which makes demands on the stability of the receptor preparation, and its tendency to artefacts, such as adsorption of the ligand, or the receptor, to the cell or dialysis membrane. Despite these disadvantages, however, equilibrium dialysis techniques can detect rapidly-dissociating components of binding, which may be lost in non-equilibrium filtration and gel-filtration assays. Thus it may be necessary to contemplate its use under some circumstances. If so, suitable low-volume cylindrical dialysis cells, plus the equipment to rotate them and so mix their contents, are commercially available, for example from Bio-Rad. A full description of the equilibrium dialysis technique is given in reference 1.

1.1 Equilibrium gel-filtration

The use of gel-filtration to separate large from small molecules is part of the everyday currency of the biochemical laboratory. The method of Hummel and Dryer employs equilibrium gel-filtration to study ligand binding to soluble proteins and has been in use for many years (2). The basis of the technique is as shown in *Protocol 1*.

Protocol 1. Equilibrium gel-filtration by the Hummel and Dryer method

1. Equilibrate gel-filtration column containing the gel-filtration matrix, for example Sephadex G-50 with a pre-determined concentration of the tracer ligand.

2. Apply a small aliquot (<0.1 bed volumes) of the binding protein, in the same solution, to the top of the column.

3. Develop the column with equilibrating buffer containing the same concentration of the tracer ligand.

4. Collect fractions, and assay them[a].

---

[a] The binding protein elutes in the excluded volume of the gel, bringing with it a peak of bound ligand. A complementary trough, in which the concentration of tracer ligand dips below the equilibrium value, is eluted in the included volume of the column (*Figure 1*). Under ideal circumstances, the area under the peak should be the same as the area of the trough. Both are equal to the amount of bound ligand. The analysis of the binding curve can then be carried out as described in Chapters 4 and 11.

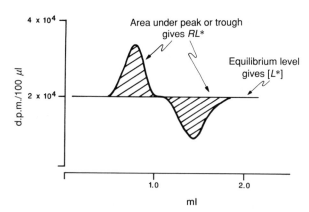

Figure 1. Measurement of ligand binding by the Hummel and Dryer method. Measurement of the area under the excluded peak gives the total amount of bound ligand, RL^*, while the baseline signal gives the free ligand concentration with which the receptor–ligand complex is in equilibrium.

In principle, the Hummel and Dryer procedure is applicable to solubilized receptors. However, it does have drawbacks, which means that it has not been widely applied to the study of ligand interactions with pharmacological receptors.

Equilibrium must be established rapidly (in much less than the gel-filtration time) if peak and trough are to be well defined and the free ligand concentrations are to be well measured (by the concentration of the tracer ligand in the equilibration buffer). However, many receptor–ligand interactions have long equilibration times, as has been discussed in Chapter 4.

As in the case of equilibrium dialysis, the technique is dependent on the existence of significant depletion of the tracer ligand concentration as a result of specific receptor binding, otherwise the peak and trough are unmeasurably small. The condition for this is the opposite of that normally applied to binding assays, i.e. $K \cdot R_t > 0.1$ (see Chapter 4). Since R_t rarely exceeds 10^{-9} M, except in the case of purified or enriched receptor preparations, the Hummel and Dryer method is limited to receptor–ligand interactions with values of $K > 10^9$/M. However, these are exactly the conditions appropriate for the application of the more sensitive non-equilibrium gel-filtration techniques.

1.2 Non-equilibrium gel-filtration

In the non-equilibrium method, the receptor–ligand complex is applied to the top of a small (approximately 2 ml) column of Sephadex and separated, as rapidly as possible, from the free ligand by gel-filtration. The separation time is 2–5 min under gravity flow, depending on the properties of the Sephadex. It can be reduced by centrifuging the columns, although this is not normally necessary. An example of a gel-filtration assay by the centrifuged-column method is given in Chapter 3 (Section 5.1). Centrifugable columns are supplied by Mobitec (agents, Scotlab).

Provided that the separation is done at low temperature (0–4 °C) many receptor–ligand complexes with $K > 10^8$/M survive gel-filtration with minimum dissociation (Chapter 4). Even when the affinity is lower than this, gel-filtration may still be worth trying. The dissociation kinetics may be unexpectedly slow, particularly if the ligand is a conformationally inflexible one (cf. the example of AF–DX 116 on muscarinic receptors, Chapter 6). Detergent solubilization of receptors may also retard binding kinetics, but the reasons for this are not well understood.

The advantages of the non-equilibrium gel-filtration assay are:

- It is very simple and cheap.

- The ratio of specific: non-specific binding is excellent.

- The throughput is high, so that it is suitable for large-scale screening assays as well as for more selective applications.

2. Equipment

2.1 Gel-filtration columns

A whole range of options is available. Numerous varieties of disposable gel-filtration columns are marketed (for example by Pierce and Bio-Rad). Gel-filtration columns can also be run in disposable 1 or 2 ml plastic syringes. We have used a particularly convenient polypropylene column marketed by Kontes (Disposaflex, Kontes; UK agents Burkard Scientific Sales) whose virtue is that it has a push-in end-piece which can be removed easily for unpacking, cleaning and repouring (see *Figure 2*). The columns can be cut down to a variety of useful sizes from 10 ml downwards. For routine assays, we use 4 ml columns containing 2 ml packed volume of gel. The columns resist soaking in 50% nitric acid for cleaning purposes.

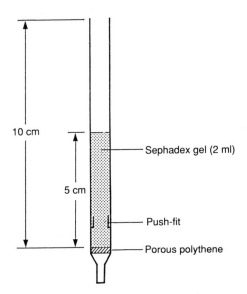

Figure 2. Disposable polypropylene columns suitable for gel-filtration binding assays.

Protocol 2. Preparing columns for gel-filtration assays

1. Obtain sinters. These can be purchased (Kontes) or can be cut out from a porous polythene disc (Amicon). This is conveniently done by using a cutting tool similar to a cork borer 6.5 mm in diameter mounted in a power drill. Place the porous polythene disc on a piece of expanded polystyrene foam (for example an ice-bucket lid) to avoid damaging the cutting edge. The cutting tool can be machined out of brass.

2. Press the sinters into the column tips with a thin brass rod or similar tool. Ensure that they fit tightly.

3. Cut the bodies of the columns to the required length. Assemble them with the endpieces.

4. Fill the columns with deionized water. Flick sharply to dislodge air bubbles. Make sure that the sinters run freely. Stand the columns in a few centi-metres of water in a beaker.

2.2 Gel-filtration media

Sephadex G-50 (fractionation range 1500–30 000; Pharmacia) turns out to be the most convenient medium for gel-filtration binding assays in which the ligand has a low molecular weight (<1 kDa), and the receptor–detergent complex high molecular weight (>100 kDa). It combines good resolution of bound from free ligand with good mechanical properties. G-50 Fine gives somewhat better resolution than G-50 Medium at the expense of a slower flow rate (approximately $0.3\times$ that of G-50 M). It should be used when the aim is complete removal of a low molecular weight substance, for example after elution of a receptor from an affinity column by means of a specific ligand, prior to assay by ligand binding or reuptake on to the affinity gel. For routine binding assays, G-50 M often seems to be adequate. Note that while G-50 M gives good resolution of bound from free ligand in the case of hydrophilic ligands, for instance quaternary amines, the removal of ligands, such as tertiary amines, which tend to partition into detergent micelles re-quires the superior resolving power of G-50 F, and, in addition, may require the lowering of the detergent concentration to a value below the critical micellar concentration (see reference 3). If the ligand has higher molecular weight, separation may still be possible using a different gel-filtration medium. The Pharmacia booklet *Gel-filtration, Theory and Practice* may use-fully be consulted for guidance.

Note that if the detergent molecules have an opposite charge to that carried by the ligand, strong binding of the ligand to detergent micelles may ensue. Since the detergent micelles will tend to be excluded from the pores of the gel, they will be eluted in the void volume of the column with the receptor–ligand–detergent complex. This can lead to high levels of non-specific bind-ing. The problem may be ameliorated by equilibrating the columns with a detergent concentration which is below the critical micellar concentration. This will promote the dissociation of detergent micelles.

Problems may also arise with adsorption of the receptor–ligand complex to the gel matrix. It may be possible to minimize such problems by tactics such as those described in Chapter 4, Section 4. In general, however, we have found that recoveries from Sephadex even of hydrophobic peptides are good in the presence of neutral or anionic detergents. Since most cell-surface receptors are negatively-charged at neutral pH, recovery may be aided by the presence of residual carboxylate and sulphate groups on the Sephadex gel, whose repulsive effect will tend to oppose adsorption of the receptor–detergent complex. In our experience, worse recoveries are obtained from media in which the beads are based on polyacrylamide (for example Bio-Gels) rather than agarose.

Protocol 3. Pouring columns for gel-filtration assays

1. Weigh out the required amount of Sephadex. The water regain of G-50 is 10 ml/g so 20 g is sufficient for 200 ml of swollen gel. Add the Sephadex slowly to 1 litre of deionized water with continuous slow stirring on a magnetic stirrer. Ensure even dispersal. Turn off the stirrer.

2. Swell the Sephadex overnight at room temperature. Pour the swollen gel into a measuring cylinder. Allow it to settle (about 2 hours). Pour off the supernatant and any fine particles. Adjust until there are equal volumes of water and settled gel.

3. Cover the measuring cylinder with Nescofilm. Invert it a number of times until the gel is evenly dispersed. Pour the gel suspension into a beaker. Add sodium azide (to 0.02% w/v) as a preservative (**care!** sodium azide is toxic).

4. Stir the gel gently with a large magnetic stirring bar to maintain an even suspension. With a 5 ml adjustable pipette, using a 5 ml disposable tip, rapidly add 4 ml of 1:1 gel suspension in water to each column. Before doing this, ensure that the column sinter is wet and free of bubbles.

5. Allow the gel to settle in the column. The column should run freely. Using a standard 2 ml column as a basis for comparison, check that each column contains 2 ml of gel. Adjust if necessary. With care, this will not be necessary. It is important to have less than 5% variation of the volume of the gel bed if reproducible results are to be obtained.

6. Wash the columns well with deionized water from a wash bottle. Before use, equilibrate them with the buffer of choice, at 4 °C.

7. The columns can be stored in deionized water at 4 °C, best in a beaker filled with water up to the height of the gel bed, to prevent drying out. If air bubbles develop, they can be removed by agitating the gel in buffer or water with a Pasteur pipette. Normally, small bubbles tend to redissolve with use.

8. The unused G-50 should be stored at 4 °C in 0.02% azide.

2.3 Performing the assays

Protocol 4. Gel-filtration assay of receptor–ligand complex

Carry out all steps at 4°C.

1. Ensure that the columns are thoroughly washed with water. Equilibrate them with the buffer of choice, at 4°C. 4 ml of buffer/column suffices for a 2 ml bed volume. When the amount of protein to be put on the column is less than about 100 μg, the appropriate detergent should be added to the buffer. However, when assaying crude supernatants containing 1% detergent (for example digitonin) and approximately 1 mg/ml protein, we have found that the addition of detergent to the buffer is not always necessary. The detergent micelles are partly or completely excluded from the gel and so tend to co-elute with the receptor. However, if the protein concentration is lower, it is advisable to pre-equilibrate the column with a detergent-containing buffer (usually 0.1% detergent suffices) otherwise recoveries of binding sites may be poor.

2. Stand the equilibrated and drained columns in scintillation vials in a suitable rack or box in a cold-room or refrigerator. The columns do not have to be completely vertical.

3. Pipette 0.1 ml of incubation mixture, containing the labelled receptor, on to the top of the gel. Allow it to run in completely.

4. Wash in with 2×0.1 ml of ice-cold buffer (containing detergent if necessary) allowing each addition to run in completely. This is conveniently done with a repetitive pipette. This washes the labelled receptor well into the gel.

5. Carefully pipette in 0.7 ml of ice-cold buffer (with detergent) down the side of the column. Do not disturb the gel bed too much. The eluate, 1 ml in total, contains the labelled receptor. Total separation time is about 2 min with Sephadex G-50 M.

6. Shake off the remaining drops of liquid from the column, and remove it. If the receptor–ligand complex is to be detected by liquid scintillation counting, add 10 ml of aqueous scintillator to the vial, cap, mix, and count. If not, γ-count or otherwise assay the eluate.

7. Regenerate the columns by washing the free ligand through with 2×2 ml of deionized water from a wash bottle. Take care not to wash out any gel. Also, rinse the outsides of the columns several times to remove any adherent radioligand. Store the columns for further use, or re-equilibrate them with buffer immediately.

8. Repour the columns (they can be used approximately 10 times before repouring). To repour, dismantle, and wash out the gel. If the sinters are blocked, soak them in 50% nitric acid for 24 h, then wash them with water.

This protocol provides good separation of bound from free ligand (*Figure 3*). A maximum of 0.2 ml of solubilized supernatant can be applied to 2 ml of gel. The final elution volume should be adjusted appropriately. The recovery of the receptor–ligand–detergent complex is normally greater than 90%.

A slightly different protocol is used for desalting or ridding a solubilized receptor of ligand, or for detergent exchange.

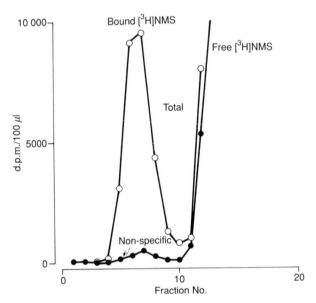

Figure 3. Non-equilibrium gel-filtration assay for solubilized mAChRs. The concentration of the radioligand, [^{3}H]-*N*-methylscopolamine was 10 pmol/ml. The mAChR concentration was 1.4 pmol/ml. 0.14 pmol of mAChR–[^{3}H]NMS complex was applied to the 2 ml of Sephadex G-50 M in 20 mM Hepes/0.1% digitonin, pH 7.5. Non-specific binding was measured by preblocking the receptor with 1 μM 3-quinuclidinyl benzilate before adding the radioligand. Elution of the column was carried out with successive 0.1 ml aliquots of Hepes/digitonin buffer. The eluted fractions were collected and counted separately in this experiment. The excellent separation, and low level of radioligand carry-through are evident.

Protocol 5. Desalting, ligand-removal, or detergent exchange by gel-filtration

1. Pipette 0.2 ml of the preparation on to a 2 ml G-50 F column equilibrated in buffer, usually containing the appropriate detergent. Allow the meniscus to run in fully.

2. Wash in with 2 × 0.2 ml aliquots of buffer (plus detergent). Discard the eluate.

3. Elute the receptor with 0.5 ml of buffer (plus detergent). Collect the eluate in a microcentrifuge tube. The desalted or de-liganded receptor is thus 2.5-fold diluted. The carry-through of a quaternary ammonium ligand is less than 0.01%, even in the presence of a detergent such as digitonin.

This protocol can be scaled up or down as appropriate. We have found it extremely useful and versatile for buffer and detergent exchange in many different procedures (see reference 3).

References

1. Klotz, I. M. (1989). In *Protein Function, a Practical Approach* (ed. T. E. Creighton), p. 25. IRL Press, Oxford.
2. Hummel, J. P. and Dreyer, W. J. (1962). *Biochim. Biophys. Acta,* **63,** 530.
3. Haga, T., Haga, K., and Hulme, E. C. (1990). In *Receptor Biochemistry, a Practical Approach* (ed. E. C. Hulme), p. 1. IRL Press, Oxford.

10

Receptor-binding kinetics

HEINO PRINZ

1. Introduction

The first steps involved in receptor–ligand interactions are always second order binding reactions. They may, or may not, involve conformational changes of the receptor protein. The velocity of binding may be extremely rapid for neuroreceptors but can be much slower for hormone receptors. This diversity leads to a variety of experimental methods employing mostly fluorescent or radioactive ligands. Likewise, the analysis of kinetic experiments is performed on very different levels depending on the type and quality of the data. Usually, and as a first approximation, simple second order association reactions are assumed, which are sufficient for the analysis of most receptor-binding kinetics (1). Possible deviations from these schemes may lead to new insights into the binding mechanism and are, therefore, the focus of this chapter. It should be remembered that disproving a given mechanism is much simpler than proving another one.

The first part of this chapter deals with experimental methods, in particular with sensitivity and speed. The section on data analysis begins with semi-logarithmic plots based on simple (second and first order) reactions. It leads to equilibrium binding and dose-response curves as a means of developing reaction schemes for the analysis of binding kinetics. A final section illustrates the detection of unique mechanistic features using the peripheral nicotinic acetylcholine receptor as an example.

2. Methods

Different receptors require different ligands and methods to study their binding kinetics. The muscarinic acetylcholine receptor, for example, can be studied with radioactive ligands in a filtration assay, whereas stopped-flow experiments with fluorescent ligands are employed for the faster nicotinic receptor. A radioactive label has the advantage that it usually does not change the pharmacological properties of a drug, but has the disadvantage that it cannot be employed for fast reactions. A fluorescent label is relatively large. Its introduction leads to a new compound with different pharmaco-

Table 1. Fluorescent probes

Fluorescent ligands

Receptor	Ligand	Chromophore	Reference
Adenosine	1,3-dialkyl-8-phenylxanthine	NBD, fluorescein	(2)
Adenosine	N6-phenyladenosine	NBD, fluorescein	(2)
Benzodiazepine	Ro 15-1788, Ro 7-1986	fluorescein	(3)
beta-Adrenergic	CGP-12177, carazolol	NBD	(4,5)
Dopamine, D1	SCH-23390, SKF 83566	boron dipyrromethene complexes, rhodamine, NBD, fluorescein	(6,7)
Dopamine, D2	N-(p-aminophenethyl)spiperone, NAPS	rhodamine, fluorescein, NBD	(7,8)
Oestrogen	2,3-diphenylindene	2,3-diphenylindene	(9)
Glucagon	glucagon	NBD, rhodamine	(10)
Nicotine	Dns-C_6-Cho, NBD-5-acylcholine	Dansyl, NBD	(11,12)
Nicotine	local anaesthetic chromophore	decidium	(13)
Opiate	enkephalins	L-1-pyrenylalanine	(14)
Oxytocin	[1-desamino,4-lysine,7-(L-3,4-dehydroprolyl)]oxytocin	fluorescein	(15)
Vasotocin	hydrin 1	fluorescein, rhodamine	(16,17)
Voltage (Na^+)	batrachotoxin	N-methylanthranylate	(18)

Membrane reporters

Function	Reporter	Reference
Membrane potential	3,3'-dipropylthiocarbocyanine iodide	(19)
Membrane potential	bis-(1,3-diethylthiobarbiturate)trimethineoxonol	(20)
Membrane fluidity	anthrylvinyl- or perylenoyl-labelled lipids	(21)
Muscarinic receptor	(1-acyl-2-[12-(9-anthryl)-11-*trans*-dodecenoyl]-sn-glycer-3-phosphocholine	(22)
NO$_2$	fluorescamine, 1-(4-trimethylaminophenyl)-6-phenyl-1,3,5-hexatriene	(23)

Ion-specific fluorochromes

Ion	Fluorochrome	Reference
Ca^{2+}	fura-2 (example: muscarinic activation)	(24)
Ca^{2+}	quin 2 (example: alpha-adrenergic activation)	(25)
Cl$^-$, I$^-$	6-methoxy-*N*-(3-sulphopropyl)quinolinium (MSQ)	(26)
Cs$^+$	anthracene-1,5-disulphonic acid	(27)
H$^+$	2,7-bis-carboxyethyl-5(6)-carboxy-fluorescein	(28)
Tl$^+$	ANTS or 1,3,6,8-pyrenetetrasulphonic acid (PTSA)	(29,30)

Selected references for fluorescent probes employed in receptor kinetics. Fluorescent ligands were used directly for kinetic studies, whereas membrane reporters detected secondary effects such as conformational changes of the receptor protein. Ion-specific fluorochromes were employed to monitor the first part of a physiological response, such as the release of Ca^{2+} as a secondary transmitter, or the opening of an ion channel. Abbreviations are used as quoted from the references.

logical properties. The spectroscopic changes observed in the course of a binding reaction cannot be predicted. This is why only a few receptors have been studied by this method. *Table 1* gives a selection of recent publications employing fluorescent ligands for receptor kinetics. The first set of references concerns ligands employed in direct binding studies, i.e. ligands whose fluorescence is either increased or decreased upon binding. The second part gives a selection of membrane reporters, i.e. alphatic molecules which are inserted in the cell membrane and monitor changes in the lipid environment. They respond to secondary effects like conformational changes of a membrane receptor *after* ligand binding. A third class of useful probes are ion-specific fluorochromes. They are employed to study the function of ligand-gated ion channels (like the nicotinic acetylcholine receptor) or the release of Ca^{2+} as a second messenger.

2.1 Filtration

Radioactive ligands are typically employed in filtration experiments, where the bound ligand is retained on the filter together with the receptor (*Figure 1*).

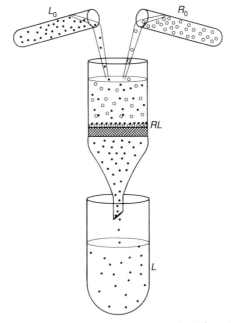

Figure 1. Filtration assay. Ligand and receptor are mixed (usually in a separate step not shown here), incubated for a given time, and filtered on a glass-fibre filter. Receptor and ligand bound to the receptor is retained on the filter. When the ligand is labelled radioactively, free ligand concentration L can be obtained from radioactivity of the filtrate, and bound ligand concentration RL either from radioactivity of the filter or from the difference $L_0 - L$. Since further association will take place during filtration, time resolution is determined by the filtration time.

Its concentration is determined from the radioactivity of the filter. The free ligand concentration is calculated directly from the radioactivity of the filtrate. This is an advantage compared to the more indirect spectroscopic methods described below. This method is, however, fraught with the problem of sensitivity. Since the concentration of bound ligand cannot be higher than the concentration of binding sites, large receptor concentrations have to be used. When low receptor concentrations are used, the concentration of bound ligand cannot necessarily be calculated from the difference in total ligand concentration (determined from the radioactivity before filtration) and free ligand concentration (determined from the radioactivity of the filtrate), because this difference will often be smaller than the absolute error of these large numbers. The bound ligand concentration also cannot be determined directly from the radioactivity of the filter, because at low receptor concentrations most of the ligand on the filter may be non-specifically bound. When slowly dissociating ligands, such as antibodies or other large proteins (toxins), are studied, specifically and non-specifically bound ligand can be separated. Washing of the filters with cold (4°C) buffer solution will only affect non-specifically bound ligand so that the remaining radioactivity corresponds to specific binding.

Washing the filter leads to erroneous results when one fraction of the bound ligand dissociates faster than another. If, for example, a receptor has two conformations, one from which the ligand dissociates rapidly and one from which it dissociates slowly, only the latter conformation will be detected. If there are two sites with such properties, the recorded stoichiometry will be wrong, and only one site will appear in an equilibrium binding curve. These errors cannot be detected within the same set of kinetic and equilibrium binding experiments, since they are self-consistent. They may, however, be detected from comparison with other experiments. For example, if the Hill coefficient (see below) obtained from physiological experiments is larger than unity, more than one ligand molecule must be involved in activation. If there is no indication of a second site after washing the filters, half of the bound ligand may have dissociated rapidly.

A general limitation of filtration experiments lies in the size of the particles retained on the filter. The commonly used glass fibre filters (Chapter 6) retain only large particles, such as membrane fragments, so that the receptor can only be studied in its membrane-bound state. Cytosolic receptors, or purified solubilized receptors, escape detection. Some of these receptors can be retained at low ionic strength on ion exchange or polyethylenimine-treated (31) (Chapter 6) filters. Time resolution is another serious limitation of filtration experiments. Mixing the reactants, application to the filter, and filtration itself require several seconds. A rapid-mixing apparatus (32), however, does not have these limitations. Since all processes are driven by syringes under remote control, reproducible reaction times in the millisecond time range can be achieved. *Note*, the term 'rapid filtration' very often means that an ordinary filtration experiment is performed manually, as rapidly as possible, with a time resolution of approximately five seconds.

Quenched-flow techniques have been applied in order to study the flux of radioactive ions into closed vesicles containing the nicotinic acetylcholine receptor (33). The time resolution of this method has been extended to the millisecond range, so that the physiological action (ion flux) could be measured with the same system in the same time range as fast-binding kinetics (stopped-flow). Similar results were obtained in a stopped-flow apparatus with fluorescent ligands as indicators for the ion flow (see below).

2.2 Stopped-flow spectrophotometry

A stopped-flow spectrophotometer (*Figure 2*) is typically used to study reaction kinetics in the millisecond time range. Receptor R and ligand L are rapidly mixed and pushed into an observation chamber. Reactions in the time between mixing and entry into the observation chamber ('dead time') cannot be observed. For dilute solutions, such as those employed in receptor-binding kinetics, a dead time of 1–5 msec is sufficient. Even very fast association reactions with a forward rate constant of $10^9/(\text{M sec})$ will proceed in the time range of seconds when the reaction partners have concentrations in the nanomolar range. The low concentrations required for fast kinetics have the additional advantage that only small amounts of precious receptor protein are used. Fluorescent ligands (*Table 1*) are ideally suited for this concentration range, but the stopped-flow spectrophotometer must be sensitive enough.

The sensitivity of a stopped-flow spectrophotometer can sometimes be enhanced by simple means. The most critical part is the observation chamber which should allow the detection of a maximal fraction of the emitted light. Polishing the wall opposite to the detection multiplier, or placing a mirror there, may help to capture a larger fraction of the emission angle. When the distance between observation chamber and photomultiplier is too large, an additional light guide placed directly on to the cuvette will further increase the amount of light reaching the detector. Another critical item for the sensitivity of a spectrophotometer is light intensity. In some cases an old lamp can be replaced with a higher intensity halogenated tungsten lamp. *Note* that a stable power supply must be used, since all fluctuations of the light intensity will be detected in the final signal. In cases where the exciting light source cannot be improved it may have to be replaced. We have found a continuous laser particularly useful as a stable, high-intensity monochromatic light source (34).

Receptor solutions are turbid when the receptor is in its membrane environment or in a detergent micelle. This leads to a considerable amount of light scattering. If the exciting light has contributions of higher wavelength, such as higher order reflections from the gratings, they will pass the emission filter and artificially increase the detected fluorescence. An additional filter on

Heino Prinz

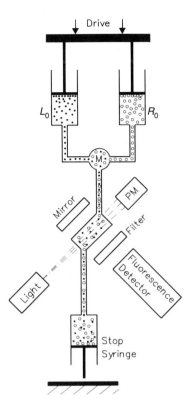

Figure 2. Stopped-flow. Ligand and receptor (initial concentrations L_0 and R_0, respectively) are injected into the mixer M and driven into an observation chamber. The flow is stopped when the piston of the stop syringe comes to a halt. Bound and free ligand are monitored continuously, together, in the observation chamber. Transmitted light is detected in the photomultiplier PM, and fluorescence is recorded at an angle of 90° after passing through a filter to block the exciting light. Very fast reactions cannot be observed in the time between mixing and reaching the observation chamber (dead time, approximately 1–5 milliseconds).

the excitation side may improve the signal to noise ratio. If the receptor is bound to large membrane particles, these particles may gradually sediment during an experiment. The effect is slow (minute time range) and can be detected by scattering experiments performed in the absence of fluorescent ligands. Particle sedimentation in the reservoir syringes may lead to apparent variations in receptor concentration. This can be avoided by stirring the membrane suspension in the syringes between kinetic runs.

Data analysis (see below) requires information additional to the stopped-flow traces (*Figure 3a*). For semi-logarithmic plots, the final amplitudes must be accurately known. Therefore, in a separate experiment the spectroscopic signal is monitored continuously until no further change is observed. The final amplitude is stored along with the original trace. For direct calculations, separate fluorescence measurements are performed of buffer, fluorescent ligand L_0, and receptor background R_0. Solutions containing membrane fragments are often slightly heterogeneous. Small variations in the background signal are, therefore, expected. Large variations in the signal itself can arise from the known artefacts of stopped-flow equipment (air bubbles, insufficient thermal equilibration of the solutions, etc.). Discrimination between large and small variations is performed by comparing a whole set of identical kinetic runs. Only similar traces are averaged and taken for further analysis.

Stopped-flow spectrophotometers have also been used to study effects *following* ligand binding to a receptor (*Table 1*). For example, changes in the intracellular calcium concentration of whole cells have been monitored using the fluorescent dye fura-2 (35, 36). Membranes containing a ligand-gated ion channel can be induced to form vesicles enclosing a fluorescent dye inside. The fluorescence of the dye is quenched upon binding of cations. For example, binding of Cs^+ to anthracene-1,5-disulphonic acid (27), of Tl^+ to 8-amino-1,3,6-naphthalenetrisulphonate (29), or to 1,3,6,8-pyrenetetrasulphonic acid (30) was employed to study the ion flux regulated by the nicotinic acetylcholine receptor.

2.3 Temperature-jump

Relaxation kinetics are the ideal method for detecting the most rapid reactions in solution. First, a solution containing all reaction partners is brought into chemical equilibrium. Then this equilibrium is disturbed, usually by a rapid increase in temperature ('T-Jump'). The ('relaxation') reactions leading from the old equilibrium to the new equilibrium (at a higher temperature) are observed by monitoring the change of some spectroscopic signal. The method has two main advantages:

- Because the temperature can be increased rapidly (for example, by capacitor discharge through a conducting solution), the reactions leading to the new equilibrium can be observed in the microsecond time range.

- Since only small perturbations of the equilibrium are performed, the signal of a relaxation experiment can be decomposed into a sum of exponential terms, with only linear powers of time in the exponent (see eqn 7).

There are, however, some disadvantages in using this method for receptors:

- The initial equilibrium of receptor and ligand leads to desensitization of some receptors.

- Only very fast reactions can be monitored, since thermal convection will disturb the signal in the time range of minutes.

- In the case of membrane-bound receptors, phase transitions of the membrane are strongly dependent on the temperature and may dominate the signal from temperature-jump experiments.

3. Data analysis

The analysis of kinetic data depends on the model applied. This is also true for the seemingly unbiased semi-logarithmic plots and the corresponding multi-exponential analysis where independent reactions are implied. This is illustrated with a few basic reaction schemes.

3.1 States and sites

The simplest reaction scheme describes the association of a ligand L to the receptor R to form a complex RL:

$$L + R \underset{k_{-1}}{\overset{k_1}{\rightleftharpoons}} RL; \tag{1}$$

(one binding site, one conformational state).

The corresponding Scatchard plot is a straight line (one site) and the Hill coefficient is equal to unity. (Compare with *Table 1* of Chapter 4). All concentrations given here are free concentrations. Total concentrations are indicated by the subscript 0 (for example R_0 or L_0). The concentrations of the three molecules involved are functions of time. Their velocities follow from the laws of first and second order reactions:

$$dL/dt = -k_1 \cdot L \cdot R + k_{-1} \cdot RL; \tag{1a}$$

$$dR/dt = -k_1 \cdot L \cdot R + k_{-1} \cdot RL; \tag{1b}$$

$$dRL/dt = +k_1 \cdot L \cdot R - k_{-1} \cdot RL. \tag{1c}$$

The differential equations show that all concentrations follow the same pattern, apart from multiplication by a factor of -1 in eqn 1c.

Receptor-binding kinetics have to consider conformational changes, since activation and desensitization must lead to different conformational *states* of the receptor. One additional (functionally active or inactive) conformation in scheme 1 will lead to the reaction scheme described in Table 4 of Chapter 4:

$$L + R \underset{k_{-1}}{\overset{k_1}{\rightleftharpoons}} R1L \underset{k_{-2}}{\overset{k_2}{\rightleftharpoons}} R2L; \tag{2}$$

(one binding site, two conformational states).

The corresponding Scatchard plot is linear, because under equilibrium conditions the ratio $R2L:R1L$ is equal to the equilibrium constant $K_2 = k_2/k_{-2}$ and does not depend on the ligand concentration. The additional complex R2L cannot be detected from the shape of the equilibrium binding curve but influences the effective equilibrium constant $K_{\text{eff}} = K_1 + K_1 \cdot K_2$. It is a common misconception that a curved Scatchard plot might be explained by multiple conformational states. Equilibrium binding data serve to yield information on the number and average affinity of *sites* (37), whereas kinetic experiments are required to obtain information on conformational *states*. Scheme 2 can be described by the following set of differential equations:

$$dL/dt \quad = - k_1 \cdot L \cdot R + k_{-1} \cdot R1L; \tag{2a}$$

$$dR/dt \quad = - k_1 \cdot L \cdot R + k_{-1} \cdot R1L; \tag{2b}$$

$$dR1L/dt = + k_1 \cdot L \cdot R - k_{-1} \cdot R1L - k_2 \cdot R1L + k_{-2} \cdot R2L; \tag{2c}$$

$$dR2L/dt = + k_2 \cdot R1L - k_{-2} \cdot R2L. \tag{2d}$$

Scheme 2 shows a sequence of second and first order reactions. The observed reaction signal depends on the molecule which is monitored in the kinetic experiment. If, for example, the conformational change from R1L to R2L is followed by changes in intrinsic protein fluorescence, an association reaction will show a 'lag phase' i.e. zero velocity at time $t = 0$ because the complex RL has not been formed. The overall velocity leading to the formation of R2L will be determined by the slowest (rate-limiting) step. For very small ligand concentrations this will be the initial association reaction, whereas for very high ligand concentrations the conformational change is rate-limiting. Therefore, the reaction leading to R2L may appear to be first or second order depending on the ligand concentration.

The following scheme also considers two different conformational states, but assumes that these states can exist in the absence of ligands.

$$
\begin{array}{ccc}
R+L & \underset{k_{-1}}{\overset{k_1}{\rightleftharpoons}} & RL \\[2mm]
k_3 \big\updownarrow k_{-3} & & k_4 \big\updownarrow k_{-4} \\[2mm]
D+L & \underset{k_{-2}}{\overset{k_2}{\rightleftharpoons}} & DL
\end{array} \tag{3}
$$

(one binding site, two conformational states).

R and D indicate two conformational states of the receptor protein. The Scatchard plot for such a scheme will be linear (37) and the Hill coefficient will be unity. Scheme 3 has been used as a basic unit of the specific allosteric model (38). Certain characteristic features of this model, such as sigmoid binding curves, are only possible when two or more subunits of R or D are present and interact allosterically. If the rate constants k_3 and k_2 become very small, scheme 3 will approach scheme 2. If the rate constants k_3 and k_4 become very small, the kinetics will show parallel-binding reactions to the different conformational states, and will not be able to be distinguished from binding kinetics of two different receptor populations. Equilibrium studies, however, will only be able to detect one population of binding sites, since the ratio R/D or RL/DL will be constant at all ligand concentrations for long incubation times.

Receptors may not only have more than one conformational state but very often more than one ligand-binding site. This must always be the case when Hill coefficients greater than unity are observed. The corresponding reaction scheme gets considerably more complicated: In the case of two binding sites generally two initial association reactions (one to each site) and two subsequent reactions leading to the saturated receptor can be expected:

$$
\begin{array}{ccc}
& RL+L & \\
\nearrow^{1} & & \searrow^{3} \\
R+2L & & LRL \qquad (4) \\
\searrow_{2} & & \nearrow_{4} \\
& LR+L &
\end{array}
$$

(two binding sites, one conformational state).

In the general case, this gives *eight* rate constants (four reactions, each in two directions), which are reflected in the overall kinetic pattern. If the sites are independent, reactions 1 and 4 and reactions 2 and 3 are identical, so that only four rate constants remain. In contrast to this, equilibrium binding curves will always give only *two* equilibrium constants (one for each site) (37).

The significance of the above reaction schemes becomes clear when they are related to function: The conformational change in scheme 2, for example, may be the activation process elicited after binding. It may also be a desensitization process, although desensitization would be more likely to follow a reaction scheme such as 3, where dissociation of ligands from the desensitized state does not necessarily pass through the active state. Reaction scheme 4, and all other schemes involving more than one binding site, are particularly useful for receptors which have to combine specificity with speed: Each of the individual rate constants in scheme 4 can be fast, yet specificity is achieved because two (or more) ligands are required for activation. This may be the reason why many neuronal receptors have more than one binding site.

3.2 Developing a model

The main task for receptor-binding kinetics is the development of a model. The initial characterization of kinetic experiments by means of semilogarithmic plots is a necessary requirement for this.

3.2.1 Semi-logarithmic plot

A first order reaction such as the dissociation of RL in scheme 1 or a conformational change such as step 2 in scheme 2 will lead to an exponential decay curve:

$$A(t) = A_0 \cdot e^{-k \cdot t}. \tag{5}$$

Experiments often do not allow the direct determination of bound or free ligand concentrations but of a spectroscopic signal $S(t)$ which is proportional to the concentration of one complex or a sum of complexes:

$$s(t) = const \cdot A_0 \cdot e^{-k \cdot t} + bkg; \tag{5a}$$

where *const* is the proportionality constant and *bkg* is the spectroscopic background. Experimentally, *bkg* is determined from the signal at equilibrium, i.e. from the signal $S(\infty)$ at infinite times, because $e^{-k \cdot t}$ then becomes zero.

$$S(t) - S(\infty) = const \cdot A_0 \cdot e^{-k \cdot t}. \tag{5b}$$

Taking the natural logarithm of both sides leads to:

$$\log_e (S(t) - S(\infty)) = \log_e (const \cdot A_0) - k \cdot t. \tag{5c}$$

This representation is called a *semi-logarithmic plot*. The slope of this plot is equal to the rate constant k of the reaction. The slope is independent of the initial concentration (A_0) or of the calibration factor (*const*). Therefore, any uncalibrated signal can be used to obtain the rate constant of a first order reaction. Second order reactions (scheme 1) can also be analysed using eqn 5c. For this, one reaction partner (typically the ligand L) is set in large excess over the other. Consequently, the concentration of free L is constant relative to the large variations in R or RL. The reaction can then be treated as a reaction of the first order with an apparent velocity constant:

$$k_{app} = k_1 \cdot L + k_{-1}; \tag{5d}$$

where k_1 and k_{-1} are association and dissociation rate constants taken from scheme 1 (see also Chapter 4).

Semi-logarithmic plots are not direct representations of experimental data, because the differences from the signal at equilibrium are used. Stopped-flow and T-jump machines typically are single-beam spectrophotometers with poor stability in an extended time range. The error in determining the signal at equilibrium may be larger than the error in an experimental trace. It leads to a systematic error influencing the curvature of the entire semi-logarithmic plot. *Figure 3b* illustrates this point. An error of 3% in the final amplitude $F(\infty)$ of *Figure 3a* leads to the two limiting curves in *Figure 3b*. This is the reason why the method of Guggenheim (39) does not depend on differences from the final amplitude; instead it uses differences of experimental points at

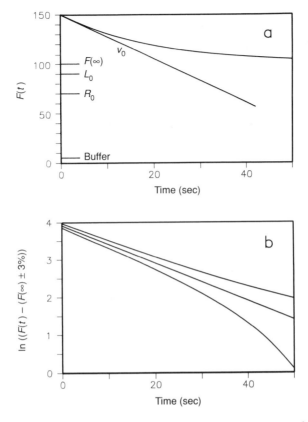

Figure 3. Analysis of a stopped-flow trace showing the time course of quenching of the free ligand fluorescence as a result of ligand binding. (a) Exponential fluorescence decay $F(t) = F_0^* \exp(-kt) - Background$. $F_0 = 50$, $k = 0.05$/sec, $Background = 100$. The final amplitude $F(\infty) = 100$ is indicated as it would appear in a stopped-flow trace. The signals corresponding to total ligand concentration L_0 and receptor concentration R_0 in the observation chamber (i.e. diluted 1:1 with buffer) are indicated. The scattering observed from the membrane suspension, R_0, is much larger than scattering of the buffer alone. The initial velocity v_0 is derived from the tangent at zero time. (b) Semi-logarithmic plot of (a) $\ln(F(t) - F(\infty))$ with $F(\infty) = 100$, 103, and 97, respectively.

finite times. It assumes, however, a single exponential curve. For receptor-binding kinetics it cannot be applied because the number of exponents (and thus of binding sites and conformational states) generally will be larger than unity.

Let us now assume that the signal from an association reaction has been recorded and that it is plotted semi-logarithmically according to eqn 5c:

- If the plot is linear, it indicates a first-order reaction. Discrimination between a first- and pseudo-first-order reaction can be obtained from a concentration dependence study. If the slope of the semi-logarithmic plot remains constant with increasing ligand concentrations, a first-order reaction (possibly a conformational change such as that shown in scheme 2) is rate-limiting. If at high ligand concentrations the slope varies according to eqn 5d the reaction is second order and might follow scheme 1:

- If a semi-logarithmic plot is curved over the whole range of an experiment and the slope decreases monotonically, no information can be obtained from the semi-logarithmic plot except that the reaction may be of a rather complex nature.
- If there is an initial increase of the slope before it decreases ('lag phase'), there must be an initial reaction which is not monitored.
- If the slope is clearly biphasic, then it may be separated into two exponential terms for initial analysis. As discussed above for the linear plot, the concentration dependence of both components will reveal whether parallel pseudo-first order reactions or rate-limiting conformational changes are involved.

The distinction between a curved and a biphasic semi-logarithmic plot sometimes is not clear at a first glance. A number of methods are available which allow the separation of a sum of exponential decay curves into their single components (40–42). The simplest of these (42) is a 'peeling' procedure which can be performed with a ruler directly from the semi-logarithmic plot. Because the fast parts of multi-exponential processes should have subsided faster than the slow parts, the curve should finally approach a straight line. The slope of this straight line will yield the rate constant k_{long} and the intersection with the y-axis will yield the amplitude A_{long} of the slow process. This slow exponential decay has to be subtracted from the total signal in eqn 5b and gives:

$$Y_{subtracted}(t) = S(t) - S(\infty) - A_{long} \cdot e^{-k_{long} \cdot t}. \tag{6}$$

If the signal had originated from a two-exponential process, the semi-logarithmic plot of $Y(t)$ should give a straight line. If a straight line is not obtained, the peeling process may be repeated. The disadvantage of the peeling procedure lies in the unreliability of the signal at the long time range. Because the difference between the signal at time t and the signal at equilibrium becomes very small, it approaches the error of the signal at equilibrium. This unreliable part of the semi-logarithmic plot then critically affects the analysis of the more reliable part. This can be avoided by using a Fourier transformation of the whole curve (40) or by applying the method of moments (41).

A good test of the reliability of such a separation into first order reactions may be found in the following procedure:

(a) Perform one experiment at one time range (for example 0–1 sec).

(b) Use one given mathematical procedure (algorithm) for the separation into rate constants and amplitudes.

(c) Perform the same experiment at a different time range (for example 0–2 sec).

(d) Apply the same algorithm as before. The derived rate constants should be independent of the time range of the experiment.

Semi-logarithmic plots are particularly useful for relaxation experiments. When small perturbations of an equilibrium are analysed, the decomposition of any complex chemical reaction into a sum of exponentials will only give linear terms:

$$Signal_{relaxation}(t) = Sum\ (A_i \cdot e^{-k_i \cdot t}). \tag{7}$$

Higher order terms can be ignored because they become very small. It should be emphasized here that eqn 7 is valid only for small perturbations. Exponential curves with only linear terms in the exponent do not form a set of orthogonal functions. Therefore, a separation of an unknown reaction according to eqn 7 is generally not unique. Moreover, the derived 'rate constants' and 'amplitudes' may well be meaningless.

3.2.2 Equilibrium binding experiments

Equilibrium binding experiments are performed in order to determine the number of binding sites and their affinities. Let us assume that they are represented as Scatchard plots, i.e. as (bound ligand)/(free ligand) versus (bound ligand). Equilibrium binding experiments are discussed in more detail in Chapter 4. If a Scatchard plot is linear, all binding sites have the same affinity. If it is curved, there are at least two binding sites with different affinities. Comparison of Scatchard plots with semi-logarithmic plots allows the distinction of four cases:

i. Linear semi-logarithmic plot, linear Scatchard plot

If the resulting rate constant k_{app} is a linear function of the ligand concentration (pseudo-first order, eqn 5d) and if the equilibrium dissociation constant is equal to k_{-1}/k_1, then reaction scheme 1 is sufficient to explain all binding data.

ii. Curved semi-logarithmic plot, linear Scatchard plot

There are at least two different possibilities for this case:

• There may be one binding site for the ligand, but more than one conformational state, as shown in schemes 2 and 3. Since conformational states cannot be distinguished in equilibrium binding studies, all information on the states must be deduced from kinetic studies.

- There may be two binding sites with identical affinities but different rate constants for complex formation. If two sites differ in their rate constants, they will probably also differ in their equilibrium constants for another ligand. Equilibrium binding studies in the presence of different classes of competing ligands may reveal the existence of different sites.

iii. Linear semi-logarithmic plot, curved Scatchard plot

This indicates that one reaction step is rate-limiting. When the rate constant is independent of the ligand concentration, the rate-limiting step is first order. Experiments at much lower ligand concentration may then reveal the rate constant for an initial association step according to eqn 9.

iv. Curved semi-logarithmic plot, curved Scatchard plot

This is the most probable case for complex receptor kinetics. The number and the stoichiometry of sites can be determined from the Scatchard plot, whereas the semi-logarithmic kinetic plot can only give an indication of the complexity of the reaction. Further analysis is required.

3.2.3 Dose-response curves

Binding of ligands to their receptors leads to the evocation or blockade of a physiological response, such as muscle contraction or an increase in blood pressure. The simplest case is that the magnitude of this response is proportional to ligand binding; in this eventuality, both curves should be equivalent. Dose-response curves are sometimes shown as Hill plots:

$$-\log \frac{(maximal\ response - response)}{(response)} \quad \text{vs. log } (concentration). \quad (8a)$$

Hill plots are linear only in the simplest case of reaction scheme 1 or of schemes where *simultaneous* binding of n ligands to a receptor is postulated:

$$n\,L + R \rightleftharpoons L_nR. \quad (8b)$$

The slope of the Hill plot (Hill coefficient n_H) is then equal to n, the number of ligand molecules required to elicit the response. Simultaneous binding, such as that postulated in scheme 8b, is unlikely in aqueous solution. Instead, intermediate steps such as those shown in scheme 4 have to be considered. If the response is initiated from a fully saturated receptor of such a complete reaction scheme (for example complex LRL in scheme 4), the Hill plot will be curved. For low ligand concentrations the slope will approach the number of bound ligand molecules, but it will always level off to unity for high concentrations (43). The experimentally-determined Hill coefficient will thus be smaller than the number of ligand molecules required for receptor activation.

For the analysis of binding curves, Hill plots (eqn 8) are not always practical. Maximal binding often cannot be determined directly from a binding curve, but has to be extrapolated from a Scatchard, or from some other secondary plot, or determined from curve fitting. Yet the shape of a Hill plot depends critically on this value. In general, a Hill plot derived from a binding curve will be different from one derived from a dose-response curve. Let us assume, for example, that all reaction steps in scheme 4 show the same affinity. In this case, the Hill plot derived from the binding curve will give a straight line with a slope $n_H = 1$. If the response is caused by LRL, the Hill coefficient, derived from the dose-response curve will be larger than unity, indicating the participation of more than one site. In this case, more significant information may be derived from the dose-response compared to the equilibrium binding curve, provided that the response is directly proportional to the formation of a particular receptor–ligand complex.

Dose-response curves need not even distantly reflect equilibrium binding curves, as documented most thoroughly for the nicotinic acetylcholine receptor. Approximately 10 μM acetylcholine is required for half-maximal response (as indicated by the number of open ion channels, by membrane depolarization, or by muscle contraction), whereas 20 nM is sufficient for half-maximal binding. Many attempts have been made in order to detect binding sites in the micromolar range, but none has led to conclusive results. Instead, the discrepancy results from a slow conformational change induced by the binding reaction itself. Equilibrium curves thus only allow detection of a final desensitized conformation of high affinity, whereas a physiological response can only be obtained from an active transient conformational state of much lower affinity. Binding to this state can only be detected in the course of kinetic experiments. This example shows that comparisons of dose-response and equilibrium binding curves can lead to the interpretation of receptor-binding kinetics.

3.2.4 Estimation of initial parameters

Once the basic features of a model, such as the number of binding sites and conformational states, have been established, a reaction scheme can be designed. This scheme has to be supplied with initial values for the rate constants. The example of scheme 3 has shown that different numerical values of rate constants can completely change the properties of a scheme. The selection of a proper set of initial parameters is essential for the success of the later fitting procedure.

The equilibrium dissociation constants of the various complexes can be obtained from equilibrium binding studies or from physiological dose-response curves. The association rate constants can only be obtained from kinetic experiments. The constants derived from semi-logarithmic plots may not be meaningful for complex reactions. If the initial association can be monitored, its rate constant can be directly obtained from the kinetic curve. At time $t = 0$, i.e. when no complex between receptor and ligand has been formed, the velocity is the product of the concentration of the reaction partners multiplied by the initial association rate constant:

$$v\big|_{t=0} = \frac{dL}{dt}\bigg|_{t=0} = -k_1 \cdot R_0 \cdot L_0; \tag{9}$$

where R_0 is the total concentration of binding sites and L_0 is the total ligand concentration.

The velocity v_0 at time $t = 0$ can be determined from a tangent through the kinetic trace at zero time (*Figure 3a*). The plot of the initial velocity versus the total concentrations of either ligand or receptor must be a straight line. If not, the time resolution is not good enough. Eqn 9 allows a direct estimation of the rate constant k_1 when R_0 and L_0 are known and when the signal is calibrated. The free ligand is calculated from the signal in the example of *Figure 3a* according to:

$$L(t) = (F(t) - F_{R0}) \cdot L_0/(F_{L0} - F_{buffer}); \tag{9b}$$

where F_{L0}, F_{R0}, and F_{buffer} are the fluorescence signals of the total ligand solution, and scattering of receptor solution and buffer, respectively.

When k_1 is determined from eqn 9b and 9, it can only serve as a lower limit for two reasons:

- The tangent cannot be determined at zero time but at some finite later time.
- If bound ligand contributes to the fluorescence signal, the reaction velocity $d(L(t))/dt$ will be faster than the value determined from eqn 9b.

3.3 Calculation of a model

Simple models, such as reaction scheme 1 or 2 can often be analysed with sufficient accuracy from semi-logarithmic plots. In these cases, the set of differential equations can also be solved analytically, so that the full course of a binding reaction can be calculated directly. The analytic solutions of these differential equations have been extended to competition binding kinetics (1, 44), so that rate constants from unlabelled ligands can be derived from competition experiments with labelled ligands. Unfortunately, analytical solutions cannot be found for all reaction schemes.

For complex reaction schemes, numerical methods must be used for the solution of the corresponding differential equations. These methods are straightforward and simple, but require a (personal) computer and a suitable program. We have used the routines given in the mathematical toolbox of Turbo Pascal (Borland International, Scotts Valley, CA 95066), but found the program FACSIMILE (Computer Science and Systems Division, Harwell Laboratory, Oxford OX11 ORA, England) more convenient. It can be obtained for personal or mainframe computers. Hopefully the manual will be improved.

Numerical methods for the solution of differential equations are simple, because the infinitely small time intervals of differential equations can be replaced with small but finite times δt. Let us assume, for example, an association reaction according to scheme 1 with the initial concentrations L_0 and R_0. Then the ligand concentration at time $t = 0 + \delta t$ will be:

$$L_{\delta t} = L_0 - k_1 \cdot L_0 \cdot R_0 \cdot \delta t; \tag{10a}$$

$$R_{\delta t} = R_0 - k_1 \cdot L_0 \cdot R_0 \cdot \delta t; \tag{10b}$$

$$RL_{\delta t} = k_1 \cdot L_0 \cdot R_0 \cdot \delta t. \tag{10c}$$

These concentrations are then used in eqn 1a–1c in order to calculate the additional changes for the next time interval. The procedure is repeated until the whole time range is covered. It is simple, but requires a computer because of the large number of time intervals δt. Various mathematical strategies (algorithms) are available for the solution of linear differential equations (see Bibliography to Chapter 11 of this book).

3.3.1 Parameter fitting

The fitting of complex reaction schemes to experimental data is not straight-forward. Different sets of rate constants may fit equally well. This may sound disappointing, but it should be kept in mind that no kinetic experiment can prove a reaction mechanism. One can only find a mechanism of minimal complexity to fit the data. If another experiment can prove that this mechanism is not sufficient (for example if scheme 3 had been used and a second binding site was found in equilibrium), the scheme must be extended to incorporate this finding. This example reflects an important feature of a fitting strategy: Many experiments (association reactions at different concentrations, dissociation reactions, and equilibrium studies) should be fitted simultaneously to the same set of data (see Chapter 11). Parameters, which do not greatly influence the shape of the theoretical curves should be kept constant at their initial values. Time points of complex kinetic reactions should be distributed in a logarithmic scale; 10–40 points will be sufficient for each. But even with these rules, a fitting strategy cannot be generalized since it depends on the experimental data available.

3.3.2 Interpretation of the results

Kinetic experiments can only show whether or not a model is sufficient to account for all data. They can never *prove* a model. For example, experiments which can be fitted to reaction scheme 1 can also be fitted to reaction scheme 2, but the resulting rate constants would be ambiguous. The main aim of kinetic experiments, therefore, lies in finding a minimal model which accounts for all data.

Rate constants derived from kinetic experiments can never be taken at face value. If for example, reaction scheme 1 had been sufficient for data analysis, but if the reaction had really followed scheme 2 or 3, then the observed rate constants would not represent the initial binding reaction of scheme 1. Or if more complicated models such as scheme 4 are necessary, then the parameters obtained are generally not independent from each other. Other completely different combinations of parameters might fit the data equally well. Choosing or fitting one rate constant will influence the whole set of remaining parameters. The solution to this problem lies in imposing a random variation on each rate constant and subsequently applying fitting procedures to find the remaining ones ('Monte Carlo methods'). This procedure requires a lot of computer time and is, therefore, generally not applied.

4. Nicotinic acetylcholine receptor-binding kinetics

The peripheral nicotinic acetylcholine receptor has been most thoroughly studied (for reviews see refs 45, 46). The binding experiments shown here (34, 47–49) required considerable amounts of material, and were possible because the receptor can be isolated and purified in large quantities from the electric organ of the electric fish *Torpedo marmorata* and *Electrophorus electricus*. A fluorescent agonist, NBD-5-acylcholine, was synthesized (12) for the purpose of these experiments.

Equilibrium binding (47) of NBD-5-acylcholine to the purified receptor from *E. electricus* indicated that there were two classes of binding sites at the receptor. The fluorescence of the dye is completely quenched upon binding. Kinetic experiments (34) performed in a stopped-flow spectrophotometer revealed complex association kinetics which could not be unambiguously separated into a sum of exponential curves. Two types of dissociation experiments gave seemingly contradictory results. When dissociation was measured after rapid dilution of the receptor–ligand complex, the kinetics were always biphasic. In fact, they were even biphasic under conditions where only one binding site of the receptor was occupied. When dissociation of the receptor–ligand complex was observed after rapid addition of another (non-fluorescent) ligand, the dissociation process was much more complex and slower. It became slower the greater the concentration of non-fluorescent ligand added (34).

These observations led to the following requirements for a reaction scheme:

- There are two binding sites (deduced from equilibrium measurements).

- There are two conformational states (deduced from the biphasic dissociation reaction).

- Ligands bind in an ordered fashion, i.e. the ligand which binds first has to dissociate last.

The latter requirement was deduced from the observation that addition of a second ligand makes the dissociation process slower. This can only be explained if the second ligand can entrap the first one. Five different reaction schemes (34) meet these requirements and can fit the data. One of these is the following:

$$R+2L \; \underset{}{\overset{1}{\rightleftharpoons}} \; LR+L \; \underset{}{\overset{2}{\rightleftharpoons}} \; L_2R$$
$$3\Big\Updownarrow \qquad\qquad 5\Big\Updownarrow$$
$$LD+L \; \underset{4}{\rightleftharpoons} \; L_2D \qquad\qquad (11)$$

Two binding sites are filled in a strictly ordered ('first on, last off') mechanism. The receptor has two conformational states denoted as R and D. Scheme 11 was used to fit a series of association and dissociation experiments simultaneously (34). The kinetic parameters derived from this fit were then used to fit a set of dissociation reactions observed after addition of competing ligand. The goodness of fit proved that this ordered mechanism was sufficient to account for the slow and complex dissociation reactions.

A second set of experiments was performed with the nicotinic acetylcholine receptor from *T. marmorata*. The tissue of this species has a much higher receptor concentration, so that membrane fragments can be used and the experimental data be compared to those for the isolated receptor. The fluorescence of NBD-5-acylcholine is only partially quenched upon binding to this receptor (48), so that the spectroscopic signal is lower. The membrane-bound receptor shows a biphasic reaction, the slow part of which (monitored in the time range of minutes) disappears after addition of detergent to the membrane, giving a similar time course as the purified receptor from *E. electricus*. The slow part can be conveniently studied with time-resolved filtration and radioactive acetylcholine (50). It corresponds quantitatively to our stopped-flow experiments performed with the fluorescent agonist NBD-5-acylcholine. The slow reaction, in the minute time range, was only detected when the ligand concentration was high enough. This led to the paradoxical situation that addition of low acetylcholine concentrations (10 nM per 100 nM receptor) was faster than the addition of high concentrations (100 nM to the same receptor). The reverse, namely increasing velocity with increasing ligand concentration, is expected for second order reactions. This observation cannot be reconciled with reaction scheme 11. An extension is required:

$$R+2L \; \underset{}{\overset{1}{\rightleftharpoons}} \; LR+L \; \underset{}{\overset{2}{\rightleftharpoons}} \; L_2R$$
$$5\Big\Updownarrow \qquad 6\Big\Updownarrow \qquad 7\Big\Updownarrow$$
$$D+2L \; \underset{3}{\rightleftharpoons} \; LD+L \; \underset{4}{\rightleftharpoons} \; L_2D \qquad (12)$$

This scheme assumes that two conformations of the receptor exist even in the absence of ligand. If conformation D has a faster association rate and a higher affinity than conformation R, and if R dominates the equilibrium in the absence of ligand, the experiments can be explained. Low concentrations of acetylcholine will bind to D rapidly. For acetylcholine concentrations larger than D, R has to be transformed to D in order to allow high affinity binding. This step is slow for the intact membrane and fast for the solubilized receptor, and is the reason why it could not be detected in the binding kinetics of *Electrophorus electricus*.

Scheme 12 considers seven molecular species (L, R, D, LR, LD, L_2R and L_2D) and 14 rate constants (seven reaction paths with one forward and one backward rate constant each). The differential equations of this complex scheme are given here to illustrate their simplicity. (They really are simple, just figure it out line by line!)

$$dR/dt = -k_1{\cdot}R{\cdot}L + k_{-1}{\cdot}LR - k_5{\cdot}R + k_{-5}{\cdot}D; \tag{12a}$$

$$dD/dt = -k_3{\cdot}D{\cdot}L + k_{-3}{\cdot}LD + k_5{\cdot}R - k_{-5}{\cdot}D; \tag{12b}$$

$$dL/dt = -k_1{\cdot}R{\cdot}L + k_{-1}{\cdot}LR - k_2{\cdot}LR{\cdot}L + k_{-2}{\cdot}L_2R$$
$$- k_3{\cdot}D{\cdot}L + k_{-3}{\cdot}LD - k_4{\cdot}LD{\cdot}L + k_{-4}{\cdot}L_2D; \tag{12c}$$

$$dLR/dt = k_1{\cdot}R{\cdot}L - k_{-1}{\cdot}LR - k_2{\cdot}LR{\cdot}L + k_{-2}{\cdot}L_2R$$
$$- k_6{\cdot}LR + k_{-6}{\cdot}LD; \tag{12d}$$

$$dLD/dt = k_3{\cdot}D{\cdot}L - k_{-3}{\cdot}LD - k_4{\cdot}LD{\cdot}L + k_{-4}{\cdot}L_2D$$
$$+ k_6{\cdot}LR - k_{-6}{\cdot}LD; \tag{12e}$$

$$dL_2R/dt = k_2{\cdot}LR{\cdot}L - k_{-2}{\cdot}L_2R - k_7{\cdot}L_2R + k_{-7}{\cdot}L_2D; \tag{12f}$$

$$dL_2D/dt = k_4{\cdot}LD{\cdot}L - k_{-4}{\cdot}L_2D + k_7{\cdot}L_2R - k_{-7}{\cdot}L_2D. \tag{12g}$$

In conclusion, I wish to point out that the purpose of receptor binding experiments is not the extraction of rate constants, but the understanding of a molecular mechanism. It is never possible to prove a mechanism, but a mechanism which does not explain the data can be disproved. This has been shown here with the example of the nicotinic acetylcholine receptor, but, in general, any reaction schemes may require further extension at some stage. For example, all schemes discussed in this chapter consider the receptor as having a few distinct conformational states and a given number of binding sites. Both may be too idealized. Proteins exhibit a whole spectrum of conformational states (51) which generally are not considered in the analysis of binding kinetics at room temperature. Moreover, the binding sites of membrane channel proteins may be charged as a result of ion flux (52). Their number might, therefore, vary with receptor activity. A spectrum of conformational states or binding sites has not yet been necessary for the analysis of receptor binding kinetics, but careful analysis of precise experiments may require such extensions in the future.

References

1. Motulsky, H. J. and Mahan, L. C. (1984). *Mol. Pharmacol.*, **25**, 1.
2. Jacobson, K. A., Ukena, D., Padgett, W., Kirk, K. L., and Daly, J. W. (1987). *Biochem. Pharmacol.*, **36**, 1697.
3. McCabe, R. T., de Costa, B. R., Miller, R. L., Havunjian, R. H., Rice, K. C., and Skolnick, P. (1990). *FASEB J.*, **4**, 2934.
4. Heithier, H., Jaeggi, K. A., Ward, L. D., Cantrill, R. C., and Helmreich, E. J. (1988). *Biochimie*, **70**, 687.
5. Tota, M. R. and Strader, C. D. (1990). *J. Biol. Chem.*, **265**, 16891.
6. Monsma, F. J., Jr, Barton, A. C., Kang, H. C., Brassard, D. L., Haugland, R. P., and Sibley, D. R. (1989). *J. Neurochem.*, **52**, 1641.
7. Madras, B. K., Canfield, D. R., Pfaelzer, C., Vittimberga, F. J., Jr, Difiglia, M., Aronin, N., Bakthavachalam, V., Baindur, N., and Neumeyer, J. L. (1990). *Mol. Pharmacol.*, **37**, 833.
8. Curcio, C. A., McNelly, N. A., and Hinds, J. W. (1985). *J. Comp. Neurol.*, **235**, 519.
9. Anstead, G. M., Altenbach, R. J., Wilson, S. R., and Katzenellenbogen, J. A. (1988). *J. Med. Chem.*, **31**, 1316.
10. Heithier, H., Ward, L. D., Cantrill, R. C., Klein, H. W., Im, M. J., Pollak, G., Freeman, B., Schiltz, E., Peters, R., and Helmreich, E. J. (1988). *Biochim. Biophys. Acta*, **971**, 298.
11. Heidmann, T., Bernhardt, J., Neumann, E., and Changeux, J. P. (1983). *Biochemistry*, **22**, 5452.
12. Meyers, H.-W., Jürss, R., Brenner, H. R., Fels, G., Prinz, H., Watzke, H., and Maelicke, A. (1983). *Eur. J. Biochem.*, **137**, 399.
13. Johnson, D. A., Brown, R. D., Herz, J. M., Berman, H. A., Andreasen, G. L., and Taylor, P. (1987). *J. Biol. Chem.*, **262**, 14022.
14. Mihara, H., Lee, S., Shimohigashi, Y., Aoyagi, H., Kato, T., Izumiya, N., and Costa, T. (1987). *Int. J. Pept. Protein. Res.*, **30**, 605.
15. Schmitz, G., Wulf, G., Bruening, T., Assmann, G., Buku, A., Yamin, N., and Gazis, D. (1988). *Peptides*, **9**, 783.
16. Buku, A., Masur, S., and Eggena, P. (1989). *Am. J. Physiol.*, **257**, E804.
17. Eggena, P., Ma, C.-L., Lu, M., and Baku, A. (1990). *Am. J. Physiol.: Endocrinol. Metab.*, **259**, E524.
18. Brown, G. B. and Bradley, R. J. (1985). *J. Neurosci. Meth.*, **13**, 119.
19. Kumazawa, T., Nomura, T., and Kurihara, K. (1988). *Biochemistry*, **27**, 1239.
20. Labrecque, G. F., Holowka, D., and Baird, B. (1989). *J. Immunol.*, **142**, 236.
21. Bergelson, L. D., Molotkovsky, J. G., and Manevich, Y. M. (1985). *Chem. Phys. Lipids.*, **37**, 165.
22. Manevich, E. M., Koeiv, A., Jaerv, J., Molotkovsky, J. G., and Bergelson, L. D. (1988). *FEBS Lett.*, **236**, 43.
23. Patel, J. M., Edwards, D. A., Block, E. R., and Raizada, M. K. (1988). *Biochem. Pharmacol.*, **37**, 1497.
24. Hughes, A. R., Takemura, H., and Putney, J. W., Jr (1988). *J. Biol. Chem.*, **263**, 10314.
25. Ambler, S. K. and Taylor, P. (1986). *J. Biol. Chem.*, **261**, 5866.

26. Dunn, S. M. J., Shelman, R. A., and Agey, M. W. (1989). *Biochemistry*, **28**, 2551.

27. Karpen, J. W. and Hess, G. P. (1986). *Biochemistry*, **25**, 1786.

28. Isom, L. L., Cragoe, E. J., Jr, and Limbird, L. E. (1987). *J. Biol. Chem.*, **262**, 17504.

29. Moore, H.-P. H. and Raftery, M. A. (1980). *Proc. Natl. Acad. Sci. USA*, **77**, 4509.

30. Donnelly, D., Mihovilovic, M., Gonzalez-Ros, J. M., Ferragut, J. A., Richman, D., and Martinez-Carrion, M. (1984). *Proc. Natl. Acad. Sci, USA*, **81**, 7999.

31. Bruns, R. F., Lawson-Wendling, K., and Pugsley, T. A. (1983). *Anal. Biochem.*, **132**, 74.

32. Boyd, N. D. and Cohen, J. B. (1980). *Biochemistry*, **19**, 5353.

33. Hess, G. P., Udgaonkar, J. B., and Olbricht, W. L. (1987). *Annu. Rev. Biophys. Biophys. Chem.*, **16**, 507.

34. Prinz, H. and Maelicke, A. (1983). *J. Biol. Chem.*, **258**, 10273.

35. Sage, S. O. and Rink, T. J. (1987). *J. Biol. Chem.*, **262**, 16364.

36. Merritt, J. E. and Rink, T. J. (1987). *J. Biol. Chem.*, **262**, 4958.

37. Prinz, H. and Maelicke, A. (1986). *J. Biol. Chem.*, **261**, 14962.

38. Monod, J., Wyman, J., and Changeux, J.-P. (1965). *J. Mol. Biol.*, **12**, 88.

39. Guggenheim, E. A. (1926). *Phil. Mag.*, **2**, 538.

40. Provencher, S. W. (1976). *Biophys. J.*, **16**, 27.

41. Dyson, R. D. and Isenberg, I. (1971). *Biochemistry*, **10**, 3233.

42. Mancini, P. and Pilo, A. (1970). *Comput. Biomed. Res.*, **3**, 1.

43. Werman, R. (1969). *Comp. Biochem. Physiol.*, **30**, 997.

44. Schreiber, G., Henis, Y. I., and Sokolovsky, M. (1985). *J. Biol. Chem.*, **260**, 8789.

45. Changeux, J.-P., Devillers-Thiéry, A., and Chemouilli, P. (1984). *Science*, **225**, 1335.

46. Maelicke, A. (1988). In *Handbook of Experimental Pharmacology*, (ed. G. V. R. Born, A. Farah, H. Herken, and A. D. Welch) Vol. 86, pp. 267–313. Springer Verlag, Heidelberg.

47. Prinz, H. and Maelicke, A. (1983). *J. Biol. Chem.*, **258**, 10263.

48. Covarrubias, M., Prinz, H., Meyers, H.-W., and Maelicke, A. (1986). *J. Biol. Chem.*, **261**, 14955.

49. Covarrubias, M., Prinz, H., and Maelicke, A. (1984). *FEBS Lett.*, **169**, 229.

50. Boyd, N. D. and Cohen, J. B. (1980). *Biochemistry*, **19**, 5344.

51. Elber, R. and Karplus, M. (1987). *Science*, **235**, 318.

52. Edmonds, D. T. (1989). *Eur. Biophys. J.*, **17**, 113.

11

Analysis and interpretation of binding at equilibrium

JAMES W. WELLS

1. Introduction

The ligand-binding properties of systems at equilibrium are an accessible and important source of information on the nature of transmembrane signalling. Levels of binding are determined by affinity and capacity, which in turn can reflect efficacy and other correlates of the response. The binding properties, therefore, report not only on the interaction between ligand and receptor, but also on the processes initiated by that interaction. Visual inspection of a binding pattern is often wasteful of the information contained therein, as the full implications of the data are rarely self-evident; rather, the information must be inferred in the context of a model, and the most useful model is one that constitutes a quantitative formulation of the hypothesis.

Despite the potential of quantitative models, much of the information from a typical binding study can remain buried within the data. The problem lies partly in ambiguity, in that the results of single experiments are seldom definitive; a unique interpretation almost inevitably requires that the system be characterized with respect to several variables. Also, all but the most naïve hypotheses can pose a formidable numerical challenge when expressed in the mathematical terms essential to quantitative assessment. The application of numerically intensive procedures to large quantities of data requires computational facilities that previously have been inaccessible to many investigators.

Inexpensive and increasingly powerful microprocessors are now a fixture in every laboratory, and the complexity of quantitative models need not be constrained by their numerical demands; rather, the limiting factor has become the ability to formulate a hypothesis and to acquire the relevant data. Biochemical data on the structure of receptors and associated proteins increasingly permit more specific and hence more practical notions of what constitutes a physically realistic model. Since quantitative analyses are no longer restricted to relatively naïve schemes, the binding of ligands is likely to become increasingly revealing as a window on transmembrane signalling. The strategy outlined in this chapter can be summarized as follows.

(a) *Build the model.* Since the model is the embodiment of the hypothesis, this really ought to be the first step. The measured signal is related to a set of parameters and independent variables via one or more equations describing the hypothesis, and derived according to physical–chemical principles.

(b) *Acquire the data.* Binding studies are carried out in a manner designed to define the parameters of the model. Should this step precede step one, or should the data prove unrecognizable in the context of the hypothesis, an empirical model may seem better than nothing.

(c) *Fit the model to the data.* Here one adjusts the parameters of the model in a stepwise and convergent manner to obtain the best possible fit to the data; at each cycle of the fitting procedure, the equations defining the model must be solved for all relevant values of the independent variable or variables. There is a clear distinction between optimizing the parameters of the model and computing the model *per se*.

(d) *Does it fit?* There ought to be reasonable agreement between the model and the experimental data. Statistical tests are in order here, but they are no substitute for inspecting a plot of the data and the fitted curve.

(e) *Can the fit be improved?* Parameters can be added up to the number of data points, when the result is no longer a fit but a solution. Each additional parameter will improve the fit, but there is a point beyond which further expansion of the model is not justified by the data; somewhat inconveniently, that point is often reached before all of the parameters known or suspected to be involved have been estimated.

(f) *Is the model sustained or consolidated by further data?* The model may be in excellent and well defined agreement between the results of experiment A on the one hand and experiment B on the other; it nevertheless may fail to accommodate both at the same time. Alternatively, a parameter may be defined by A and B taken together but by neither taken separately. This is the stuff of simultaneous analyses.

In the strictest sense, all models are either empirical or mechanistic. The former is a description of the phenomenon; the latter is a description of the factors and processes that govern the phenomenon. A purely empirical model is a useful tool for the investigator who wishes to interpolate from available data but has little interest in the underlying events. Typical examples include the polynomials used to describe the results of a Lowry assay, or the data obtained with quenched standards and the external source in a liquid scintillation spectrometer; the shape of the curve is seldom of concern in either case. When the underlying events are the issue, however, the properties of a functional relationship become important for their mechanistic implications.

The distinction between empirical and mechanistic models is blurred in practice, since the latter often tend to be mechanistic descriptions based upon a limited knowledge of the system. It may be clear that the model is inadequate in many respects, but the parameters are nevertheless interpreted in mechanistic terms.

2. General approach to mechanistic models

2.1 General comments

For systems at equilibrium, concentration is the independent variable in the majority of investigations. The interaction of a probe with another species, such as a receptor, will be influenced not only by the concentration of each but also by that of any additional component that perturbs the interaction. Experiments are typically designed to study the concentration-dependent distribution of the probe among multiple phases or states, which, therefore, must be identified along with the various processes whereby the distribution is established and maintained. Such an exercise involves many assumptions, some of which will introduce errors or inappropriate simplifications.

For the lucky or shrewd investigator, unwarranted assumptions will be revealed as a discrepancy between the model and the data. Somewhat perversely, however, profoundly different mechanistic schemes can predict similar results. If an error or omission is not reflected in the predicted behaviour of the system for the experimental protocol being used, the problem will remain hidden until revealed by new data of a different nature. Consistency between a model and the data does not indicate that the model is correct; it indicates only that the model *may* be correct. In many biological systems, the investigator is limited in the range of experimental approaches that can be applied. Accordingly, there is a tendency for essentially the same experiment to be repeated in various guises and in many laboratories. Such repetition may confirm the reproducibility of the phenomenon; it does little to establish the validity of a model that happens to describe the data.

The chemical potential of a component within a heterogeneous system is related to the logarithm of its local concentration; therefore, it is important to account for all factors that reduce the concentration in the phase or state of interest. The aqueous concentration of a radiolabelled probe may be depleted through dissolution in the hydrophobic interior of a membrane or through binding to the container, to charged groups on the surface of the membrane, to other receptors, or to the receptor of interest; similarly, the effective concentration of unliganded receptor within the membrane may be depleted through binding of the probe, through association with other proteins, or through enzymatic processes in appropriately constituted systems. If the aqueous concentration of the probe is believed important to the phenomenon under investigation, it must be measured directly or inferred from the total concentration and a model that accounts for all forms of depletion; similarly, if binding of the probe to a particulate fraction of a homogenate is to be measured, the contribution arising from the specific interaction with the receptor of interest must be differentiated from all others.

Any component within a reaction mixture can exist in at least two states: complexed specifically with other components and dissolved in one or another phase of the system. The observed signal in the presence of a labelled probe will typically reflect both states, and often multiple forms or substates thereof, as well as a background contribution which is independent of the probe and arises from the measuring equipment. For a mechanistic model, the aim is to define all of the states and substates believed to exist, and to formulate an explicit expression for the distribution of the probe at the time of the measurement. Such formulations need not include time if the measurements are made at equilibrium or steady state with respect to all processes within the system; otherwise, time must appear explicitly if at least some of the parameters estimated from the model are not to be wrong or uninterpretable.

The observed signal will reflect not only the distribution of the probe among the possible states or substates of the system but also the specific signal of the probe in each. For radioligands, that signal depends only upon the specific radioactivity, which in turn depends only upon the initial value and the intrinsic rate of decay for the nuclide in question. The specific signal emitted by the probe is thus the same and is known in advance for all states and substates within the system. On the one hand, this simplifies the expressions that constitute an explicit description of the model and can reduce the number and complexity of experiments required to define the system; on the other hand, it usually obliges the investigator to separate the state of interest from other states before the signal is measured. Physical separation is generally tedious; also, it may be incomplete or may initiate time-dependent events that violate kinetic assumptions. In experiments with concentration as the independent variable, it precludes adding successive aliquots of a ligand to the same sample of receptor.

An important advantage of radiolabelled probes is the high sensitivity achievable with several radionuclides of biological interest. At an isotopic abundance of 100%, the specific activity of tritium is 28.8 Ci/matom. Hydrogenation of a double bond with carrier-free $^3H_2$ will thus yield a probe with a specific activity of 57.6 Ci/mmol; similarly, methylation of a tertiary amino group with C^3H_3I will yield a specific activity of 86.4 Ci/mmol. At a counting efficiency of 50%, these levels of activity yield a signal of 100 counts/min with only 1.0–1.6 fmol of the probe. Much higher sensitivity is obtainable with sulphur-35 (1494 Ci/matom), iodine-125 (2176 Ci/matom) and phosphorus-32 (9131 Ci/matom), although the comparatively rapid decay of these nuclides complicates the determination of absolute count rates when the counting efficiency is appreciably less than 100%.

As purified receptors and systems for enhanced expression become available, probes of lower sensitivity than radioligands can be employed. Fluorescent and spin-labelled compounds offer a number of technical advantages over radioligands, but the specific signal generally depends upon the environment and must be measured for each state if the assessment of the data is to be fully quantitative. Failure to do so can reduce the information gained from the experiment and may confound the interpretation of the data. Nevertheless, differences in the specific signal can simplify experimental procedures: for example, the quantum yield of a fluorescent probe may differ tenfold or more between the polar environment of bulk water and the non-polar environment of a specific site on a receptor; the difference can preclude the need to separate bound from unbound probe prior to measurement.

2.2 Distribution of the probe

For any system, the total concentration of a probe P with respect to total volume can be described according to eqn 1:

$$[P]_t = \sum_{i=1}^{m} f_i([P_i] + \sum_{j=1}^{n} [B_{ij}]), \tag{1}$$

where $\sum_{i=1}^{m} f_i = 1$.

The quantity f_i represents the fractional volume occupied by each phase i, and m is the total number of phases. Concentrations within phase i are designated as $[P_i]$ for the unbound probe and $[B_{ij}]$ for the probe in a specific complex j with another component of the system ($j = 1, 2, \ldots, n$). Levels of B_{ij} in any phase i are determined by the value of $[P_i]$; values of $[P_x]$ ($x \neq i$) are relevant only insofar as they influence $[P_i]$, provided that interactions leading to the formation of B_{ij} occur exclusively within phase i. It is assumed throughout this chapter that dimers and higher aggregates of the probe do not occur to any appreciable extent; similarly, multiple conformations of the probe are assumed to interconvert spontaneously and rapidly on the time scale of the binding assay and not to affect the pH or the free concentration of any species not explicit in the model. The concentration can thus be represented by a single, time-averaged value of $[P_i]$; evidence to the contrary would require that eqn 1 be expanded accordingly.

The simplest, specific example of eqn 1 is perhaps an aqueous system in which all constituents are water soluble, and in which no constituent, apart from water, contributes appreciably to the total volume (i.e. $m = 1$). Eqn 1 then reduces to eqn 2, in which the subscript A denotes the aqueous phase (i.e. $f_A = 1$); the probe is distributed between bulk solvent P_A on the one hand and the various complexes B_{Aj} on the other:

$$[P]_t = [P_A] + \sum_{j=1}^{n} [B_{Aj}]. \tag{2}$$

If a purified, globular protein R with a single binding site is the only constituent that forms a specific complex with the probe, the expression can be refined further as in eqn 3; the complex PR_{Aj} represents one of n states differentiated by factors such as conformation, degree of protonation, or degree of complexation with metal ions:

$$[P]_t = [P_A] + \sum_{j=1}^{n} [PR_{Aj}].$$ (3)

It is noteworthy that eqn 3 describes the distribution of the probe only if the protein does not contribute appreciably to the total volume; at sufficient concentrations, the protein itself will constitute a separate phase into which a sufficiently hydrophobic probe may partition to an appreciable extent.

Typically, at least two phases are present with proteins that, in their native state, are associated with a membrane. Eqn 4 describes the distribution of a probe in a system comprising an aqueous and a non-aqueous phase, differentiated by the subscripts A and M respectively:

$$[P]_t = f_A([P_A] + \sum_{j=1}^{n} [B_{Aj}]) + f_M([P_M] + \sum_{j=1}^{o} [B_{Mj}]).$$ (4)

In a homogenate of disrupted tissue or reconstituted membranes, the non-aqueous phase is the membrane itself; the local concentrations of dissolved ($[P_M]$) and specifically bound probe ($\sum_{j=1}^{o} [B_{mj}]$) relate to the volume of the membrane, which in turn constitutes the fraction f_M of the total volume of the suspension. The membrane is generally dispersed as fragments throughout the aqueous phase, and there is likely to be restricted mobility of the binding sites from one fragment to another; also, membrane-bound components that contribute either directly or indirectly to B_{Mj} may not be distributed randomly among the fragments. The volume corresponding to f_M in eqn 4 thus represents only an upper limit on that accessible to individual components. A similar qualification pertains to f_A if the aqueous phase is compartmentalized owing to vesicles or similar formations. The quantity $\sum_{j=1}^{o} [B_{Mj}]$ represents binding to sites accessible only from the non-aqueous phase, and the relevant concentration of probe is therefore $[P_M]$; similarly, the quantity $\sum_{j=1}^{n} [B_{Aj}]$ represents binding to sites accessible only from the aqueous phase, and the relevant concentration of the probe is therefore $[P_A]$. For membrane-bound receptors accessible from the aqueous phase, the relevant fractional volume is f_A with respect to the probe, but f_M with respect to other components of the membrane.

The solubilization of membrane-bound receptors and other proteins typically requires a detergent to substitute for lipid at the hydrophobic surfaces that normally interact with the membrane. If the concentration of detergent exceeds the critical micelle concentration, the distribution of the probe between the resulting micellar phase M and the aqueous phase A may also be approximated by eqn 4 or a similar expression. In systems that form lamellar micelles and related aggregates, the interpretation of f_A and f_M may be qualified as noted above.

The implications of eqn 4 can be demonstrated by considering the example of a membrane-bound receptor R in a suspension of homogenized tissue. If saturable binding of the probe is exclusive to the receptor, and if access is exclusively via either the membrane or the aqueous phase, then the distribution of the probe can be described by eqn 5; the subscript X denotes either M or A, depending upon whether $[P_M]$ or $[P_A]$ is the relevant concentration with respect to formation of the complex:

$$[P]_t = f_M[P_M] + f_A[P_A] + f_X \sum_{j=1}^{n} [B_{Xj}]. \tag{5}$$

The expression can be written as shown in eqn 6 if the stoichiometry of the probe and receptor in the complex is $1:1$ (i.e. PR) and if oligomers of the receptor either do not occur or are without effect on specific binding:

$$[P]_t = f_M[P_M] + f_A[P_A] + f_X \sum_{j=1}^{n} [PR_{Xj}]. \tag{6}$$

The distribution of unbound probe between the membrane and the aqueous phase can be described by a partition coefficient $\mathcal{P}$ as in eqn 7:

$$\mathcal{P} = [P_M]/[P_A]. \tag{7}$$

The overall distribution of the probe is discussed below for specific models in which multiple states of the complex arise from distinct and mutually independent sites and from various forms of co-operativity. In these examples, however, independent variables are limited to the probe, ligands that compete with the probe, and specific instances of ligands that act allosterically. It is important to recognize that the binding properties of most receptors are sensitive to many constituents common to most experiments, including protons, metal ions, lipids, and other proteins. Such variables must be included in the formulation of a mechanistic model should they undergo any change, either explicitly or as a consequence of changes in another variable. Although not reflected in eqns 1–6, the state of the probe may similarly depend upon extraneous factors that might require inclusion in the model.

Typically, the measured or dependent variable is a membrane-bound probe P_B, which in the reaction mixture consists of at least two components, as described by eqn 8:

$$[P_B] = f_M[P_M] + f_X \sum_{j=1}^{n} [PR_{Xj}]. \tag{8}$$

Since the specific activity of a radioligand is the same in all states, eqn 8 is only a useful expression when $f_X\sum_{j=1}^{n}[PR_{Xj}]$ is an appreciable fraction of $[P]_t$ (eqn 6). Since $f_X\sum_{j=1}^{n}[PR_{Xj}]$ is typically exceeded by $f_A[P_A]$ and perhaps even by $f_M[P_M]$, it is usually necessary to achieve a physical separation of PR_{Xj} from

P_A. For membranes and membrane fragments, bound radioligand is separated together with the insoluble phase when the reaction mixture is either filtered or centrifuged. For proteins in solution, or for fragments of microsomal size, several options are available: the reaction mixture can be chromatographed on a gel of appropriate porosity to separate free and bound ligand; the protein together with bound ligand can be precipitated with polyethyleneglycol and separated by filtration or centrifugation; free ligand can be adsorbed to charcoal and removed by centrifugation; and free ligand can be separated by centrifugation through semi-permeable membranes of appropriate porosity.

All the foregoing techniques can reduce binding from the levels present in the reaction mixture, and can augment the observed signal to the extent that the removal of P_A is incomplete. Apparent binding following the separation of free and bound probe is thus described by eqn 9:

$$[P_B] = C_{PM}f_M[P_M] + C_{PA}f_A[P_A] + f_X\sum_{j=1}^{n}C_{PRj}[PR_{Xj}]. \qquad (9)$$

The constants C_{PM}, C_{PA} and C_{PRj} represent the fractions of P_M, P_A and PR_{Xj}, respectively, that appear as bound radioligand (i.e. $0 < C < 1$). For P_M and PR_{Xj}, this constitutes a loss of membrane-bound probe; the result is a lower level of noise on the one hand (P_M) but an artefactually low signal on the other (PR_{Xj}). For P_A, it constitutes a source of noise that is only introduced at the time of separation. It is important to note that both $f_M[P_M]$ and $f_X\sum_{j=1}^{n}[PR_{Xj}]$ affect $C_{PA}f_A[P_A]$, but the converse is not true; that is, $C_{PA}f_A[P_A]$ affects neither $f_M[P_M]$ nor $f_X\sum_{j=1}^{n}[PR_{Xj}]$.

Neither $[P_M]$ nor $[P_A]$ can be known without measurements that are generally tedious but may be dictated by the complex effects of C_{PM}, C_{PA} and C_{PRj} on total binding (eqn 9). The total probe is usually known or readily measured, however, and it is useful to express the total binding as a function of $[P]_t$. Substituting either $[P_M]/\wp$ for $[P_A]$, or $\wp[P_A]$ for $[P_M]$ in eqn 6 yields eqns 10 and 11 respectively; rearrangement leads, in turn, to the corresponding expressions for $[P_M]$ and $[P_A]$ (eqns 12 and 13):

$$[P]_t = f_M[P_M] + f_A[P_M]/\wp + f_X\sum_{j=1}^{n}[PR_{Xj}] \qquad (10)$$

$$[P]_t = \wp f_M[P_A] + f_A[P_A] + f_X\sum_{j=1}^{n}[PR_{Xj}] \qquad (11)$$

$$[P_M] = \frac{\wp([P]_t - f_X\sum_{j=1}^{n}[PR_{Xj}])}{f_A + \wp f_M} \qquad (12)$$

$$[P_A] = \frac{[P]_t - f_X\sum_{j=1}^{n}[PR_{Xj}]}{f_A + \wp f_M} \qquad (13)$$

Substituting for $[P_M]$ (eqn 12) and $[P_A]$ (eqn 13) in eqn 9 yields eqn 14, in which $\Sigma_{j=1}^{n}[PR_{Xj}]$ is the only variable that cannot be measured independently or is not known *a priori*:

$$[P_B] = \frac{C_{PM}\mathscr{P}f_M([P]_t - f_X\sum_{j=1}^{n}[PR_{Xj}])}{f_A + \mathscr{P}f_M} \qquad (14)$$

$$+ \frac{C_{PA}f_A([P]_t - f_X\sum_{j=1}^{n}[PR_{Xj}])}{f_A + \mathscr{P}f_M}$$

$$+ f_X\sum_{j=1}^{n}C_{PRj}[PR_{Xj}].$$

Eqn 14 is a generic function that encompasses many of the contributions to total binding that are typically encountered in assays with radioligands. Analogous expressions can be derived to include effects disregarded in the present discussion, such as binding of the probe to the container or the contribution of radiolabelled enantiomers differing in affinity for the receptor. If it is known that saturable binding constitutes a negligible fraction of the total probe (i.e. $f_X\Sigma_{j=1}^{n}[PR_{Xj}] \ll [P]_t$), the first two terms of eqn 14 can be simplified accordingly; the same simplification can be made in the calculation of PR_{Xj}, as described in Section 3.

Practical applications of eqn 14 require that specific binding be related to a set of variables (x) and a set of parameters (a) as in eqn 15:

$$\sum_{j=1}^{n}[PR_{Xj}] = f(x,a). \qquad (15)$$

This represents the saturable component of the signal and thus affects the first two terms of eqn 14 only insofar as it reduces $[P_M]$ and $[P_A]$ (cf. eqn 9); it is assumed here to be without effect on the fractional volume of the membrane f_M, the partition coefficient $\mathscr{P}$, or the fractions of P_M and P_A that eventually appear as bound probe (C_{PM}, C_{PA}). The implications of eqn 14, some of which are considered below, are thus independent of all x (eqn 15) except the total concentration of the probe. More complex relationships will emerge, however, if x includes variables that directly affect the amount of probe in either of the non-saturable compartments. With an ionizable ligand, for example, changes in pH will alter the overall partition coefficient $\mathscr{P}$. Also, metal ions can reduce the size of the pellet obtained when bound probe is obtained by centrifugation, and can thereby reduce the quantity of entrapped P_A (C_{PA}). Finally, different concentrations of receptor may imply differences in f_M, as in studies involving native membranes; total receptor usually is considered a parameter (i.e. $[R]_t$), but it effectively becomes a variable when two or more values are required in the analysis. These and other effects can be accommodated by extensions of eqn 14 in which the relevant parameter $(C_{PX}, \mathscr{P}, f_X)$ is expressed as a function of the offending variable.

It is important to note that eqn 15 does not contain the parameter C_{PRj} (cf. eqn 14), which, therefore, is assumed to be the same for all PR_{Xj}. The separation of free and bound probe is often accompanied by a loss of specific binding (i.e., $C_{PRj} < 1$), particularly when membranes are harvested by filtration: either the membranes are not retained quantitatively by the filter, or the probe dissociates from the receptor during washing, or both. A selective loss of one or another complex would be difficult to assess and could render specific binding almost uninterpretable in mechanistic terms.

2.3 Specific activity of the probe

Also required for practical purposes is an expression that relates $[P_B]$ in eqn 14 to the measured signal. In studies with radioligands, the latter comprises emission from the probe itself and a background signal due to cosmic radiation, local sources of radioactivity other than the probe, static electricity, chemiluminescence, and noise from the photomultipliers and associated circuitry. The basic unit of measurement is counts per minute (c.p.m.), while the basic unit of radioactive emission is given variously as disintegrations per minute (d.p.m.), disintegrations per second (becquerel, Bq) or curies (Ci; 1 Ci $= 2.22 \times 10^{12}$ d.p.m.).

For the signal originating from the probe, the ratio of measurement (c.p.m.) to emission (d.p.m.) is the counting efficiency, which in turn depends upon the level of quench; the efficiency is readily measured for each sample with the help of an external standard and quenched samples of known emission. The notion of efficiency is vague as regards the background, which consists of quenchable and unquenchable contributions; also, any dependence on quench is likely to differ from that of the signal from the probe. Background rates can be measured with samples lacking the probe but otherwise identical in composition and volume to those of the assay. If the level of quench undergoes only minor fluctuations from sample to sample, or if the background rate is small relative to the signal due to the probe, the latter can generally be taken as the difference (Δ c.p.m.) between the total counts per minute c.p.m.$_{obsd}$ and the background rate c.p.m.$_{bgrd}$ taken as measured. The absolute rate of emission d.p.m. for the probe can then be calculated from Δc.p.m. and the efficiency E according to eqn 16:

$$\text{d.p.m.} = (\text{c.p.m.}_{obsd} - \text{c.p.m.}_{bgrd})/E. \tag{16}$$

It has been found convenient in the author's laboratory to express all data as d.p.m. per ml of reaction mixture; values from replicate assays then are averaged to obtain the mean and standard error which in turn are used in subsequent manipulations. Raw data, including a measure of quench for each sample, are transferred from the counting machine via a local network to a central computer, where a simple programme is used to compute d.p.m. (eqn 16) and d.p.m./ml for each sample and mean d.p.m./ml (± SEM) for replicates. The standard error is expressed both in absolute units (d.p.m./ml) and as a percentage of the mean; the former is required for subsequent weighting of the data in some procedures, and the latter is useful in scanning the results for anomalous statistics. The efficiency is calculated from the coefficients of a cubic polynomial previously fitted to data obtained from a series of quenched standards. All counting machines will perform at least some of these operations, and some machines will perform all; thus, the requirement for a separate computer will vary from situation to situation. The quantity d.p.m./ml is an absolute measure of radioactivity and, once obtained, can serve as the dependent variable B_{obsd} for either empirical or mechanistic models. For mechanistic models, the function fitted to the data is eqn 17, in which $[P_B]$ is as defined in eqns 14 and 15 (mol per litre of reaction mixture), and SA is the specific activity of the radioligand in Ci/mmol:

$$B_{obsd} = [P_B] \cdot SA \cdot 2.22 \times 10^{12}. \tag{17}$$

The procedure described above departs from the preferred practice of analysing the data taken as measured; corrections for quench and background are made prior to the analysis, and neither quantity appears explicitly in the model being fitted. The consequences of prior manipulation relate primarily to the distortions that occur if the relative weight of individual members within a set of data is changed without a corresponding change in the weighting function. Subtraction of the background rate can distort the analysis in at least two ways. Firstly, the background rate is a measurement like any other, but it is assumed to be without error; if the estimate is wrong, the resulting fit will be biased accordingly. Secondly, the value of $c.p.m._{obsd}$ (eqn 16) is presumably dependent upon some variable, while $c.p.m._{bgrd}$ is generally assumed to be independent; thus the subtraction will alter, to some extent, the relationship between the measured error and what is taken as the dependent variable. An additional source of error relates to the possibility that estimates of $c.p.m._{bgrd}$ differ from sample to sample, perhaps as a consequence of different levels of quench.

The adverse consequences of subtracting $c.p.m._{bgrd}$ are generally negligible in practice, since the background rate is relatively small and is the same for all samples; moreover, the alternative would involve fitting a more complex function that incorporates eqn 16 and requires the inclusion of efficiency as an additional, independent variable. In the author's experience, however, erroneous estimates of $c.p.m._{bgrd}$ can lead to appreciable distortions when the total signal is comparable in magnitude with the background rate. This can be accommodated by fitting eqn 18, which differs from eqn 17 in that an empirical parameter $(B_{[P]_t=0})$ is included to reflect the discrepancy in $c.p.m._{bgrd}$

between the measured value and that consistent with the model as fitted to the rest of the data:

$$B_{obsd} = [P_B] \cdot SA \cdot 2.22 \times 10^{12} + B_{[P]_t=0}. \quad (18)$$

A final word here concerns the advantage gained when the specific signal of the probe is independent of its environment. The specific activity SA of a radioligand is the same for P_M, P_A and all PR_{Xj} in eqn 9, and thus for all three terms of eqn 14; accordingly, SA is a common factor that appears only in eqns 17 and 18. This simplification is likely to be inappropriate in analogous expressions for the binding of fluorescent ligands, whose quantum yield can be highly sensitive to the environment.

2.4 Properties of hydrophilic and lipophilic ligands

2.4.1 Non-specific and apparent specific binding

It is of interest at this point to consider the implications of eqn 14 and the circumstances in which it can be simplified further. Total binding tends to be uninformative unless non-specific binding from all sources is negligible at all desired concentrations of the probe. This is seldom achieved, and the investigator is generally obliged to measure total binding at least twice: once under the conditions of interest with respect to PR_{Xj} and once under conditions that prelude PR_{Xj}. The former is described by eqn 14; the latter is described by eqn 14 with $\sum_{j=1}^{n}[PR_{Xj}]$ equal to zero, as shown in eqn 19, and represents non-specific binding $P_{B,ns}$ from all sources:

$$[P_{B,ns}] = \frac{C_{PM}\wp f_M[P]_t}{f_A + \wp f_M} + \frac{C_{PA}f_A[P]_t}{f_A + \wp f_M}. \quad (19)$$

If the two measurements are made at the same total concentration of radioligand, the difference $P'_{B,sp}$ is related to specific binding $(f_X\sum_{j=1}^{n}[PR_{Xj}])$ according to eqns 20 and 21:

$$[P'_{B,sp}] = [P_B] - [P_{B,ns}] \quad (20)$$

$$[P'_{B,sp}] = f_X\sum_{j=1}^{n} C_{PRj}[PR_{Xj}] - \frac{f_X\sum_{j=1}^{n}[PR_{Xj}](C_{PA}f_A + C_{PM}\wp f_M)}{f_A + \wp f_M}. \quad (21)$$

If C_{PRj} is the same for all j, eqn 21 can be written as follows:

$$[P'_{B,sp}] = f_X\sum_{j=1}^{n}[PR_{Xj}]\left\{C_{PR} - \frac{C_{PA}f_A + C_{PM}\wp f_M}{f_A + \wp f_M}\right\}. \quad (22)$$

Eqn 22 indicates that the apparent and true levels of specific binding in such an experiment differ by a factor that depends upon the nature of the experiment. Perhaps the most favourable situation is one in which the probe is very hydrophilic ($\wp = 0$), binds via the aqueous phase and dissociates slowly from a receptor that is retained quantitatively during separation ($C_{PR} = 1$). The simplified form of eqn 22 then is

$$[P'_{B,sp}] = f_A \sum_{j=1}^{n} [PR_{Aj}](1 - C_{PA}), \tag{23}$$

and the true level of specific binding is only underestimated by the extent to which P_A is retained during separation. With a lipophilic probe, however, the discrepancy is determined by the partition coefficient $\wp$ and the fraction of the total volume occupied by the hydrophobic compartment (f_M). In the worst scenario, the product $\wp f_M$ is large relative to f_A, and the separation removes virtually all of P_A ($C_{PA} = 0$) without affecting P_M ($C_{PM} = 1$). Equation 22 then becomes

$$[P'_{B,sp}] = f_X \sum_{j=1}^{n} [PR_{Xj}](C_{PR} - C_{PM}). \tag{24}$$

Equation 24 demonstrates that apparent specific binding can be zero when most of the probe is either bound to the receptor or dissolved in the membrane. Any change in $\sum_{j=1}^{n}[PR_{Xj}]$ at constant $[P]_t$ will be offset by a comparable but opposite change in $f_M[P_M]$; total binding in the reaction mixture will thus be virtually unchanged, since the perturbation has merely redistributed the probe between the receptor and the membrane. Indeed, conditions expected to block binding to the receptor could lead to an increase in total binding if the probe were removed more readily from the receptor than from the membrane during separation of the aqueous and particulate phases (i.e. $C_{PR} < C_{PM}$).

Equations 19–24 demonstrate that it can be misleading to estimate specific binding simply as the difference between two estimates of total binding at the same total concentration of the probe. Unrecognized discrepancies between $[P'_{B,sp}]$ and $f_X\sum_{j=1}^{n}[PR_{Xj}]$ (eqn 21) will lead to erroneous estimates of maximal binding irrespective of whether free or total concentration is taken as the independent variable; similarly, errors can be introduced into estimates of affinity and other parameters, as described below. The preferred practice is therefore to avoid such manipulations and to analyse the data taken as measured; this requires eqn 14 or a similar expression that accounts for all states of the probe to the extent that they can be identified or surmized.

The accounting provided by eqn 14 is most accurate when separation of the aqueous phase is efficient ($C_{PA} < 0.01$) but exclusive ($C_{PM} = 1$, $C_{PR} = 1$). Also, the specific *signal* in any binding assay is in effect $[P'_{B,sp}]$, irrespective of whether specific *binding* is inferred from eqn 14 or obtained by calculating explicitly $[P'_{B,sp}]$ (e.g. eqns 20–24); accordingly, it is always beneficial to minimize the difference between $[P'_{B,sp}]$ and $f_X\Sigma^n_{j=1}[PR_{Xj}]$. Such differences reflect the direct loss of specifically bound probe (i.e. $C_{PR} < 1$) and the difference in non-specific binding that accompanies appreciable depletion of the probe through binding to the receptor. The latter is exacerbated if the ligand partitions into the membrane.

The best approach is to avoid depletion altogether, as illustrated in the example described below. When depletion is unavoidable, the most advantageous conditions are secured with slowly dissociating, hydrophilic ligands and separation by filtration. If dissociation is too rapid or if there is some leakage of the membranes through the filter (i.e. $C_{PR} < 1$), the value of C_{PR} can be determined independently and fixed accordingly in eqn 14. However, such corrections are often unnecessary when the membranes are separated by centrifugation; values of C_{PA} tend to be higher than with filtration, but the reaction mixture is maintained at equilibrium.

Similarly optimal conditions may not be practicable with lipophilic probes. When the partition coefficient is large, the quantity of probe dissolved in the membrane may approximate or exceed that bound specifically to the receptor. If receptor-related depletion is unavoidable, the investigator is faced with a dilemma: the difference between $[P'_{B,sp}]$ and $f_X\Sigma^n_{j=1}[PR_{Xj}]$ is small when C_{PM} is small (i.e. $C_{PM}/\wp f_M \ll C_{PR}$), as occurs typically when filters are washed extensively following application of the sample; in contrast, the accounting provided by eqn 14 is most accurate when C_{PM} equals unity, as when bound probe is separated by centrifugation.

Analyses in terms of eqn 14 require that the efficacy of separation (C_{PA}, C_{PM}, C_{PRj}), the relative volumes of distribution f_X and the relative solubility of the probe $\wp$ be known or implicit in the data. Some of this information is not available from experiments typically performed to define the parameters of specific binding (i.e. **a** in eqn 15). Either $\wp$ or f_X will generally have to be determined independently, but the other can then be calculated from estimates of non-specific binding, provided that the relative contribution of $C_{PM}\wp f_M$ and $C_{PA}f_A$ is known. If the measured levels of non-specific binding cannot be related to levels in the reaction mixture, the analysis will require independent estimates of both $\wp$ and f_X. These complications might suggest that the best approach is to minimize C_{PM} and thereby achieve the best estimate of $f_X\Sigma^n_{j=1}[PR_{Xj}]$. The problem is that $\wp$ and f_X are required to define at least some of the parameters in most models for the interaction of the ligand with the receptor (i.e. **a** in eqn 15), as illustrated below.

2.4.2 Parameters inferred from specific binding

In the example of eqn 14, it is assumed that access to the receptor is from either the aqueous phase or the membrane but not from both. Any tendency of the probe to localize in the phase from which access is denied will reduce the effective concentration with respect to the receptor. Failure to account for such effects can result in erroneous estimates of affinity and related parameters in whatever model is supplied as eqn 15. Experimental procedures are typically devised in part to achieve levels of non-specific binding below those present in the reaction mixture (i.e. $C_{PM} \ll 1$ and $C_{PA} \ll 1$ in eqn 19). The accompanying loss of information may preclude a complete accounting of the probe and thereby prevent the estimation of any parameters other than capacity.

These problems can be illustrated by considering the simple case of binding to a population of identical and mutually independent sites, as illustrated in Scheme I.

Scheme I

$$PR \underset{[P]/K_P}{\rightleftharpoons} R \overset{[A]/K_A}{\rightleftharpoons} AR$$

The receptor R can bind either the probe P or an unlabelled ligand A, and the complex PR corresponds to the quantity $\Sigma_{j=1}^{n}[PR_{Xj}]$ in eqn 14. Other parameters or variables are defined in eqns 25–29, where the subscript X denotes access from the aqueous phase A or the membrane M:

$$K_P = [P_X][R_X]/[PR_X] \tag{25}$$

$$K_A = [A_X][R_X]/[AR_X] \tag{26}$$

$$[P]_t = f_X[PR_X] + f_A[P_A] + f_M[P_M] \tag{27}$$

$$[A]_t = f_X[AR_X] + f_A[A_A] + f_M[A_M] \tag{28}$$

$$[R]_t = f_X[PR_X] + f_X[AR_X] + f_X[R_X]. \tag{29}$$

If one assumes for simplicity that P and A are identical with respect to partition coefficient and affinity (i.e. $K_P = K_A$), as might occur with tritiated and unlabelled analogues of the same ligand, the expression corresponding to eqn 15 is as follows (eqn 30):

$$f_X[PR_X] = f([P]_t, [A]_t, [R]_t, K_P, f_A, f_M, \wp). \tag{30}$$

From eqns 25–28, it can be shown that

$$[R_X] = [PR_X]K_P/[P_X] \tag{31}$$

$$[AR_X] = [PR_X][A]_t/[P]_t. \tag{32}$$

The value of $[P_X]$ in eqn 31 is given by eqn 33 or 34, depending upon whether access to the receptor is from the aqueous phase ($X \equiv A$) or the membrane ($X \equiv M$) respectively:

$$[P_A] = \frac{[P]_t - f_A[PR_A]}{f_A + \wp f_M} \tag{33}$$

$$[P_M] = \frac{w([P]_t - f_M[PR_M])}{f_A + \wp f_M}. \tag{34}$$

Substituting for $[R_X]$ (eqn 31) and $[AR_X]$ (eqn 32) in eqn 29 yields the quadratic polynomial shown as eqn 35, and the root corresponding to $f_X[PR_X]$ can be evaluated numerically as described in Section 3.2:

$$f_X^2[PR_X]^2([A]_t + [P]_t) - f_X[PR_X][P]_t([R]_t + [A]_t + [P]_t + K_P') \\ + [P]_t^2[R]_t = 0. \tag{35}$$

The parameter K_P' (eqn 35) represents the apparent affinity in units of total concentration of the ligand; it is related to K_P (eqn 25) according to eqn 36 or 37, depending upon whether access to the receptor is from the aqueous phase or the membrane:

$$K_P' = K_P(f_A + \wp f_M) \qquad \text{when } X \equiv A \tag{36}$$

$$K_P' = K_P(f_A + \wp f_M)/\wp \qquad \text{when } X \equiv M. \tag{37}$$

The relationships between K_P' and K_P illustrate the complications that can arise when only the total concentration of the ligand is known. Since K_P' depends upon $\wp$ and f_M, the apparent affinity will depend upon the quantity of membrane whenever appreciable amounts of the ligand are found in both the aqueous and non-aqueous phases. This effect may underlie the common observation that affinities tend to be higher at lower concentrations of protein (see, for example, ref. 1).

The parameters $\wp$ and f_M can be estimated independently from non-specific binding under appropriate conditions (eqn 19), and the values can then be substituted in eqns 14 and 35 during subsequent analyses. This approach only succeeds to the extent that non-specific binding following separation of the bound probe can be related quantitatively to that in the reaction mixture; in particular, reliable estimates of parameters such as C_{PM} and C_{PA} (eqn 19) must be available for each experiment. Eqn 14 avoids the explicit subtraction of non-specific binding (e.g. eqn 21 or 22), but does not in itself address the difference between K_P' and K_P (eqns 36 and 37). The latter must be incorporated explicitly in the model used to describe specific binding (eqn 15 or, in the present example, eqn 35) if the ligand is to enter as total concentration.

It is clear from the foregoing that a quantitative description of the data is easier if one knows the concentration of unbound ligand in the phase that gives access to the receptor. When access is from the aqueous phase, the simplest approach is to use hydrophilic ligands. Eqn 14 is much simplified when the partition coefficient is zero and f_A is essentially unity; similarly, the parameters of eqn 15 can be taken at face value. When lipophilic ligands are unavoidable, the best strategy is generally to work at relatively low concentrations of membrane and receptor. If $f_M[P_M]$ and $f_A\sum_{j=1}^{n}[PR_{Aj}]$ together constitute a negligible fraction of $[P]_t$, the latter can be taken as equal to $[P_A]$. Total binding is then described by eqn 38 (cf. eqn 9) or, in the example of Scheme I, by eqn 39 ($K_A = K_P$):

$$[P_B] = C_{PM}\wp f_M[P_A] + C_{PA}f_A[P_A] + f_A \sum_{j=1}^{n} C_{PRj}[PR_{Aj}] \tag{38}$$

$$[P_B] = C_{PM}\mathcal{P}f_M[P_A] + C_{PA}f_A[P_A] + f_A C_{PR} \frac{[R]_t[P_A]}{[P_A] + [A_A] + K_P}. \quad (39)$$

Since probe in the aqueous phase has direct access to the membrane on the one hand and to the receptor on the other, the equations need not reflect the mutual depletion between $f_M[P_M]$ and $f_A\Sigma_{j=1}^n[PR_{Aj}]$ in the reaction mixture. Non-specific binding is therefore a function of $[P_A]$ and the quantity $C_{PM}\mathcal{P}f_M + C_{PA}f_A$ but not of $\Sigma_{j=1}^n[PR_{Aj}]$ (eqn 40); conversely, specific binding is a function of $\mathbf{x}$ and $\mathbf{a}$ as defined by the model (eqn 15) but not of $f_A + \mathcal{P}f_M$ (eqn 41):

$$[P_{B,ns}] = [P_A](C_{PM}\mathcal{P}f_M + C_{PA}f_A) \quad (40)$$

$$[P_{B,sp}] = f_A C_{PR} \frac{[R]_t[P_A]}{[P_A] + [A_A] + K_P}. \quad (41)$$

An appreciable difference between $[P]_t$ and $[P_A]$ may be unavoidable in some systems. If the ratio of receptor to lipid cannot be increased, as in native membranes, depletion due to $f_M[P_M]$ may only be negligible at unacceptably low concentrations of receptor; in any system, attempts to minimize depletion due to $f_A\Sigma_{j=1}^n[PR_{Aj}]$ may similarly be restricted by a lower limit on $[R]_t$. If the reaction mixture can be of any volume, the lower limit on $[R]_t$ is determined by the requirement that the reaction time be both practicable and short relative to processes such as proteolysis and thermal denaturation. In most assays, however, there is an upper limit on the volume of liquid that can be processed; the lower limit on $[R]_t$ is then probably determined by the specific radioactivity of the probe.

If depletion of the probe is unavoidable and eqn 14 impractical, the free concentration of the probe must be measured directly. Such measurements are technically easier when the method of separation does not perturb the distribution of the probe among the different states and when the aqueous phase is readily accessible. With centrifugation of membranes, for example, the value of $[P_A]$ can be estimated by counting an aliquot of the supernatant fraction. When high levels of non-specific binding require that the reaction mixture be filtered, the estimation of $[P_A]$ is complicated by any failure to retain 100% of the receptors and by the need to collect the initial filtrate separately from that of subsequent washes; in such systems, it may be preferable to estimate bound probe by filtration and $[P_A]$ by centrifugation in a parallel assay. Even in ostensibly well behaved systems, when depletion seems to be negligible or when binding can be expressed as a function of $[P]_t$ (eqn 14), it is advisable to measure $[P_A]$ at least once and to confirm that all states of the probe have indeed been recognized.

Direct measurement of the unbound probe is tedious but relatively straightforward; the free concentration of an unlabelled analogue can be calculated from the same data provided that the two ligands are otherwise identical.

Other ligands are presumably without effect on the non-specific terms of eqn 14, but appreciable dissolution in the membrane will be reflected in at least some of the parameters pertaining to specific binding (i.e. **a** in eqn 15). Since direct measurement is likely to involve analytical procedures such as HPLC, appropriate adjustments to the model may be the only practical means of dealing with the depletion of unlabelled ligands.

3. Examples of mechanistic models

3.1 General comments

A mechanistic model is a quantitative formulation of a hypothesis. The simulated behaviour of such models can be used both to suggest experiments and to test the hypothesis; the latter is typically achieved by assessing a best fit of the model to experimentally acquired data. If the fit is acceptable, the hypothesis remains tenable until further notice; if the fit is unacceptable, it is the business of the investigator to identify the source of the discrepancy. It is important to recognize that the hypothesis is not limited to interactions with receptor (i.e. eqn 15) but includes all contributions to the measured signal (e.g. eqn 14). Problems attributable unequivocally to the former raise the central question of whether the proposed mechanism of specific binding is fundamentally flawed or merely incomplete. Lack of agreement may also be artefactual, however, in the sense that the conditions of the experiment violate assumptions in a model that is otherwise correct; common examples include the relatively steep binding curves that occur when depletion of a ligand goes unrecognized or when the system is falsely assumed to be at thermodynamic equilibrium (see Section 4.1.2).

Models based on different mechanisms can predict similar behaviour, and some artefacts can mimic such mechanistic properties as co-operativity. The quality of the data is therefore crucial to the success of model-based interpretations. Moments of truth are lost through uncertainty: binding data have little discriminatory potential when the standard error on multiple determinations is of the order of 10%; in contrast, very subtle differences can be discerned at a standard error of 1%. If the mechanistic hypothesis is indeed correct, capricious data may only result in parameters of dubious numerical value. If the hypothesis is wrong, however, tell-tale discrepancies with the data can be obscured by experimental error; moreover, parameters may have a physical significance quite different from that ascribed to them in the model.

No mechanism can be considered final; for some receptors, even the broad principles that govern binding of the agonist and appearance of the response remain unclear. Hypotheses and their corresponding models are thus subject to continuous modification and refinement, and the investigator ought to be wary of prepackaged schemes that purport to account for the properties of his receptor. At the very least, he wants to be fully aware of the properties and limitations of such schemes, particularly as they relate to the conditions of his own experiments; at best, he will build a model exactly to his specifications and then will modify it from time to time as required.

With eqn 14 or a similar expression in hand, the practical requirement is for a function or set of functions that permit one to compute $\sum_{j=1}^{n}[PR_{Xj}]$ according to the putative mechanism (i.e. eqn 15). The nature of the problem relates to whether or not a single equation of state is sufficient to describe the system. Formulation of the model is relatively straightforward when specific interactions account for an appreciable fraction of only one species, usually the receptor; the binding function is linear in $\sum_{j=1}^{n}[PR_{Xj}]$, and an explicit expression can therefore be obtained for eqn 15. When specific interactions deplete two or more species, the computation of $\sum_{j=1}^{n}[PR_{Xj}]$ generally involves evaluating the roots of one or more polynomials. The models described below are representative of the mechanistic schemes most widely used to rationalize the binding properties of receptors. It is assumed throughout that the system is at thermodynamic equilibrium, and time therefore is irrelevant.

With the mechanistic schemes typically of interest, higher degree or multiple polynomials invariably arise when an appreciable but unknown fraction of one or more ligands is bound to the receptor. The extent of such depletion is often in the hands of the investigator, but the situation is only avoidable when practicable concentrations of the receptor can be small relative to its equilibrium dissociation constant for the ligand. With schemes that involve reversible interactions among components of the membrane, the depletion of unbound species may be outside experimental control. Higher degree or multiple polynomials may then arise irrespective of efforts to avoid the depletion of unbound ligands.

When two or more equations of state are required, explicit expressions for eqn 15 tend to be unsatisfactory or unavailable. Numerical methods are therefore in order, and the problem can be approached in essentially two ways: the quantity $\sum_{j=1}^{n}[PR_{Xj}]$ can be computed numerically from appropriate expressions in which depleted species appear as total concentration; alternatively, numerical methods can be used to compute the free concentrations from the total concentrations, and $\sum_{j=1}^{n}[PR_{Xj}]$ in turn can be calculated from explicit expressions. Since the desired values must lie within known limits, the Newton–Raphson procedure is usually effective in providing the right answer with reasonable dispatch. That procedure thus constitutes the basis of numerical solutions suggested for the models described below.

Several examples are presented in which the relevant quantity is one root of a single polynomial. Explicit solutions may be available, but they are not generally recommended owing to the errors that can arise from loss of significance. With quadratic polynomials, for example, the formula for the roots includes a subtraction; the results can be meaningless when the difference is comparable with or less than the precision of the computation. Numerical methods are not without their own problems, as illustrated below, but they can succeed when explicit solutions are unreliable; also, the formulae for the roots of cubic and quartic polynomials are cumbersome. There is no general expression for the roots of quintic and higher degree polynomials, and numerical methods are unavoidable in those situations.

Explicit and numerical solutions are shown for all the mechanistic schemes described below. The explicit solutions are based on the premise that the receptor is the only species to undergo an appreciable change in free concentration over the course of the binding curve. Two or more algebraically interchangeable forms are shown for each scheme; some are more computationally efficient, while others provide further insight into the nature and properties of the model. Numerical methods generally allow for considerable scope in the approach to a problem, but they can also fail for reasons that may initially be obscure. That versatility and some of the problems are illustrated with various examples, particularly in the case of the ever popular, multi-site model. It is assumed throughout that all ligands are found only in the aqueous phase of the reaction mixture and that the non-aqueous phase contributes negligibly to the total volume (i.e. $f_A = 1$ and $\wp = 0$ in eqn 14).

A final comment here concerns scaling factors and their usefulness when the analysis involves multiple sets of data. It often happens that a parametric value is to be constrained in relative but not absolute terms. Different concentrations of protein and hence receptor may have been used in different experiments; if the membranes were from the same preparation throughout, it may be appropriate to assign a common value of $[R]_t$ to all of the data and then to scale it appropriately for the data from each experiment. Similarly, the investigator may know or may wish to explore the relative values of a dissociation constant or any other parameter among data from different experiments. Any parameter in eqn 14, the models described below, or the empirical models described in Section 4 might be associated with a scaling factor for such purposes.

3.2 Root-finding

The Newton–Raphson procedure, or Newton's method, is an iterative process that begins with an initial value and converges upon the root via a series of intermediate approximations. In the example of Scheme I, specific binding of the probe is one root (x_r) of a quadratic polynomial (eqn 42) (cf. eqn 35, $[A]_t = 0$, $\wp = 0$, $f_A \approx 1$):

$$f(x) = \varphi_0 + \varphi_1 x + \varphi_2 x^2, \tag{42}$$

where $\varphi_0 = [P]_t[R]_t$

$\varphi_1 = -([P]_t + [R]_t + K_P)$

$\varphi_2 = 1.$

The function is illustrated in *Figure 1* for conditions under which [PR$_A$] is 0.8 nM and represents 80% of both [P]$_t$ and [R]$_t$. The determination of x_r by Newton's method begins with an estimate such as 0.2 nM (i.e. x_0 in *Figure 1*).

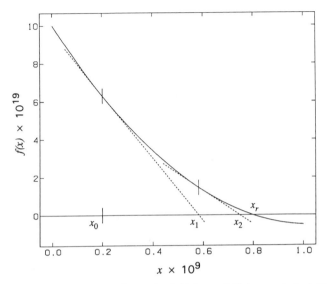

Figure 1. Root-finding by Newton's method. Values of $f(x)$ were calculated according to eqn 42 to obtain the curve shown in the figure; the contributions to ϕ_i were as follows: [P]$_t$ = 1.0 nM, [R]$_t$ = 1.0 nM and K_p = 0.05 nM. The dotted lines illustrate the tangents to the curve at $x = x_0$ and $x = x_1$. Labelled values on the abscissa are as follows: x_0 = 0.20, x_1 = 0.58, x_2 = 0.75, x_r = 0.80.

One then constructs the tangent to the curve at $x = x_0$ and finds the x-intercept (i.e. $x_1 = 0.58$ nM in *Figure 1*), which is used in turn to obtain the tangent and corresponding intercept at $x = x_1$. The process is repeated until the difference between two successive values of x_i is acceptably small. In practice, each value is calculated from its predecessor as follows:

$$x_{i+1} = x_i + \delta x.$$

An expression for the increment δx is obtained from Taylor's series by taking only the linear terms and setting $f(x + \delta)$ equal to zero:

$$\delta x = -\frac{f(x_i)}{f'(x_i)}. \tag{43}$$

When $f'(x)$ is well behaved at all physically relevant x, the investigator will not be inconvenienced by first derivatives equal to zero. Such behaviour is not guaranteed, however, and it is prudent either to design the algorithm such that the problem is avoided or to test $f'(x)$ before performing the division.

Left on its own, Newton's method will continue indefinitely, eventually reaching a point after which the value of δx from each cycle is effectively zero. The process therefore must be stopped when δx is negligible according to some predetermined criterion. For non-zero roots, convergence can usually be assessed by computing the relative change ε in successive values of x_i (eqn 44); a conservative practice would be to stop iterating when ε is less than 10^{-5}:

$$\varepsilon = \left| \frac{x_{i+1} - x_i}{x_{i+1} + x_i} \right|. \tag{44}$$

Convergence to zero is not a trivial matter, owing to increasing uncertainty over the value of x as it approaches the precision of the processor. The resolution of the problem depends in part upon whether one is solving for a complex, as in the example of *Figure 1*, or for an unbound species (e.g. eqns 72–74 below). A root of zero is inevitable in the absence of receptor or the relevant ligand; such conditions can be identified in advance, and Newton's method can be avoided. In other situations, the proximity of the root to zero will depend upon the parameters and variables (i.e. x and a in eqn 15, φ_j in eqn 42). If one is solving for a complex in a model that includes an unlabelled ligand (e.g. $f_X[PR_X]$ in eqn 35), the root will approach zero as the concentration of that ligand becomes saturating; if one is solving for the free concentrations of one or more ligands, the root approaches zero as depletion of the ligand becomes complete. Provided that the process is indeed converging to zero, these and similar situations can be handled by taking the root as zero at the first negative value of x_i.

Absolute convergence criteria are more general than relative criteria, but the tolerance must be specified in units relevant to the expected value of the root. In binding studies, any root must lie between zero (x_{min}) and the total concentration of either receptor or ligand (x_{max}). The tolerance for convergence therefore can be taken as $\varepsilon'(x_{min} + x_{max})/2$, where ε' is the precision of the processor (2).

When there is one polynomial in one variable, as in the example of *Figure 1*, the problem is said to be one-dimensional. In practice, however, the investigator is often required to evaluate several variables. When the probe forms more than one complex, it may not be desirable or even possible to obtain a single polynomial in which x is total specific binding and all ligands appear in units of total concentration; the concentrations of n complexes must then be evaluated simultaneously from a set of n polynomials. If the investigator has opted to solve for the free concentrations of interacting species, rather than for their corresponding complexes, there will be up to one polynomial for every species to undergo depletion.

Newton's method is readily applied to multiple equations of the sort encountered in binding studies, and a system of n functions in n variables is said to pose a problem in n dimensions. Values of δx_j for the set of n variables at each iteration are computed from a system of linear equations defined by the functions and their partial derivatives with respect to x_j. The solution can generally be obtained by standard analytical methods such as Gaussian elimination with pivoting, Gauss–Jordan elimination or LU decomposition. In extreme circumstances, however, direct methods can fail through loss of significance; such failure must be recognized and trapped by the programme, which can then back up and try again. In the author's experience, the iterative Gauss–Seidel procedure is almost always successful when other methods fail.

3.3 The multi-site model

In this frequently cited model, one or more ligands (L_i, where $i = 1, 2, \ldots, m$) compete for distinct and mutually independent sites (R_j, where $j = 1, 2, \ldots, n$) differentiated by their equilibrium dissociation constant for the ligand ($K_{ij} = [L_i][R_j]/[L_iR_j]$). The various states of R for two ligands and n classes of sites are shown in Scheme II.

Scheme II

$$L_1R_1 \underset{[L_1]/K_{11}}{\rightleftharpoons} R_1 \xrightarrow{[L_i]/K_{i1}} L_iR_1$$

$$L_1R_2 \underset{[L_1]/K_{12}}{\rightleftharpoons} R_2 \xrightarrow{[L_i]/K_{i2}} L_iR_2$$

$$\vdots \qquad\qquad \vdots \qquad\qquad \vdots$$

$$L_1R_j \underset{[L_1]/K_{1j}}{\rightleftharpoons} R_j \xrightarrow{[L_i]/K_{ij}} L_iR_j$$

$$\vdots \qquad\qquad \vdots \qquad\qquad \vdots$$

$$L_1R_n \underset{[L_1]/K_{1n}}{\rightleftharpoons} R_n \xrightarrow{[L_i]/K_{in}} L_iR_n$$

3.3.1 Explicit expressions

Total specific binding of a radioligand or other probe, shown here as L_1, is described in eqns 45–47 as a function of the free concentrations of all ligands present. The three expressions are equivalent and have been rearranged to illustrate the functional form with respect to the concentration of the probe (L_1, eqn 45) and an unlabelled ligand (L_2, eqns 46 and 47):

$$\sum_{j=1}^{n} [L_1R_j] = \sum_{j=1}^{n} \frac{[R_j]_t[L_1]}{[L_1] + K_{1j}\{1 + [L_2]/K_{2j} + \sum_{i=3}^{m} [L_i]/K_{ij}\}} \tag{45}$$

$$\sum_{j=1}^{n} [L_1R_j] = \sum_{j=1}^{n} \left\{ \frac{[R_j]_t[L_1]}{[L_1] + K'_{1j}} \cdot \frac{K'_{2j}}{[L_2] + K'_{2j}} \right\} \tag{46}$$

$$\sum_{j=1}^{n} [L_1 R_j] = \left\{ \sum_{j=1}^{n} \frac{[R_j]_t [L_1]}{[L_1] + K'_{1j}} \right\} \left\{ \sum_{j=1}^{n} \frac{F'_j K'_{2j}}{[L_2] + K'_{2j}} \right\}. \tag{47}$$

where

$$K'_{1j} = K_{1j} \left\{ 1 + \sum_{i=3}^{m} \frac{[L_i]}{K_{ij}} \right\}$$

$$K'_{2j} = K_{2j} \left\{ 1 + \frac{[L_1]}{K_{1j}} + \sum_{i=3}^{m} \frac{[L_i]}{K_{ij}} \right\}$$

$$F'_j = \frac{[R_j]_t [L_1]/([L_1] + K'_{1j})}{\sum_{j=1}^{n} [R_j]_t [L_1]/([L_1] + K'_{1j})}$$

$$[R_j]_t = [R_j] + \sum_{j=1}^{m} [L_i R_j].$$

If only the concentration of the radioligand is varied, the expression for specific binding is a rational function as in eqn 48; the coefficients φ'_j and φ_j are defined by the free concentrations of all ligands except the probe ($[L_i]$, $i = 2$, $3, \ldots, m$), by the total concentrations of all sites ($[R_j]_t$, $j = 1, 2, \ldots, n$) and by the dissociation constants for the various equilibria K_{ij}:

$$\sum_{j=1}^{n} [L_1 R_j] = \frac{\sum_{j=1}^{n} \varphi'_j [L_1]^j}{1 + \sum_{j=1}^{n} \varphi_j [L_1]^j}. \tag{48}$$

Equation 49 is the corresponding expression when the concentration of an unlabelled ligand L_2 is varied at fixed concentrations of the probe and all other ligands; the coefficients ψ'_j and ψ_j are defined by the free concentrations of the probe ($[L_1]$) and unlabelled ligands other than L_2 ($[L_i]$, $i = 3, 4, \cdots$, m), by the total concentrations of all sites ($[R_j]_t$, $j = 1, 2, \cdots, n$) and by the equilibrium dissociation constants K_{ij}:

$$\sum_{j=1}^{n} [L_1 R_j] = \frac{\sum_{j=0}^{n-1} \psi'_j [L_2]^j}{1 + \sum_{j=1}^{n} \psi_j [L_2]^j}. \tag{49}$$

Equations 50–53 and 54–57 describe the coefficients for the specific example in which a labelled and unlabelled ligand compete for two classes of sites (i.e. $m = 2$ and $n = 2$):

$$\varphi'_1 = \frac{[R_1]_t/K_{11}}{1 + [L_2]/K_{21}} + \frac{[R_2]_t/K_{12}}{1 + [L_2]/K_{22}} \tag{50}$$

$$\varphi_2' = \frac{([R_1]_t + [R_2]_t)/K_{11}K_{12}}{\{1 + [L_2]/K_{21}\}\{1 + [L_2]/K_{22}\}} \tag{51}$$

$$\varphi_1 = \frac{1/K_{11}}{1 + [L_2]/K_{21}} + \frac{1/K_{12}}{1 + [L_2]/K_{22}} \tag{52}$$

$$\varphi_2 = \frac{1/K_{11}K_{12}}{\{1 + [L_2]/K_{21}\}\{1 + [L_2]/K_{22}\}} \tag{53}$$

$$\psi_0' = [L_1]\left\{\frac{[R_1]_t/K_{11}}{1 + [L_1]/K_{11}} + \frac{[R_2]_t/K_{12}}{1 + [L_1]/K_{12}}\right\} \tag{54}$$

$$\psi_1' = \frac{[R_1]_t[L_1]/K_{11}K_{21} + [R_2]_t[L_1]/K_{12}K_{22}}{\{1 + [L_1]/K_{11}\}\{1 + [L_1]/K_{12}\}} \tag{55}$$

$$\psi_1 = \frac{1/K_{21}}{1 + [L_1]/K_{11}} + \frac{1/K_{22}}{1 + [L_1]/K_{12}} \tag{56}$$

$$\psi_2 = \frac{1/K_{21}K_{22}}{\{1 + [L_1]/K_{11}\}\{1 + [L_1]/K_{12}\}}. \tag{57}$$

Although eqns 45–49 are interchangeable, the last two are less satisfactory for practical purposes. Considerable algebraic manipulation is required to obtain the appropriate coefficients, and the computation of $\Sigma_{j=1}^n[L_1R_j]$ involves a relatively large number of floating point operations. In contrast, eqn 45 or 46 is readily obtained for any number of ligands and sites, and the number of computational operations is minimal. Other algebraic forms of the model tend to pose the problems found with eqns 48 and 49, and some may yield intermediate values that exceed the limits imposed by the architecture of the computer. An example of the latter pitfall is illustrated by eqn 58, where the values of products such as $\Pi_{i=1}^m K_{ij}$ (i.e. $K_{1j}K_{2j} \cdots K_{mj}$) may result in numerical underflow if the individual parameters are not scaled appropriately prior to the multiplication:

$$\sum_{j=1}^n [L_1R_j] = \sum_{j=1}^n \frac{[R_j]_t[L_1]\prod_{i=2}^m K_{ij}}{\prod_{i=1}^m K_{ij} + [L_1]\prod_{i=2}^m K_{ij} + \sum_{i=2}^m [L_i]\prod_{\substack{k=1\\k\neq i}}^m K_{kj}}. \tag{58}$$

A special case of Scheme II is that in which two of the ligands exhibit the same affinity, as is found typically with the protonated and tritiated analogues of the same compound (i.e. cold versus hot). Total specific binding then can be described by eqn 59, which is identical with eqn 45 except for the restriction that K_{1j} and K_{2j} both equal K_j:

$$\sum_{j=1}^{n} [L_1 R_j] = \sum_{j=1}^{n} \frac{[R_j]_t [L_1]}{[L_1] + [L_2] + K_j \{1 + \sum_{i=3}^{m} [L_i]/K_{ij}\}}. \qquad (59)$$

If L_1 and L_2 are the only ligands present, eqn 45 and 46 reduce to eqns 60 and 61 respectively:

$$\sum_{j=1}^{n} [L_1 R_j] = \sum_{j=1}^{n} \frac{[R_j]_t [L_1]}{[L_1] + EC_{50(j)}} \qquad (60)$$

$$\sum_{j=1}^{n} [L_1 R_j] = \sum_{j=1}^{n} \left\{ \frac{[R_j]_t [L_1]}{[L_1] + K_j} \cdot \frac{IC_{50(j)}}{[L_2] + IC_{50(j)}} \right\}, \qquad (61)$$

where $EC_{50(j)} = [L_2] + K_j$

   $IC_{50(j)} = [L_1] + K_j$.

It is noteworthy that eqn 60 or 61 permits the simultaneous estimation of both $[R_j]_t$ and K_j from experiments at a single concentration of the probe $[L_1]$ and graded concentrations of the unlabelled ligand $[L_2]$. The two parameters can be highly correlated, however, and reliable estimates may require data at multiple concentrations of both ligands; also, compromises may be unavoidable when there are two or more classes of sites. The value of K_j is better defined than that of $[R_j]_t$ when $[L_1]$ is small (i.e. $[L_1] \ll K_j$); in contrast, $[R_j]_t$ is better defined than K_j when $[L_1]$ is large (i.e. $[L_1] \gg K_j$). At low concentrations of the probe, the value of $[R_j]_t$ will be linearly dependent upon K_j when the concentration $[L_2]$ is varied over a range sufficient to define inhibitory potency; since $IC_{50(j)}$ and the attendant error are defined on a geometric scale in $[L_2]$, estimates of $[R_j]_t$ can be highly uncertain for practical purposes. At higher concentrations of the probe, the inhibitory potency approaches $[L_1]$ in value; K_j is estimated from the difference and thus becomes increasingly ill defined. A further problem at higher concentrations of the probe relates to the inexorable increase in the relative contribution of non-specific binding to the total signal (cf. eqn 14). A final comment regarding eqns 59–61 relates to the requirement that the inhibitory potency of the unlabelled ligand exceed the concentration of the radioligand; the fitting procedure may fail if the data and the proffered value of $[L_1]$ are inconsistent on this score.

There is an important difference between distinct sites as illustrated in Scheme II and multiple states of the same site as illustrated below in Scheme III. In the latter model, the sites interconvert spontaneously from one state to another; there is no interconversion in the former, at least on the time-scale of the binding assays.

Scheme III

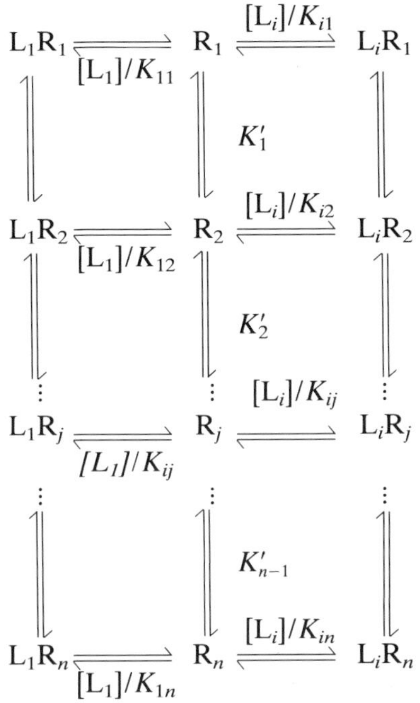

$$
\begin{array}{ccccc}
L_1R_1 \;\underset{[L_1]/K_{11}}{\rightleftharpoons}\; R_1 & \xrightleftharpoons{[L_i]/K_{i1}} & L_iR_1 \\[1.5em]
\Big\updownarrow & \Big\updownarrow\; K_1' & \Big\updownarrow \\[1.5em]
L_1R_2 \;\underset{[L_1]/K_{12}}{\rightleftharpoons}\; R_2 & \xrightleftharpoons{[L_i]/K_{i2}} & L_iR_2 \\[1.5em]
\Big\updownarrow & \Big\updownarrow\; K_2' & \Big\updownarrow \\[1.5em]
\vdots & \vdots & \vdots \\[0.5em]
L_1R_j \;\underset{[L_1]/K_{ij}}{\rightleftharpoons}\; R_j & \xrightleftharpoons{[L_i]/K_{ij}} & L_iR_j \\[1.5em]
\vdots & \vdots & \vdots \\[0.5em]
\Big\updownarrow & \Big\updownarrow\; K_{n-1}' & \Big\updownarrow \\[1.5em]
L_1R_n \;\underset{[L_1]/K_{1n}}{\rightleftharpoons}\; R_n & \xrightleftharpoons{[L_i]/K_{in}} & L_iR_n
\end{array}
$$

Total specific binding of the probe is described by eqn 62, in which $K_j' = [R_j]/[R_{j+1}]$, $K_{ij} = [L_i][R_j]/[L_iR_j]$, $[R]_t = \Sigma_{j=1}^n [R_j]_t$ and $[R_j]_t = [R_j] + \Sigma_{i=1}^m [L_iR_j]$:

$$
\sum_{j=1}^{n} [L_1R_j] = \frac{[R]_t[L_1]}{[L_1] + K_{obsd}}, \tag{62}
$$

where

$$
K_{obsd} = \frac{\displaystyle\sum_{j=1}^{n}\left\{1+\sum_{i=2}^{m}[L_i]/K_{ij}\right\}\prod_{k=1}^{j} 1/K_{k-1}'}{\displaystyle\sum_{j=1}^{n} 1/K_{1j}\sum_{k=1}^{j} 1/K_{k-1}'}
$$

$$
K_o' = 1.
$$

Eqn 62 is a rectangular hyperbola for any value of n or m, provided that the free concentration of only one ligand is varied over the course of the experiment; the same result is obtained with more highly branched schemes, which differ only in the definition of K_{obsd}. A population of interconvertible sites will thus appear homogeneous irrespective of the number of conformational equilibria that may occur. The differences among R_j may be intrinsic to the receptor or may reflect interactions with protons, metal ions and other ingredients present at constant concentrations; conformity with eqn 62 requires only that all reactions occur spontaneously and rapidly on the time scale of a binding assay. When two or more R_j do not interconvert or interconvert slowly, the functional form is that of eqn 48 (Scheme II).

3.3.2 One-dimensional numerical solutions

Equations 45–62 require the free concentrations of all ligands, which must therefore be measured if reduced appreciably through binding to the receptor. This inconvenience can be avoided if the model is formulated in terms of total concentrations, but the resulting equations are no longer linear in $\Sigma_{j=1}^{n}[L_1R_j]$. The binding function is a single polynomial for relatively simple variants of Scheme II, and some common examples are listed below. The derivations involve substituting from the equilibrium ratio (i.e. $K_{ij} = [L_i][R_j]/[L_iR_j]$) to eliminate the concentrations of free ligand (i.e. $[L_i] = [L_i]_t - \Sigma_{j=1}^{n}[L_iR_j]$) and unlabelled complexes (i.e. $[L_iR_j]$, where L_i is not the probe) from the equation of state for each class of sites (i.e. $[R_j]_t = [R_j] + \Sigma_{i=1}^{m}[L_iR_j]$). Subsequent expansion of the expression for $[R]_t$ (i.e. $[R]_t = \Sigma_{j=1}^{n}[R_j]_t$) and collection of the terms in $\Sigma_{j=1}^{n}[L_1R_j]$, where L_1 is the probe, yields the desired polynomial. These manipulations can be tedious, as illustrated by the 54 terms that constitute the expanded form of eqn 67 (see below), and errors tend to occur as the investigator's patience flags. Commercially available programmes such as *Mathematica* (Wolfram Research, Inc., Champaign, IL) and *Macsyma* (Symbolics, Inc. Burlington, MA) are thus recommended for manipulating the symbols in all but the most trivial derivations.

With one ligand L_1 and a homogeneous population of sites R_1, the concentration of the complex L_1R_1 is obtained as one root of a quadratic:

$$[R_1]_t[L_1]_t - ([R_1]_t + [L_1]_t + K_{11})[L_1R_1] + [L_1R_1]^2 = 0, \qquad (63)$$

where $\quad [R_1]_t = [R_1] + [L_1R_1]$

$$[L_1]_t = [L_1] + [L_1R_1]$$

$$K_{11} = [L_1][R_1]/[L_1R_1].$$

With one ligand and two classes of sites (i.e. $n = 2$), the polynomial is cubic in $\Sigma_{j=1}^{n}[L_1R_j]$ (eqn 64); analogous expressions can be derived for three or more classes:

$$\varphi_0 + \varphi_1[L_1]_b + \varphi_2[L_1]_b^2 + \varphi_3[L_1]_b^3 = 0, \qquad (64)$$

where

$$\varphi_0 = -[L_1]_t\{[R_1]_t([L_1]_t + K_{12}) + [R_2]_t([L_1]_t + K_{11})\}$$

$$\varphi_1 = ([L_1]_t + K_{11})([L_1]_t + K_{12}) + [R_1]_t(2[L_1]_t + K_{12}) + [R_2]_t(2[L_1]_t + K_{11})$$

$$\varphi_2 = -([R_1]_t + [R_2]_t + 2[L_1]_t + K_{11} + K_{12})$$

$$\varphi_3 = 1$$

$$[L_1]_b = [L_1R_1] + [L_1R_2].$$

Expressions that describe the binding of a single ligand are of limited usefulness in practice. Most experiments involve one or more ligands in addition to the probe; the unlabelled ligand itself may be studied at various concentrations, or it may be added only to define the level of non-specific binding. If a probe L_1 and an unlabelled ligand L_2 compete for a single class of sites R_1, the corresponding polynomial is cubic in $[L_1R_1]$ (eqn 65):

$$\varphi_0 + \varphi_1[L_1R_1] + \varphi_2[L_1R_1]^2 + \varphi_3[L_1R_1]^3 = 0, \tag{65}$$

where $\varphi_0 = [R_1]_t[L_1]_t^2 K_{21}$

$$\varphi_1 = [L_1]_t\{[R_1]_t(K_{11} - 2K_{21}) - [L_2]_t K_{11} - [L_1]_t K_{21} - K_{11}K_{21}\}$$

$$\varphi_2 = -\{(K_{11} - K_{21})([R_1]_t + K_{11}) + [L_1]_t(K_{11} - 2K_{21}) - [L_2]_t K_{11}\}$$

$$\varphi_3 = K_{11} - K_{21}$$

$$K_{11} = [L_1][R_1]/[L_1R_1]$$

$$K_{21} = [L_2][R_1]/[L_2R_1]$$

$$[R_1]_t = [R_1] + [L_1R_1] + [L_2R_1]$$

$$[L_1]_t = [L_1] + [L_1R_1]$$

$$[L_2]_t = [L_2] + [L_2R_1].$$

When the two ligands are of equal affinity (i.e. both K_{11} and K_{12} equal K_1), the expression becomes a quadratic (cf. eqn 35):

$$[R_1]_t[L_1]_t^2 - ([R_1]_t + [L_1]_t + [L_2]_t + K_1)[L_1]_t[L_1R_1]$$

$$+ ([L_1]_t + [L_2]_t)[L_1R_1]^2 = 0. \tag{66}$$

With three classes of sites and two ligands of equal affinity (i.e. both K_{1j} and K_{2j} equal K_j), the polynomial in $\Sigma_{j=1}^{n}[L_1R_j]$ is a quartic:

$$\varphi_0 + \varphi_1[L_1]_b + \varphi_2[L_1]_b^2 + \varphi_3[L_1]_b^3 + \varphi_4[L_1]_b^4 = 0, \tag{67}$$

where $\varphi_0 = -[L_1]_t^4\{[L]_t([R]_t[L]_t + d) + c\}$

$$\varphi_1 = +[L_1]_t^3\{[L]_t([L]_t([L]_t + a + 3[R]_t) + b + 2d) + c + K_1K_2K_3\}$$

$$\varphi_2 = -[L_1]_t^2[L]_t\{[L]_t(3[L]_t + 2a + 3[R]_t) + b + d\}$$

$$\varphi_3 = +[L_1]_t[L]_t^2\{a + [R]_t + 3[L]_t\}$$

$$\varphi_4 = -[L]_t^3$$

$$a = K_1 + K_2 + K_3$$

$$b = K_1K_2 + K_1K_3 + K_2K_3$$

$$c = [R_1]_t K_2K_3 + [R_2]_t K_1K_3 + [R_3]_t K_1K_2$$

$$d = [R_1]_t(K_2 + K_3) + [R_2]_t(K_1 + K_3) + [R_3]_t(K_1 + K_2)$$

$$K_j = \frac{[L_1][R_j]}{[L_1R_j]} = \frac{[L_2][R_j]}{[L_2R_j]}$$

$$[L_1]_b = [L_1R_1] + [L_1R_2] + [L_1R_3]$$

$$[L]_t = [L_1]_t + [L_2]_t$$

$$[L_i]_t = [L_i] + [L_iR_1] + [L_iR_2] + [L_iR_3]$$

$$[R]_t = [R_1]_t + [R_2]_t + [R_3]_t$$

$$[R_j]_t = [R_j] + [L_1R_j] + [L_2R_j].$$

When all ligands enter in units of total concentration, single polynomials can be obtained for one ligand and multiple classes of sites (e.g. eqn 64), for multiple ligands and one class of sites (e.g. eqn 65), and for the special case of multiple ligands and multiple classes of sites in which all ligands bind with the same affinity (e.g. eqn 67). More complex schemes require two or more expressions as described below. In some instances, however, the conditions may permit a compromise in which an n-dimensional system can be modelled by a single expression. One example is the family of G protein-linked receptors, where the binding sites often appear to be homogeneous with respect to at least some antagonists but heterogeneous with respect to agonists; moreover, affinities and capacities are such that binding tends to result in appreciable depletion of the former but not of the latter. If the data are to be analysed in the context of Scheme II, and an appropriate antagonist is selected as the probe L_1, it can be assumed that K_{1j} is the same for all j; also, the free and total concentrations of agonist are effectively the same (i.e. $[L_2] \approx [L_2]_t$), at least at the levels of receptor typically found in homogenates of native membranes. With three classes of sites, the polynomial is quartic in $\Sigma_{j=1}^n[L_1R_j]$:

$$\varphi_0 + \varphi_1[L_1]_b + \varphi_2[L_1]_b^2 + \varphi_3[L_1]_b^3 + \varphi_4[L_1]_b^4 = 0, \tag{68}$$

where

$$\varphi_0 = -[L_1]_t[R]_t\{a(K_{11} + [L_1]_t)^2 + e[L_2]K_{11}(K_{11} + [L_1]_t) + d[L_2]^2K_{11}^2\}$$

$$\varphi_1 = a\{(K_{11} + [L_1]_t)^3 + [R]_t(K_{11} + [L_1]_t)(K_{11} + 3[L_1]_t)\}$$
$$+ [L_2]K_{11}\{b(K_{11} + [L_1]_t)^2 + e[R]_t(K_{11} + 2[L_1]_t)\}$$
$$+ [L_2]^2K_{11}^2\{c(K_{11} + [L_1]_t) + d[R]_t\} + [L_2]^3K_{11}^3$$

$$\varphi_2 = -a\{3(K_{11} + [L_1]_t)^2 + [R]_t(2K_{11} + 3[L_1]_t)\}$$
$$- [L_2]K_{11}\{2b(K_{11} + [L_1]_t) + e[R]_t\} - c[L_2]^2K_{11}^2$$

$$\varphi_3 = a\{3(K_{11} + [L_1]_t) + [R]_t\} + b[L_2]K_{11}$$

$$\varphi_4 = -a$$

$$a = K_{21}K_{22}K_{23}$$

$$b = K_{21}K_{22} + K_{21}K_{23} + K_{22}K_{23}$$

$$c = K_{21} + K_{22} + K_{23}$$

$$d = [R_1]_t K_{21} + [R_2]_t K_{22} + [R_3]_t K_{23}$$

$$e = ([R_1]_t + [R_2]_t)K_{21}K_{22} + ([R_1]_t + [R_3]_t)K_{21}K_{23}$$
$$\quad + ([R_2]_t + [R_3]_t)K_{22}K_{23}$$

$$K_{11} = \frac{[L_1][R_1]}{[L_1R_1]} = \frac{[L_1][R_2]}{[L_1R_2]} = \frac{[L_1][R_3]}{[L_1R_3]}$$

$$[L_1]_b = [L_1R_1] + [L_1R_2] + [L_1R_3].$$

It ought to be noted in passing that G-linked receptors are well known to interconvert between states. The interconversion is promoted not only by guanyl nucleotides, sulphydryl reagents and some monovalent cations, but apparently also by agonists. These effects are not understood but seem to violate a principal assumption underlying Scheme II. It follows that the mechanistic relevance of the model is at least questionable, and the physical significance of the parameters is therefore unclear.

3.3.3 Numerical pitfalls

It is generally convenient to evaluate $\sum_{j=1}^{n}[L_1R_j]$ by means of the Newton–Raphson procedure, which converges upon the root from a starting value supplied by the investigator. Convergence is efficient when in the region of the root, but success depends upon the correct root being found. With an unfortunate starting value, the procedure will converge to the wrong root or perhaps enter a non-convergent cycle. Since the equation that defines the model is solved hundreds and even thousands of times in a typical analysis, the starting value at each stage must be computed from the information available rather than by direct intervention on the part of the user. The situation thus differs from that encountered in the optimization of parameters to achieve the best fit of the model to the data; starting values of the parameters are only required once per analysis and are probably best supplied by the investigator following an inspection of the plotted data (see Section 5).

For the multi-site model, the value of $\sum_{j=1}^{n}[L_1R_j]$ must lie between zero and the smaller of $\sum_{j=1}^{n}[R_j]_t$ and $[L_1]_t$. When the depletion of all ligands is small, a good approximation can be obtained from the explicit expression derived by assuming that the free and total concentrations are equal (e.g. eqn 45 or 46); that value then can be refined by the Newton–Raphson procedure. Such an approach may lead to the wrong root under extreme conditions, however, and it is helpful to investigate the properties of the polynomial if one requires initial estimates that are infallible.

The nature of the problem can be illustrated for the simple case of eqn 67 with no unlabelled ligand (i.e. $[L_2]_t = 0$). The curves in *Figure 2* represent the polynomial for values of $[R]_t$ between 0 and 2074 units, as listed in *Table 1*; there is an equal distribution of receptors among the three classes of sites, and the values of K_1, K_2 and K_3 are 0.1, 1.0 and 10 respectively. Since $[L_1]_t$ is 1000 throughout, most or virtually all of the receptor is occupied when $[L_1]_t$ exceeds $[R]_t$; similarly, virtually all of the probe is bound when $[R]_t$ exceeds $[L_1]_t$.

For the examples in which $[R]_t$ is 900 or less (curves a–j), the polynomial is concave upward and the first derivative is consistently negative at values of x between 0 and $[R]_t$; the Newton–Raphson procedure will therefore converge to the required root from any starting value within that range. At values of $[R]_t$ greater than 920, the first derivative is zero at some value of x within the permitted range as bounded by either $[R]_t$ (curves l–o) or $[L_1]_t$ (curves o–u) (*Table 1*). An injudicious starting value will thus result in convergence either to 1000, which is $[L_1]_t$, or to a value that approaches $[R]_t$ as $[R]_t$ becomes large relative to K_j and $[L_1]_t$. Finite values of $[L_2]_t$ reduce the value of x at which the first derivative of the function equals zero, and the range of starting

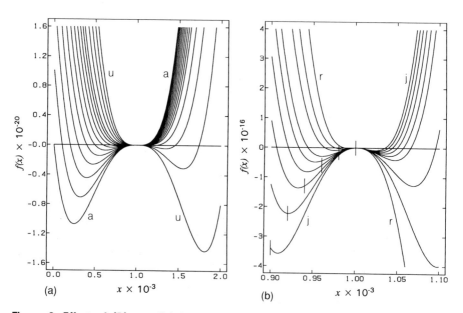

Figure 2. Effect of $[R]_t$ on $f(x)$ in eqn 67. The polynomial was evaluated for the parametric values listed in *Table 1* and at values of x over the range shown on the abscissa. The vertical lines in (b) indicate $[R]_t$ for curves j–o inclusive. The calculations for (a) and (b) were performed with single and double precision respectively.

Table 1. Effect of $[R]_t$ on $f(x)$ in eqn 67 [a]

Curve	$[R]_t$	x		$[L_1]_b/[R]_t$ [b]
		$f(x) = 0$	$f'(x) = 0$	%
Figure 2(a)				
a	0	0	251	0
b	100	99.6	326	99.6
c	200	199	400	99.5
d	300	298	475	99.5
e	400	398	549	99.4
f	500	496	624	99.3
g	600	595	697	99.1
h	700	692	770	98.8
i	800	787	842	98.3
j	900	875	909	97.2
o	1000	943	960	94.3
r	1200	983	992	81.9
s	1440	986	998	68.5
t	1728	990	999	57.3
u	2074	993	1000	47.9
Figure 2(b)				
j	900	875	909	97.2
k	920	891	921	96.9
l	940	906	932	96.4
m	960	920	943	95.9
n	980	932	952	95.1
o	1000	943	960	94.3
p	1050	965	975	91.9
q	1100	975	984	88.7
r	1200	983	992	81.9

[a] The polynomial was evaluated to obtain the appropriate root ($f(x) = 0$) and the value of x corresponding to the minimum between zero and $[L_1]_t$ ($f'(x) = 0$). Values of $[R]_t$ are listed in the table, and other parameters were as follows: $K_1 = 0.1$, $K_2 = 1.0$, $K_3 = 10$ and $[L_1]_t = 1000$. The data are illustrated in *Figure 2*.
[b] $x = [L_1]_b$ when $f(x) = 0$.

values that yields the correct root is correspondingly narrower. When the root is out of bounds, the process must be repeated with a new starting value selected according to a standard procedure, such as bisection, or in some manner specific to the function. Intermediate estimates that fall out of bounds do not necessarily lead to the wrong root. In the author's experience, the nature of the problem determines whether it is better to restart immediately or to wait for convergence; such questions can be particularly important in *n*-dimensional systems, including those described below.

Numerical problems such as that illustrated in *Figure 2* occur infrequently with preparations derived from native membranes, at least with the multi-site model; they are likely to become more common as genes are manipulated to produce controlled and ever larger quantities of receptors. Bolder experimental design and the likely requirement for more complex models can lead to considerable difficulty in the selection of starting values, even when the expression can be reduced to a single polynomial. Some of the limitations of the Newton–Raphson procedure can be overcome by the inclusion of bracketing and bisection as suggested by Press *et al.* (3). Alternatives to Newton–Raphson may be preferable in some circumstances, and the reader is referred to books on numerical methods for descriptions of available techniques. Excellent presentations are to be found in refs 2 and 3.

While numerical failure may be relatively infrequent in the absence of depletion, it is not wholly precluded. The properties of eqn 67 or any other function are determined by the parametric values supplied initially by the investigator or subsequently by the optimizing routine (i.e. **a** in eqn 15). Those values can be independent of the data, and one may therefore be faced with a situation, such as depletion, that never existed in the reaction mixture. Common causes of failure include outlandish initial estimates and parameters that are poorly defined by the data. The latter are troublesome owing to a lack of control during the fitting procedure. Since the parameter is without effect on chi-square, the selection of successive estimates is essentially random; moreover, instability in one parameter can destabilize others. Undefined parameters can be avoided by incorporating additional data acquired under more appropriate conditions; alternatively, they can be considered surplus and removed from the matrix used to minimize chi-square. Surplus parameters are generally correlated and can therefore be combined, as with b_1 and b_2 in the example $y = mx + b_1 + b_2$; alternatively, one or more values can be fixed arbitrarily and left unchanged during successive iterations of the fitting procedure.

A complication that arises with higher order polynomials is the increased risk of overflow or, more likely, of underflow in the floating-point variables that store the products of intermediate operations. The smallest power that can be represented in a computer that conforms to the IEEE standard is approximately -2^5 or -32 in single precision and -2^7 or -128 in double precision; the largest power is $(2^5 - 1)$ or 31 and $(2^7 - 1)$ or 127 respectively. Values of K_{ij}, $[L_i]_t$ and $[R_j]_t$ are often in the nanomolar range or less. The exponents of the coefficients in eqns 64–68 and similar expressions may thus require more than the number of bits allotted for a real variable, at least in single precision. The problem can be avoided by scaling all variables and parameters such that the mean exponent is zero; the coefficients can then be calculated with little fear of encountering an underflow, and the value of $\sum_{j=1}^{n}[L_1 R_j]$ obtained from the Newton–Raphson procedure can be converted back to the original units.

A final comment regarding polynomials relates to the order of the calculation. The most CPU-intensive, floating point operation is exponentiation, and the least efficient procedure is therefore to compute eqn 67 as shown above. The calculation is much faster if the expression is rearranged as follows:

$$f([L_1]_b) = \varphi_0 + [L_1]_b\{\varphi_1 + [L_1]_b(\varphi_2 + [L_1]_b(\varphi_3 + [L_1]_b\varphi_4))\}.$$

3.3.4 *n*-Dimensional numerical solutions

When depletion of two or more ligands occurs with multiple classes of sites, the system of equations that describes Scheme II can be solved by applying the Newton–Raphson method in n dimensions. There is some flexibility here, depending upon the extent to which the number of equations is reduced algebraically. Since the method proceeds via a system of linear equations, the number of dimensions is a major determinant of the programme's requirement for CPU time. A further consideration relates to the problem illustrated in *Figure 2*. The Newton–Raphson method succeeds handily when the starting value is well chosen; it can fail insidiously when that value is not in the neighbourhood of the desired root. Providing a first approximation of the root is relatively straightforward with a single polynomial, which can be investigated in detail if untoward complications arise. Locating roots in n-dimensional space is rather more difficult, especially when multiple roots are clustered, and the difficulties may increase with the number of dimensions.

The approach analogous to that described above for eqns 63–68 is to compute the values of $[L_iR_j]$ from $m \times n$ implicit equations of the form shown in eqn 69 ($i = 1, 2, \ldots, m; j = 1, 2, \ldots, n$):

$$\frac{[L_i][R_j]}{K_{ij}} - [L_iR_j] = 0. \tag{69}$$

The expansion of partial derivatives required by Newton–Raphson leads to the simultaneous linear equations shown below (eqn 70, $x_{ij} = [L_iR_j]$); **a** is the vector of initial or current values of **x**, and successive approximations are obtained according to the expression $x_{ij} = a_{ij} + \delta x_{ij}$ (cf. eqn 43):

$$0 = f_{11}\,(\mathbf{a}) + \frac{\partial f_{11}}{\partial x_{11}}\delta x_{11} + \cdots + \frac{\partial f_{11}}{\partial x_{m1}}\delta x_{m1} + \frac{\partial f_{11}}{\partial x_{12}}\delta x_{12} + \cdots + \frac{\partial f_{11}}{\partial x_{m2}}\delta x_{m2} + \cdots + \frac{\partial f_{11}}{\partial x_{mn}}\delta x_{mn}$$

$$0 = f_{m1}\,(\mathbf{a}) + \frac{\partial f_{m1}}{\partial x_{11}}\delta x_{11} + \cdots + \frac{\partial f_{m1}}{\partial x_{m1}}\delta x_{m1} + \frac{\partial f_{m1}}{\partial x_{12}}\delta x_{12} + \cdots + \frac{\partial f_{m1}}{\partial x_{m2}}\delta x_{m2} + \cdots + \frac{\partial f_{m1}}{\partial x_{mn}}\delta x_{mn}$$

$$0 = f_{12}\,(\mathbf{a}) + \frac{\partial f_{12}}{\partial x_{11}}\delta x_{11} + \cdots + \frac{\partial f_{12}}{\partial x_{m1}}\delta x_{m1} + \frac{\partial f_{12}}{\partial x_{12}}\delta x_{12} + \cdots + \frac{\partial f_{12}}{\partial x_{m2}}\delta x_{m2} + \cdots + \frac{\partial f_{12}}{\partial x_{mn}}\delta x_{mn}$$

$$0 = f_{m2}\,(\mathbf{a}) + \frac{\partial f_{m2}}{\partial x_{11}}\delta x_{11} + \cdots + \frac{\partial f_{m2}}{\partial x_{m1}}\delta x_{m1} + \frac{\partial f_{m2}}{\partial x_{12}}\delta x_{12} + \cdots + \frac{\partial f_{m2}}{\partial x_{m2}}\delta x_{m2} + \cdots + \frac{\partial f_{m2}}{\partial x_{mn}}\delta x_{mn}$$

$$0 = f_{mn}\,(\mathbf{a}) + \frac{\partial f_{mn}}{\partial x_{11}}\delta x_{11} + \cdots + \frac{\partial f_{mn}}{\partial x_{m1}}\delta x_{m1} + \frac{\partial f_{mn}}{\partial x_{12}}\delta x_{12} + \cdots + \frac{\partial f_{mn}}{\partial x_{m2}}\delta x_{m2} + \cdots + \frac{\partial f_{mn}}{\partial x_{mn}}\delta x_{mn}$$

$$\tag{70}$$

Values of $f_{ij}(\mathbf{x})$ at successive iterations are calculated according to eqn 71:

$$f_{ij}(\mathbf{x}) = \frac{([L_i]_t - \sum_{a=1}^{n} [L_iR_a])([R_j]_t - \sum_{a=1}^{m} [L_aR_j])}{K_{ij}} - [L_iR_j]. \tag{71}$$

The diagonal and non-diagonal elements of the matrix of partial derivatives are given by eqns 72 and 73–75 respectively:

$$\frac{\partial f_{ij}}{\partial [L_iR_j]} = -\frac{([L_i]_t - \sum_{a=1}^{n} [L_iR_a]) + ([R_j]_t - \sum_{a=1}^{m} [L_aR_j])}{K_{ij}} - 1 \tag{72}$$

$$\frac{\partial f_{ij}}{\partial [L_xR_j]} = -\frac{[L_i]_t - \sum_{a=1}^{n} [L_iR_a]}{K_{ij}} \quad (x \neq i) \tag{73}$$

$$\frac{\partial f_{ij}}{\partial [L_iR_y]} = -\frac{[R_j]_t - \sum_{a=1}^{m} [L_aR_j]}{K_{ij}} \quad (y \neq j) \tag{74}$$

$$\frac{\partial f_{ij}}{\partial [L_xR_y]} = 0 \quad (x \neq i, y \neq j). \tag{75}$$

An alternative to solving for $[L_iR_j]$ is to determine $[L_i]$ and $[R_j]$ by the Newton–Raphson procedure and then to compute the concentration of each complex according to the expression $[L_iR_j] = [L_i][R_j]/K_{ij}$. The solution involves $m + n$ equations of the form shown in eqns 76 and 77:

$$-[L_i]_t + [L_i] + \sum_{j=1}^{n} [L_iR_j] = 0 \tag{76}$$

$$-[R_j]_t + [R_j] + \sum_{i=1}^{m} [L_iR_j] = 0. \tag{77}$$

The set of linear equations to be solved for $\delta[L_i]$ and $\delta[R_j]$ in successive cycles of Newton–Raphson is as follows:

$$0 = f_{L1}(\mathbf{a}) + \frac{\partial f_{L1}}{\partial [R_1]} \delta[R_1] + \cdots + \frac{\partial f_{L1}}{\partial [R_n]} \delta[R_n] + \frac{\partial f_{L1}}{\partial [L_1]} \delta[L_1] + \cdots + \frac{\partial f_{L1}}{\partial [L_m]} \delta[L_m]$$

$$\vdots \qquad \vdots \qquad \vdots \qquad \vdots \qquad \vdots$$

$$0 = f_{Lm}(\mathbf{a}) + \frac{\partial f_{Lm}}{\partial [R_1]} \delta[R_1] + \cdots + \frac{\partial f_{Lm}}{\partial [R_n]} \delta[R_n] + \frac{\partial f_{Lm}}{\partial [L_1]} \delta[L_1] + \cdots + \frac{\partial f_{Lm}}{\partial [L_m]} \delta[L_m]$$

$$\tag{78}$$

$$0 = f_{R1}(\mathbf{a}) + \frac{\partial f_{R1}}{\partial [R_1]} \delta[R_1] + \cdots + \frac{\partial f_{R1}}{\partial [R_n]} \delta[R_n] + \frac{\partial f_{R1}}{\partial [L_1]} \delta[L_1] + \cdots + \frac{\partial f_{R1}}{\partial [L_m]} \delta[L_m]$$

$$\vdots \qquad \vdots \qquad \vdots \qquad \vdots \qquad \vdots$$

$$0 = f_{Rn}(\mathbf{a}) + \frac{\partial f_{Rn}}{\partial [R_1]} \delta[R_1] + \cdots + \frac{\partial f_{Rn}}{\partial [R_n]} \delta[R_n] + \frac{\partial f_{Rn}}{\partial [L_1]} \delta[L_1] + \cdots + \frac{\partial f_{Rn}}{\partial [L_m]} \delta[L_m].$$

The expansion of eqns 76 and 77 in terms of **x** is given by eqns 79 and 80 respectively; the partial derivatives are given by eqns 81–86:

$$f_{Li}(\mathbf{x}) = - [L_i]_t + [L_i] \left\{ 1 + \sum_{j=1}^{n} \frac{[R_j]}{K_{ij}} \right\} \tag{79}$$

$$f_{Rj}(\mathbf{x}) = - [R_j]_t + [R_j] \left\{ 1 + \sum_{i=1}^{m} \frac{[L_i]}{K_{ij}} \right\}, \tag{80}$$

$$\frac{\partial f_{Li}}{\partial [L_i]} = 1 + \sum_{j=1}^{n} \frac{[R_j]}{K_{ij}} \tag{81}$$

$$\frac{\partial f_{Rj}}{\partial [R_j]} = 1 + \sum_{i=1}^{m} \frac{[L_i]}{K_{ij}} \tag{82}$$

$$\frac{\partial f_{Li}}{\partial [R_j]} = \frac{[L_i]}{K_{ij}} \tag{83}$$

$$\frac{\partial f_{Rj}}{\partial [L_i]} = \frac{[R_j]}{K_{ij}} \tag{84}$$

$$\frac{\partial f_{Li}}{\partial [L_x]_{(x \neq i)}} = 0 \tag{85}$$

$$\frac{\partial f_{Rj}}{\partial [R_y]_{(y \neq j)}} = 0. \tag{86}$$

The matrix of partial derivatives is sparse in eqn 70 (see eqn 75) and somewhat less so in eqn 78 (see eqns 85 and 86). Zero-elements can be avoided altogether if Newton–Raphson is used to compute $[L_i]$ or $[R_j]$ but not both. Substitution of $[R_j]$ from eqn 80 ($f_{Rj}(\mathbf{x}) = 0$) into eqn 79 ($f_{Lj}(\mathbf{x}) = 0$) yields eqn 87, which can be solved numerically for all L_i. This is the procedure followed by Feldman *et al.* (4). The corresponding values of $[R_j]$ can be calculated from eqn 88; $[L_iR_j]$ can be obtained in turn from the expression $[L_iR_j] = [L_i][R_j]/K_{ij}$ or can be calculated from eqn 89:

$$-[L_i]_t + [L_i] \left\{ 1 + \sum_{j=1}^{n} \frac{[R_j]_t}{K_{ij} \left(1 + \sum_{a=1}^{m} [L_a]/K_{aj} \right)} \right\} = 0 \tag{87}$$

$$[R_j] = \frac{[R_j]_t}{1 + \sum_{i=1}^{m} [L_i]/K_{ij}} \tag{88}$$

$$[L_iR_j] = [R_j]_t \frac{[L_i]/K_{ij}}{1 + \sum_{a=1}^{m} [L_a]/K_{aj}} \tag{89}$$

The set of m equations to be solved for $\delta[L_i]$ in successive cycles of Newton–Raphson is as follows:

$$0 = f_{L1}(\mathbf{a}) + \frac{\partial f_{L1}}{\partial[L_1]} \delta[L_1] + \cdots + \frac{\partial f_{L1}}{\partial[L_m]} \delta[L_m]$$
$$\vdots \qquad \vdots \qquad \vdots \tag{90}$$
$$0 = f_{Lm}(\mathbf{a}) + \frac{\partial f_{Lm}}{\partial[L_1]} \delta[L_1] + \cdots + \frac{\partial f_{Lm}}{\partial[L_m]} \delta[L_m],$$

where

$$f_{Li}(\mathbf{x}) = -[L_i]_t + [L_i]\left\{1 + \sum_{j=1}^{n} \frac{[R_j]_t}{K_{ij}\left(1 + \sum_{a=1}^{m} [L_a]/K_{aj}\right)}\right\} \tag{91}$$

$$\frac{\partial f_{Li}}{\partial[L_i]} = 1 + \sum_{j=1}^{n} \frac{[R_j]_t\{1 + \sum_{\substack{a=1 \\ a \neq i}}^{m} [L_a]/K_{aj}\}}{K_{ij}\{1 + \sum_{a=1}^{m} [L_a]/K_{aj}\}^2} \tag{92}$$

$$\frac{\partial f_{Li}}{\partial[L_x]_{(x \neq i)}} = -[L_i] \sum_{j=1}^{n} \frac{[R_j]_t}{K_{ij}K_{xj}\{1 + \sum_{a=1}^{m} [L_a]/K_{aj}\}^2}. \tag{93}$$

For a system involving three ligands and three sites, the matrix of partial derivatives will contain 81 elements if based on eqn 69, 36 elements for eqns 76–77 and nine elements for eqn 87. Since the preparation and manipulation of the matrix is the most CPU intensive part of the overall procedure, one can expect a solution to be obtained most rapidly with eqn 87. Eqn 69 is unnecessarily tedious for most applications, particularly when m and n are large. The concentration of each complex is computed directly, but much of the information is irrelevant when only those complexes containing the radioligand contribute to the measured signal; also, 36 of the 81 elements are zero by definition. Convergence to $[L_iR_j]$ may be advantageous in some situations where convergence to $[L_i]$ (eqn 87) or to $[L_i]$ and $[R_j]$ (eqns 76–77) is accompanied by particularly intractable initial value problems.

3.3.5 General comments
Several descriptions of the multi-site model have been presented, and the investigator is faced with the question of which to use in a particular situation. If memory and CPU time are unlimited, the most general solution is to install

eqn 87 or perhaps eqns 76–77. A dimensioned capacity of three ligands and four sites ought to be sufficient to handle the data from most experiments or series of experiments. Systems with fewer ligands or fewer sites could be contrived by setting the values of the appropriate parameters or variables to zero; alternatively, the programme could be written to recognize the number of each and to handle the arrays accordingly. The latter approach is more efficient in that it avoids computing levels of binding for ligands and sites that do not exist; such operations would account for most of the CPU time if a model with three ligands and four sites were used to describe the binding of one ligand to one site. Refinements such as the equivalence of K_{ij} for labelled and unlabelled analogues of L_i at a particular R_j (e.g. eqns 66 and 67) or for a particular L_i at two or more R_j (e.g. eqn 68) can be incorporated by means of appropriate switches built into the programme.

With slower processors, it may be impractical to perform a large number of unnecessary operations with an overly complex model. Even relatively fast machines, such as those based on 80386 or 68030 chips, can seem to plod when asked to deal with the very large quantities of data that can be involved when the results of several experiments are analysed simultaneously. In such cases, the investigator can speed things up considerably by using the equations that describe the mechanism precisely. As noted above, there are several advantages to be gained by reducing the number of dimensions in the Newton–Raphson procedure. A single polynomial is likely to be solved most expeditiously, although very complex coefficients may require that the variables be stored in memory as double rather than single precision. As always, the major gain is realized when the free and total concentrations are effectively the same for all ligands; the system can then be described explicitly, and numerical methods can be abandoned insofar as the solution of the model is concerned.

If the various ligands are taken as independent variables (i.e. x in eqn 15), the parameters of the multi-site model comprise the affinity K_{ij} and the capacity $[R_j]_t$ corresponding to the sites of each class (i.e. a in eqn 15). Capacities can be expressed either explicitly (i.e. $[R_j]_t$) or as a fraction of the total (i.e. $F_j = [R_j]_t/[R_t]$, where $[R]_t = \Sigma_{j=1}^n[R_j]_t$), and the choice will depend upon the requirements of the analysis. It is largely a matter of taste in analyses involving data from a single experiment, although it may be convenient to estimate the range of uncertainty in terms of one parameter rather than another (see Section 5.2.2). With multiple sets of data, however, the definition of capacities will determine the constraints that can be applied in simultaneous analyses. If the fitting programme optimizes the values of F_j and $[R]_t$, the investigator can impose relative values on the fraction of all sites comprising each class or on the total number of sites in different experiments. Alternatively, optimization of $[R_j]_t$ allows one to constrain the absolute number of sites within one class or another. The equations presented here have all been formulated in terms of $[R_j]_t$, which can be obtained directly from the fitting routine or computed from values of F_j and $[R]_t$.

Practical applications of the multi-site model as formulated above require a predetermined value of n. In the absence of other information, the decision is made by testing successive increments for their effect on chi-square and the distribution of the data about the fitted curve (see Section 5.4). When there is only one independent variable $[L_i]$, each increment in j adds two parameters to the model (i.e. K_{ij} and $[R_j]_t$); with m variables ($[L_i]$, $i = 1, 2, \ldots, m$), each increment adds a minimum of two and a maximum of $m + 1$ parameters, depending upon the number of ligands that recognize the additional sites as distinct from any of the others. Statistically justifiable values of n rarely exceed three with data of the resolution and quality typical of binding studies; also, there is often a grey area in which n is clearly better than $n - 1$, but two or more parameters are highly correlated during successive iterations of the fitting procedure. Useful analyses in which n is four or more are likely to require pooled data from several experiments performed under different and carefully chosen conditions. It follows that the approach illustrated above is of practical relevance only when n is comparatively small.

Schemes in which n is large can be handled by assuming a dispersion in which affinities are distributed according to some function that can be integrated over the range of ligand concentration required to define the binding curve (5–7). Alternatively, one can assume an arbitrary value of n, such as 100, and values of K_{ij} equally distributed on a logarithmic scale over the required range; since the model is linear in $[R_j]_t$, it is relatively straightforward to find a set of values that correspond to the minimum in chi-square (8).

3.4 Strictly heterotropic co-operativity

The simplest example of a co-operative system involves two sites (Scheme IV), each with exclusive specificity for one or the other of two ligands (L and X).

Scheme IV

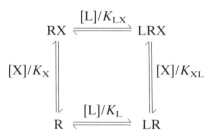

If R is a receptor and L a radioligand that recognizes the agonist-specific site, X might represent various species such as a proton, metal ions, organic ligands specific for an allosteric site, or a macromolecule other than the receptor. The last possibility constitutes the basis of mechanistic schemes typically put forward to rationalize the binding patterns observed for agonists at G protein-linked receptors (see below). There are four equilibrium ratios (eqns 94–97), any three of which are sufficient to define the system at equilibrium (eqn 98):

$$\frac{[L][R]}{[LR]} = K_L \tag{94}$$

$$\frac{[R][X]}{[RX]} = K_X \tag{95}$$

$$\frac{[L][RX]}{[LRX]} = K_{LX} \tag{96}$$

$$\frac{[LR][X]}{[LRX]} = K_{XL} \tag{97}$$

$$\frac{K_X}{K_{XL}} = \frac{K_L}{K_{LX}}. \tag{98}$$

3.4.1 Explicit expressions

If the free concentrations of both ligands are known, total binding of the probe $[L]_b$ is described by eqn 99:

$$[L]_b = [R]_t \frac{[L]/K_L + ([X]/K_X)([L]/K_{LX})}{1 + [X]/K_X + [L]/K_L + ([X]/K_X)([L]/K_{LX})}, \tag{99}$$

where $[L]_b = [LR] + [LRX]$

$[R]_t = [R] + [LR] + [RX] + [LRX]$

The functional dependence of total binding on $[L]$ and $[X]$ is illustrated by eqns 100 and 101 respectively, which are obtained by rearrangement of eqn 99:

$$[L]_b = \frac{[R]_t[L]}{[L] + K'_L} \tag{100}$$

$$[L]_b = [L]_{b,[X]\to\infty} + ([L]_{b,[X]=0} - [L]_{b,[X]\to\infty})\frac{K'_X}{[X] + K'_X}, \tag{101}$$

where $K'_L = \dfrac{1 + [X]/K_X}{1/K_L + ([X]/K_X)(1/K_{LX})}$

$K'_X = \dfrac{1 + [L]/K_L}{1/K_X\{1 + [L]/K_{LX}\}}$

$$[L]_{b,[X]=0} = \frac{[R]_t[L]}{[L] + K_L}$$

$$[L]_{b,[X]\to\infty} = \frac{[R]_t[L]}{[L] + K_{LX}}.$$

Both expressions describe a rectangular hyperbola provided that [X] is constant at all values of [L] (eqn 100) or that [L] is constant at all values of [X] (eqn 101). If [L] and [X] both change, one variable is disregarded in effect when the system is mapped onto two-dimensional coordinates, and the Hill coefficient will differ from unity.

A useful extension of Scheme IV is illustrated in Scheme V, where the probe L_1 and one or more unlabelled ligands (L_i, $i = 2, 3, \ldots, m$) compete for the L-specific site.

Scheme V

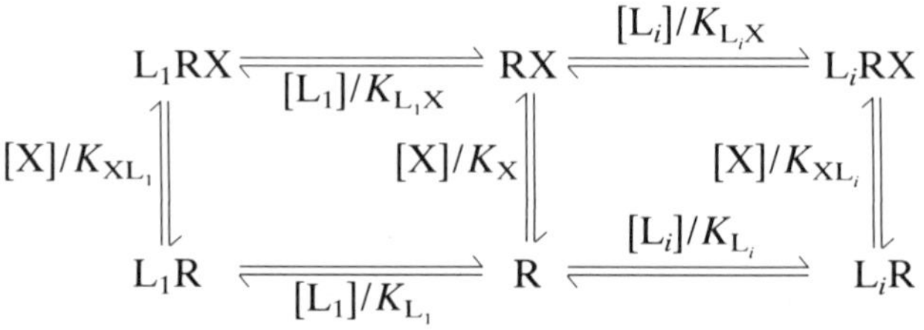

$$
\begin{array}{ccccc}
L_1RX & \underset{[L_1]/K_{L_1X}}{\overset{}{\rightleftharpoons}} & RX & \underset{}{\overset{[L_i]/K_{L_iX}}{\rightleftharpoons}} & L_iRX \\[2pt]
[X]/K_{XL_1}\updownarrow & & [X]/K_X\updownarrow & & [X]/K_{XL_i}\updownarrow \\[2pt]
 & & & [L_i]/K_{L_i} & \\[2pt]
L_1R & \underset{[L_1]/K_{L_1}}{\overset{}{\rightleftharpoons}} & R & \rightleftharpoons & L_iR
\end{array}
$$

Total binding of the probe is described by eqn 102, which in turn can be rearranged to obtain eqns 103–105 for those situations in which only one of the three or more ligands is varied during the experiment (cf. eqns 100 and 101):

$$[L_1]_b = [R]_t \frac{[L_1]/K_{L_1} + ([X]/K_X)([L_1]/K_{L_1X})}{1 + [X]/K_X + \sum_{i=1}^{m} \{[L_i]/K_{L_i} + ([X]/K_X)([L_i]/K_{L_iX})\}} \tag{102}$$

$$[L_1]_b = \frac{[R]_t[L_1]}{[L_1] + K'_{L_1}} \tag{103}$$

$$[L_1]_b = \left\{ \frac{[R]_t[L_1]}{[L_1] + K''_{L_1}} \right\} \left\{ \frac{K''_{L_2}}{[L_2] + K''_{L_2}} \right\} \tag{104}$$

$$[L_1]_b = [L_1]_{b,[X]\to\infty} + ([L_1]_{b,[X]=0} - [L_1]_{b,[X]\to\infty}) \left\{ \frac{K'_X}{[X] + K'_X} \right\}, \tag{105}$$

where $\quad [L_1]_b = [LR] + [LRX]$

$$[R]_t = [R] + [RX] + \sum_{i=1}^{m} ([L_iR] + [L_iRX])$$

$$K'_{L_1} = \frac{1 + [X]/K_X + \sum_{i=2}^{m} \{[L_i]/K_{L_i} + ([X]/K_X)([L_i]/K_{L_i X})\}}{1/K_{L_1} + ([X]/K_X)(1/K_{L_1 X})}$$

$$K''_{L_1} = \frac{1 + [X]/K_X + \sum_{i=3}^{m} \{[L_i]/K_{L_i} + ([X]/K_X)([L_i]/K_{L_i X})\}}{1/K_{L_1} + ([X]/K_X)(1/K_{L_1 X})}$$

$$K''_{L_2} = \frac{1 + [X]/K_X + \sum_{\substack{i=1 \\ i\neq 2}}^{m} \{[L_i]/K_{L_i} + ([X]/K_X)([L_i]/K_{L_i X})\}}{1/K_{L_2} + ([X]/K_X)(1/K_{L_2 X})}$$

$$K'_X = \frac{1 + \sum_{i=1}^{m} [L_i]/K_{L_i}}{(1/K_X) \{1 + \sum_{i=1}^{m} [L_i]/K_{L_i X}\}}$$

$$[L_1]_{b,[X]=0} = \frac{[R]_t[L_1]}{[L_1] + K_{L_1}\{1 + \sum_{i=2}^{m} [L_i]/K_{L_i}\}}$$

$$[L_1]_{b,[X]\to\infty} = \frac{[R]_t[L_1]}{[L_1] + K_{L_1 X}\{1 + \sum_{i=2}^{m} [L_i]/K_{L_i X}\}}.$$

It is noteworthy that eqns 104 and 105 are of the same form if the probe binds more tightly to R than to RX (i.e. $K_{L_1} < K_{L_1 X}$) and if levels of the ternary complex L_1RX are negligible (i.e. $[L_1] \ll K_{L_1 X}$, $[L_1]_{b,[X]\to\infty} \approx 0$). In such an event, competitive (eqn 104) and non-competitive (eqn 105) inhibition will be indistinguishable with data from a single experiment conducted at a fixed concentration of the probe and graded concentrations of an unlabelled ligand (L_2 or X). Similarly, competitive and non-competitive ligands are identical in the effect of the probe on inhibitory potency (K''_{L_2}, eqn 104; K'_X, eqn 105) if the former are studied in the absence of the latter (eqn 104, [X] = 0) and if the latter sustain only negligible levels of the ternary complex L_1RX at even the highest concentration of the probe (eqn 105, $[L_1] \ll K_{L_1 X}$). These considerations illustrate the ambiguity inherent in such data and highlight the need for caution when drawing mechanistic conclusions.

3.4.2 Numerical solutions and the mobile or floating receptor model

When the total concentrations of both R and X are taken into account, the equations that describe Scheme V are no longer linear in $[L_i]_b$. An important consequence of non-linearity is that the labelled sites can appear heterogeneous with respect to L_i whenever R binds an appreciable fraction of X. The model can therefore account for a dispersion of affinities in a population of mutually independent and intrinsically identical sites. It has been widely adopted, at least implicitly, to rationalize the binding properties of receptors that activate G proteins (i.e. $X \equiv G$). The relevance of the model to such systems is typically based upon two premises: first, that pools of rapidly diffusing proteins coexist within the membrane; second, that there is a random and spontaneous exchange between free elements on the one hand (e.g. R and X) and one or more complexes on the other (e.g. RX). It is implicit in the equations presented here that the normal assumptions of solution kinetics are not violated by the presence of other proteins, interactions with the cytoskeleton, or the likelihood that diffusion is in two dimensions rather than three.

The simplest form of the model for most applications is that in which the probe L_1 and an unlabelled ligand L_2 compete for the receptor. If there is mutual depletion of R and X but no depletion of either L_1 or L_2, the values of $[L_1R]$ and $[L_1RX]$ can be calculated independently from two polynomials (eqns 106 and 107). Both functions are quadratic, and the roots can therefore be obtained explicitly; those estimates can then be polished by using them as starting values in the Newton–Raphson procedure:

$$\varphi_0 + \varphi_1[L_1R] + \varphi_2[L_1R]^2 = 0 \tag{106}$$

$$\psi_0 + \psi_1[L_1RX] + \psi_2[L_1RX]^2 = 0, \tag{107}$$

where $\varphi_0 = -\dfrac{[L_1]^2}{K_{L_1}^2}[R]_t$

$$\varphi_1 = \frac{[L_1]}{K_{L_1}}\left\{1 + \frac{[L_1]}{K_{L_1}} + \frac{[L_2]}{K_{L_2}} + \left(\frac{[X]_t}{K_X} - \frac{[R]_t}{K_X}\right)\right.$$
$$\left. \times \left(1 + \frac{[L_1]}{K_{L_1}X} + \frac{[L_2]}{K_{L_2}X}\right)\right\}$$

$$\varphi_2 = \frac{1}{K_X}\left\{1 + \frac{[L_1]}{K_{L_1}} + \frac{[L_2]}{K_{L_2}}\right\}\left\{1 + \frac{[L_1]}{K_{L_1}X} + \frac{[L_2]}{K_{L_2}X}\right\}$$

$$\psi_0 = -\frac{[L_1]^2 [X]_t}{K_{L_1X}^2 K_X}[R]_t$$

$$\psi_1 = \frac{[L_1]}{K_{L_1}X}\left\{1 + \frac{[L_1]}{K_{L_1}} + \frac{[L_2]}{K_{L_2}} + \left(\frac{[X]_t}{K_X} + \frac{[R]_t}{K_X}\right)\right.$$
$$\left. \times \left(1 + \frac{[L_1]}{K_{L_1}X} + \frac{[L_2]}{K_{L_2}X}\right)\right\}$$

$$\psi_2 = -\frac{1}{K_X}\left\{1 + \frac{[L_1]}{K_{L_1X}} + \frac{[L_2]}{K_{L_2X}}\right\}^2$$

$$[R]_t = [R] + [RX] + [L_1RX] + [L_2RX]$$

$$[X]_t = [X] + [RX] + [L_1RX] + [L_2RX].$$

The total concentrations of all reactants are considered in eqns 108–111, which encompass both Scheme V and Scheme I. The latter is included to address the possibility that some receptors (S in eqn 109) may bind L_i but not the allosteric ligand X; such an addition to the model can be useful when evaluating data from native membranes or other systems where there is evidence for distinct populations of sites. If Scheme V is relevant to G-mediated systems, for example, some receptors may be sequestered in a G-free compartment or otherwise unable to interact with the G protein. It is assumed here that the affinity of L_i is the same for R and for S (i.e. $[L_i][S]/[L_iS] = K_{L_i}$, $[L_i][R]/[L_iR] = K_{L_i}$):

$$- [R]_t + [R] + [RX] + \sum_{i=1}^{m} [L_iR] + \sum_{i=1}^{m} [L_iRX] = 0 \qquad (108)$$

$$- [S]_t + [S] + \sum_{i=1}^{m} [L_iS] = 0 \qquad (109)$$

$$- [X]_t + [X] + [RX] + \sum_{i=1}^{m} [L_iRX] = 0 \qquad (110)$$

$$- [L_i]_t + [L_i] + [L_iR] + [L_iRX] + [L_iS] = 0. \qquad (111)$$

Values of [R], [S], [X] and $[L_i]$ can be calculated from eqns 108–111 by applying the Newton–Raphson method in $3 + m$ dimensions ($i = 1, 2, \ldots, m$). The expansion for the calculation of successive increments is shown below (eqn 112). The individual functions are defined in eqns 113–116, and the elements of the matrix of partial derivatives are defined in eqns 117–128; partial derivatives not shown are equal to zero. The quantity $\mathbf{x}$ is the vector of $3 + m$ variables, and $\mathbf{a}$ is the current point at each iteration:

$$0 = f_R(\mathbf{a}) + \frac{\partial f_R}{\partial[R]}\delta[R] + \frac{\partial f_R}{\partial[S]}\delta[S] + \frac{\partial f_R}{\partial[X]}\delta[X] + \frac{\partial f_R}{\partial[L_1]}\delta[L_1] + \cdots + \frac{\partial f_R}{\partial[L_m]}\delta[L_m]$$

$$0 = f_S(\mathbf{a}) + \frac{\partial f_S}{\partial[R]}\delta[R] + \frac{\partial f_S}{\partial[S]}\delta[S] + \frac{\partial f_S}{\partial[X]}\delta[X] + \frac{\partial f_S}{\partial[L_1]}\delta[L_1] + \cdots + \frac{\partial f_S}{\partial[L_m]}\delta[L_m]$$

$$0 = f_X(\mathbf{a}) + \frac{\partial f_X}{\partial[R]}\delta[R] + \frac{\partial f_X}{\partial[S]}\delta[S] + \frac{\partial f_X}{\partial[X]}\delta[X] + \frac{\partial f_X}{\partial[L_1]}\delta[L_1] + \cdots + \frac{\partial f_X}{\partial[L_m]}\delta[L_m]$$

$$(112)$$

$$0 = f_{L1}\ (\mathbf{a}) + \frac{\partial f_{L1}}{\partial[R]}\ \delta[R] + \frac{\partial f_{L1}}{\partial[S]}\ \delta[S] + \frac{\partial f_{L1}}{\partial[X]}\ \delta[X] + \frac{\partial f_{L1}}{\partial[L_1]}\ \delta[L_1] + \cdots + \frac{\partial f_{L1}}{\partial[L_m]}\ \delta[L_m]$$

$$\vdots \qquad\qquad \vdots \qquad\qquad \vdots \qquad\qquad \vdots \qquad\qquad \vdots \qquad\qquad \vdots$$

$$0 = f_{Lm}\ (\mathbf{a}) + \frac{\partial f_{LM}}{\partial[R]}\ \delta[R] + \frac{\partial f_{Lm}}{\partial[S]}\ \delta[S] + \frac{\partial f_{Lm}}{\partial[X]}\ \delta[X] + \frac{\partial f_{Lm}}{\partial[L_1]}\ \delta[L_1] + \cdots + \frac{\partial f_{Lm}}{\partial[L_m]}\ \delta[L_m]$$

where

$$f_R(\mathbf{x}) = -[R]_t + [R]\left\{1 + \frac{[X]}{K_X} + \sum_{i=1}^{m} \frac{[L_i]}{K_{L_i}}\left(1 + \frac{[X]}{K_{XL_i}}\right)\right\} \tag{113}$$

$$f_S(\mathbf{x}) = -[S]_t + [S]\left\{1 + a\right\} \tag{114}$$

$$f_X(\mathbf{x}) = -[X]_t + [X]\left\{1 + [R]\left(\frac{1}{K_X} + b\right)\right\} \tag{115}$$

$$f_{Li}(\mathbf{x}) = -[L_i]_t + [L_i]\left\{1 + \frac{[S]}{K_{L_i}} + \frac{[R]}{K_{L_i}}\left(1 + \frac{[X]}{K_{XL_i}}\right)\right\} \tag{116}$$

$$\frac{\partial f_R}{\partial[R]} = 1 + \frac{[X]}{K_X} + \sum_{i=1}^{m} \frac{[L_i]}{K_{L_i}}\left\{1 + \frac{[X]}{K_{XL_i}}\right\} \tag{117}$$

$$\frac{\partial f_R}{\partial[X]} = [R]\left\{\frac{1}{K_X} + b\right\} \tag{118}$$

$$\frac{\partial f_R}{\partial[L_i]} = \frac{[R]}{K_{L_i}}\left\{1 + \frac{[X]}{K_{XL_i}}\right\} \tag{119}$$

$$\frac{\partial f_S}{\partial[S]} = 1 + a \tag{120}$$

$$\frac{\partial f_S}{\partial[L_i]} = \frac{[S]}{K_{L_i}} \tag{121}$$

$$\frac{\partial f_X}{\partial[R]} = [X]\left\{\frac{1}{K_X} + b\right\} \tag{122}$$

$$\frac{\partial f_X}{\partial[X]} = 1 + [R]\left\{\frac{1}{K_X} + b\right\} \tag{123}$$

$$\frac{\partial f_X}{\partial [L_i]} = \frac{[R][X]}{K_{L_i} K_{XL_i}} \qquad (124)$$

$$\frac{\partial f_{Li}}{\partial [R]} = \frac{[L_i]}{K_{L_i}} \left\{ 1 + \frac{[X]}{K_{XL_i}} \right\} \qquad (125)$$

$$\frac{\partial f_{Li}}{\partial [S]} = \frac{[L_i]}{K_{L_i}} \qquad (126)$$

$$\frac{\partial f_{Li}}{\partial [X]} = \frac{[L_i][R]}{K_{L_i} K_{XL_i}} \qquad (127)$$

$$\frac{\partial f_{Li}}{\partial [L_i]} = 1 + \frac{[S]}{K_{L_i}} + \frac{[R]}{K_{L_i}} \left\{ 1 + \frac{[X]}{K_{XL_i}} \right\}, \qquad (128)$$

and where

$$a = \sum_{i=1}^{m} \frac{[L_i]}{K_{L_i}}$$

$$b = \sum_{i=1}^{m} \frac{[L_i]}{K_{L_i} K_{XL_i}}.$$

In the author's experience, successful convergence with eqn 112 often requires starting values that are close to the desired roots. Excellent estimates can be obtained from an iterative procedure based on eqns 129–132:

$$\varphi_0 + \varphi_1[R] + \varphi_2[R]^2 = 0 \qquad (129)$$

$$\psi_0 + \psi_1[X] + \psi_2[X]^2 = 0 \qquad (130)$$

$$[S] = \frac{[S]_t}{1 + a} \qquad (131)$$

$$[L_i] = \frac{[L_i]_t}{1 + (1/K_{L_i})\{[R] + [S] + [X][R]/K_{XL_i}\}}, \qquad (132)$$

where

$$\varphi_0 = -[R]_t$$

$$\varphi_1 = 1 + a + ([X]_t - [R]_t) \left\{ \frac{1}{K_X} + b \right\}$$

335

$$\varphi_2 = (1 + a)\left\{\frac{1}{K_X} + b\right\}$$

$$\psi_0 = -[X]_t(1 + a)$$

$$\psi_1 = 1 + a + ([R]_t - [X]_t)\left\{\frac{1}{K_X} + b\right\}$$

$$\psi_2 = \frac{1}{K_X} + b.$$

Values of $[L_i]_t$ are initially substituted for $[L_i]$ in eqns 129–131 to obtain values of $[R]$, $[S]$ and $[X]$; the latter are then substituted in eqn 132 to obtain revised values of $[L_i]$ for the next cycle. This process generally converges in less than ten iterations, although it is occasionally necessary to trap oscillations and try again. The resulting values of $[R]$, $[S]$, $[X]$ and $[L_i]$ can be polished with the Newton–Raphson procedure, which usually converges within one or two iterations to yield virtually the same numbers. Complexes involving the probe or any other ligand then can be calculated from eqns 133–135:

$$[L_iR] = \frac{[L_i][R]}{K_{L_i}} \tag{133}$$

$$[L_iRX] = \frac{[L_i][R][X]}{K_{L_iX}K_X} \tag{134}$$

$$[L_iS] = \frac{[L_i][S]}{K_{L_i}}. \tag{135}$$

Models in which X is a protein distinct from the receptor suggest an extension of Scheme V that includes the binding of ligands specific for X (N_j, $j = 1, 2, \ldots, n$). The system is described by $3 + m + n$ equations as shown by eqns 136–140:

$$-[R]_t + [R] + [RX] + \sum_{i=1}^{m} [L_iR] + \sum_{j=1}^{n} [RXN_j] + \sum_{i=1}^{m} [L_iRX]$$

$$+ \sum_{i=1}^{m}\sum_{j=1}^{n} [L_iRXN_j] = 0 \tag{136}$$

$$-[S]_t + [S] + \sum_{i=1}^{m} [L_iS] = 0 \tag{137}$$

$$-[X]_t + [X] + [RX] + \sum_{j=1}^{n} [XN_j] + \sum_{i=1}^{m} [L_iRX]$$

$$+ \sum_{j=1}^{n} [RXN_j] + \sum_{i=1}^{m}\sum_{j=1}^{n} [L_iRXN_j] = 0 \tag{138}$$

$$-[L_i]_t + [L_i] + [L_iR] + [L_iRX] + \sum_{j=1}^{n} [L_iRXN_j]$$
$$+ [L_iS] = 0 \tag{139}$$

$$-[N_j]_t + [N_j] + [XN_j] + [RXN_j] + \sum_{i=1}^{m} [L_iRXN_j] = 0. \tag{140}$$

Such schemes are numerically intensive but not as intractable as they might appear. They present the interesting possibility of combining data from multiple experiments on the binding of radioligands specific for the receptor on the one hand and to protein X on the other.

If o species of X compete for R, eqns 110 and 138 can be expressed as in eqns 141 and 142 respectively; eqns 108 and 136 can be expanded to obtain eqns 143 and 144, and other expressions can be revised in a similar manner:

$$-[X_k]_t + [X_k] + [RX_k] + \sum_{i=1}^{m} [L_iRX_k] = 0 \tag{141}$$

$$-[X_k]_t + [X_k] + [RX_k] + \sum_{j=1}^{n} [X_kN_j] + \sum_{i=1}^{m} [L_iRX_k]$$
$$+ \sum_{j=1}^{n} [RX_kN_j] + \sum_{i=1}^{m} \sum_{j=1}^{n} [L_iRX_kN_j] = 0 \tag{142}$$

$$-[R]_t + [R] + \sum_{k=1}^{o} [RX_k] + \sum_{i=1}^{m} [L_iR] + \sum_{i=1}^{m} \sum_{k=1}^{o} [L_iRX_k] = 0 \tag{143}$$

$$-[R]_t + [R] + \sum_{k=1}^{o} [RX_k] + \sum_{i=1}^{m} [L_iR] + \sum_{j=1}^{n} \sum_{k=1}^{o} [RX_kN_j]$$
$$+ \sum_{i=1}^{m} \sum_{k=1}^{o} [L_iRX_k] + \sum_{i=1}^{m} \sum_{j=1}^{n} \sum_{k=1}^{o} [L_iRX_kN_j] = 0. \tag{144}$$

If the assumptions underlying Scheme V are valid for G-linked receptors, the extension described in eqns 136–140 permits an explicit description of the interaction between agonists L_i and guanyl nucleotides N_j. Evidence for subtypes of G proteins raises the possibility that a particular pool of receptors may interact with more than one subtype X_k (eqns 141–144). Other variants of Scheme V are suggested by the heterotrimeric nature of G proteins and the observation that, at least in solution, guanyl nucleotides can promote the dissociation of holo G into an α subunit on the one hand and a $\beta\gamma$ dimer on the other. Additional possibilities include the interaction of R with two or more equivalents of the G protein to form ternary complexes (i.e. RX_2) and higher oligomers. With more complicated models, unique solutions are likely to require pooled data from multiple experiments designed to define specific parameters. It is particularly important that internal consistency be demonstrated by means of labelled probes for both the receptor and the G protein, at least until the validity of the model is established.

Systems in which both R and X are localized to the membrane present a dilemma, in that the effective total concentrations for mutual interactions cannot be less than $[R]_t/f_M$ and $[X]_t/f_M$ respectively (cf. eqn 4); the actual values may be unknowable, given the likelihood of uneven distribution among and restricted diffusion within the membrane fragments. Total concentrations of R and X are typically estimated from the binding of specific ligands; moreover, it is generally assumed that access to the binding site is via the aqueous phase, which in turn is assumed to represent virtually 100% of the total volume of the reaction mixture. It follows that estimates of K_X and similar parameters are a function of f_M in the absence of an explicit correction (cf. eqns 36 and 37). That dependence can probably be disregarded in analyses of data acquired under identical conditions with respect to the properties and concentration of the membrane fragements, although the units of K_X cannot be interpreted literally; should those conditions differ, however, estimates of K_X cannot be compared without accounting for differences in f_M.

A cautionary word is in order here regarding models in which there are alternative pathways from one point to another. In Scheme IV, for example, the receptor can go from R to LRX via LR or RX. An important consequence of this arrangement is that $K_L/K_{LX} = K_X/K_{XL}$. Only three of the four parameters can thus be defined explicitly when setting up the various equations; inconsistent redundancies present in the initial estimates of the parameters or generated during the fitting procedure constitute a violation of the principle of microscopic reversibility.

3.5 Unrestricted co-operativity

If the specificity of the two sites in Schemes IV and V is not exclusive, the receptor is bivalent with respect to any particular ligand L_i, as illustrated in Scheme VI. The structural composition of R is not defined, but the functional unit responsible for the binding properties is likely to be a homo- or hetero-oligomer. It is important to note that the oligomeric nature of R must be retained, or the model becomes a variant of that illustrated in Scheme IV; specifically, there must be no exchange of subunits among oligomers, at least under the conditions of the experiment.

When the probe is the only ligand present (L_1), the system is defined by three microscopic dissociation constants as shown in eqns 145–147:

$$\frac{[R][L_1]}{[RL_1]} = K_{R1} \tag{145}$$

$$\frac{[L_1][R]}{[L_1R]} = K_{1R} \tag{146}$$

$$\frac{[L_1][RL_1]}{[L_1RL_1]} = K_{1(R1)}. \tag{147}$$

Scheme VI

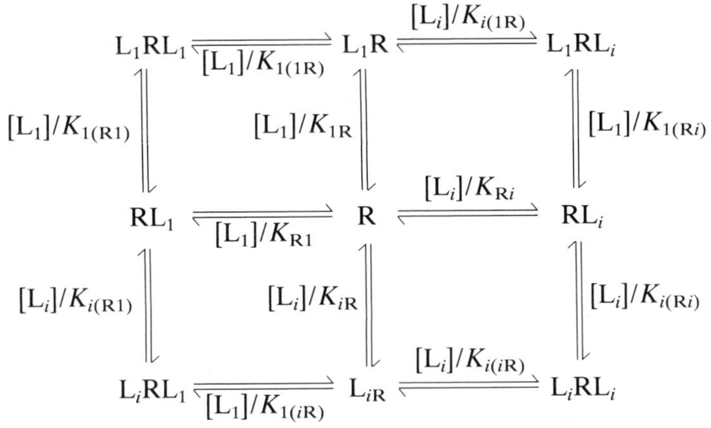

$$
\begin{array}{ccccc}
L_1RL_1 & \underset{[L_1]/K_{1(1R)}}{\overset{}{\rightleftharpoons}} & L_1R & \underset{}{\overset{[L_i]/K_{i(1R)}}{\rightleftharpoons}} & L_1RL_i \\[4pt]
{\scriptstyle [L_1]/K_{1(R1)}}\updownarrow & & {\scriptstyle [L_1]/K_{1R}}\updownarrow & & {\scriptstyle [L_1]/K_{1(Ri)}}\updownarrow \\[4pt]
RL_1 & \underset{[L_1]/K_{R1}}{\overset{}{\rightleftharpoons}} & R & \underset{}{\overset{[L_i]/K_{Ri}}{\rightleftharpoons}} & RL_i \\[4pt]
{\scriptstyle [L_i]/K_{i(R1)}}\updownarrow & & {\scriptstyle [L_i]/K_{iR}}\updownarrow & & {\scriptstyle [L_i]/K_{i(Ri)}}\updownarrow \\[4pt]
L_iRL_1 & \underset{[L_1]/K_{1(iR)}}{\overset{}{\rightleftharpoons}} & L_iR & \underset{}{\overset{[L_i]/K_{i(iR)}}{\rightleftharpoons}} & L_iRL_i \\
\end{array}
$$

A second ligand (L_i) introduces five additional constants: three analogous to those shown above for the probe, and two that reflect hybrid occupancy of the receptor by one equivalent of each ligand:

$$\frac{[L_i][RL_1]}{[L_iRL_1]} = K_{i(R1)} \tag{148}$$

$$\frac{[L_1][RL_i]}{[L_1RL_i]} = K_{1(Ri)} . \tag{149}$$

Eight binding constants plus $[R]_t$ are therefore required to define a system with two ligands; in contrast, the corresponding multi-site model requires only four binding constants plus $[R_1]_t$ and $[R_2]_t$ (Scheme II, $n = 2$). Each additional ligand introduces five more constants, as in eqns 145–149, plus two constants for each of all possible hybrids not involving the probe (eqns 150–151):

$$\frac{[L_x][RL_y]}{[L_xRL_y]} = K_{x(Ry)} \tag{150}$$

$$\frac{[L_y][RL_x]}{[L_yRL_x]} = K_{y(Rx)} . \tag{151}$$

For the present purposes, all expressions have been derived in terms of the parameters defined above, which represent eight of the twelve microscopic equilibria in the example of Scheme VI; the remaining parameters are implicit (e.g. eqn 152):

$$\frac{K_{1R}}{K_{1(R1)}} = \frac{K_{R1}}{K_{1(1R)}} . \tag{152}$$

3.5.1 Explicit expressions

Total binding of the radioligand $[L_1]_b$ in the presence of $m - 1$ unlabelled ligands (L_i, $i = 2, 3, \ldots, m$) is given by eqn 153:

$$[L_1]_b = [L_1R] + [RL_1] + 2[L_1RL_1] + \sum_{i=2}^{m} ([L_1RL_i] + [L_iRL_1]). \quad (153)$$

If the radioligand is the only variable and the free concentrations of all ligands are known, the expression for total binding is a rational function as in eqn 154:

$$[L_1]_b = [R]_t \frac{\varphi_1[L_1] + 2\varphi_2[L_1]^2}{1 + \varphi_1[L_1] + \varphi_2[L_1]^2}, \quad (154)$$

where $\quad \varphi_1 = \dfrac{1}{a}\left\{ \dfrac{1}{K_{R1}} + \dfrac{1}{K_{1R}} + \sum_{i=2}^{m}\left(\dfrac{1}{K_{R1}} \cdot \dfrac{[L_i]}{K_{i(R1)}} + \dfrac{[L_i]}{K_{Ri}} \cdot \dfrac{1}{K_{1(Ri)}} \right) \right\} \quad (155)$

$$\varphi_2 = \frac{1}{a} \cdot \frac{1}{K_{R1}K_{1(R1)}} \quad (156)$$

$$a = 1 + \sum_{i=2}^{m}\left\{ \frac{1}{K_{Ri}} + \frac{1}{K_{iR}} + \frac{[L_i]}{K_{Ri}} \sum_{j=2}^{m} \frac{[L_j]}{K_{j(Ri)}} \right\}. \quad (157)$$

Equation 158 is the corresponding expression when the concentration of an unlabelled ligand $[L_2]$ is varied at fixed concentrations of the probe $[L_1]$ and any other ligands ($[L_i]$, $i = 3, \ldots, m$):

where $$[L_1]_b = [R]_t \frac{\psi_0' + \psi_1'[L_2]}{1 + \psi_1[L_2] + \psi_2[L_2]^2}. \quad (158)$$

$$\psi_0' = \frac{1}{a}\left\{ \frac{[L_1]}{K_{R1}} + \frac{[L_1]}{K_{1R}} + \frac{2[L_1]^2}{K_{R1}K_{1(R1)}} + \sum_{i=3}^{m}\left(\frac{[L_1]}{K_{R1}} \cdot \frac{[L_i]}{K_{i(R1)}} + \frac{[L_i]}{K_{Ri}} \cdot \frac{[L_1]}{K_{1(Ri)}} \right) \right\} \quad (159)$$

$$\psi_1' = \frac{1}{a}\left\{ \frac{[L_1]}{K_{R1}} \cdot \frac{1}{K_{2(R1)}} + \frac{1}{K_{R2}} \cdot \frac{[L_1]}{K_{1(R2)}} \right\} \quad (160)$$

$$\psi_1 = \frac{1}{a}\left\{ \frac{1}{K_{R2}} + \frac{1}{K_{2R}} + \sum_{\substack{i=1 \\ i\neq 2}}^{m}\left(\frac{[L_i]}{K_{Ri}} \cdot \frac{1}{K_{2(R1)}} + \frac{1}{K_{R2}} \cdot \frac{[L_i]}{K_{i(R2)}} \right) \right\} \quad (161)$$

$$\psi_2 = \frac{1}{a} \cdot \frac{1}{K_{R2}K_{2(R2)}} \quad (162)$$

$$a = 1 + \sum_{\substack{i=1 \\ i \neq 2}}^{m} \left\{ \frac{[L_i]}{K_{Ri}} + \frac{[L_i]}{K_{iR}} + \frac{[L_i]}{K_{Ri}} \sum_{\substack{j=1 \\ j \neq 2}}^{m} \frac{[L_j]}{K_{j(Ri)}} \right\}. \tag{163}$$

Equations 154 and 158 describe binding to a bivalent receptor as in Scheme VI; the corresponding expressions for binding to a multivalent receptor with n sites for L_i are given by eqns 164 and 165 respectively:

$$[L_1]_b = [R]_t \frac{\sum_{j=1}^{n} j\varphi_j[L_1]^j}{1 + \sum_{j=1}^{n} \varphi_j[L_1]^j} \tag{164}$$

$$[L_1]_b = [R]_t \frac{\sum_{j=0}^{n-1} \psi_j'[L_2]^j}{1 + \sum_{j=1}^{n} \psi_j[L_2]^j}. \tag{165}$$

Equation 164 is the Adair equation, and the same parameters (φ_j) appear in numerator and denominator. A practical consequence of this property arises in the case of negatively cooperative systems, where each inflection (i.e. $f''(x) = 0$) in plots of $[L_1]_b$ versus $\log [L_1]$ is associated with a/n of the total signal ($a = 1, 2, \ldots$, or $n - 1$). Negative co-operativity is therefore indistinguishable from the special case of the multi-site model in which the capacity of each subclass is some multiple of a basic unit (e.g. $[R_j]_t/[R]_t = 1/n$ for all j in Scheme II; cf. eqn 48). The data are ambiguous, but the choice is limited when n equals 2; the possibilities increase as the oligomer becomes more complex. A tetramer would be indistinguishable from 1, 2, 3 or 4 classes of mutually independent sites, depending upon the degree to which each equivalent of L_i affected the affinity of the next.

The coefficients of the numerator and denominator differ in eqn 165; accordingly, there is no restriction on the fractional signal associated with each point of inflection defined by $\log[L_2]$. The value will depend upon the concentration of the probe and upon the nature and magnitude of the co-operativity between the probe and the unlabelled ligand. Data from experiments involving different unlabelled ligands at multiple concentrations of the probe can thus be pooled to define the relatively large number of parameters associated with larger oligomers. Such experiments can also help to determine whether multiple points of inflection reflect co-operativity or mutually independent sites, since the coefficients are defined differently in eqn 165 (Scheme VI) and eqn 49 (Scheme II).

Equations 154–165 describe the total binding of one ligand, typically the probe, but it is sometimes of interest to know the concentrations of individual complexes. If L_x and L_y represent two of m ligands present in the reaction mixture (i.e. L_i, where $i = 1, 2, \ldots, m$), all possible complexes can be calculated as follows: (eqns 166–170).

341

$$[RL_x] = \frac{[R]_t}{1 + \sum_{i=1}^{m} [L_i]/K_{i(Rx)} + (K_{Rx}/[L_x])\{1 + \sum_{i=1}^{m} [L_i]/K_{iR} + \sum_{\substack{i=1 \\ i \neq x}}^{m} ([L_i]/K_{Ri})(1 + \sum_{j=1}^{m} [L_j]/K_{j(Ri)})\}}$$

$$(166)$$

$$[L_xR] = \frac{[R]_t}{1 + (K_{xR}/[L_x])\{1 + \sum_{\substack{i=1 \\ i \neq x}}^{m} [L_i]/K_{iR} + \sum_{i=1}^{m} ([L_i]/K_{Ri})(1 + \sum_{j=1}^{m} [L_j]/K_{j(Ri)})\}}$$

$$(167)$$

$$[L_xRL_y] = \frac{[R]_t}{1 + (K_{x(Ry)}/[L_x])[1 + \sum_{\substack{i=1 \\ i \neq x}}^{m} [L_i]/K_{i(Ry)} + (K_{Ry}/[L_y])\{1 + \sum_{i=1}^{m} [L_i]/K_{iR} + \sum_{\substack{i=1 \\ i \neq y}}^{m} ([L_i]/K_{Ri})(1 + \sum_{j=1}^{m} [L_j]/K_{j(Ri)})\}]}$$

$$(168)$$

$$[L_yRL_x] = \frac{[R]_t}{1 + (K_{y(Rx)}/[L_y])[1 + \sum_{\substack{i=1 \\ i \neq y}}^{m} [L_i]/K_{i(Rx)} + (K_{Rx}/[L_x])\{1 + \sum_{i=1}^{m} [L_i]/K_{iR} + \sum_{\substack{i=1 \\ i \neq x}}^{m} ([L_i]/K_{Ri})(1 + \sum_{j=1}^{m} [L_j]/K_{j(Ri)})\}]}$$

$$(169)$$

$$[L_xRL_x] = \frac{[R]_t}{1 + (K_{x(Rx)}/[L_x])[1 + \sum_{\substack{i=1 \\ i \neq x}}^{m} [L_i]/K_{i(Rx)} + (K_{Rx}/[L_x])\{1 + \sum_{i=1}^{m} [L_i]/K_{iR} + \sum_{\substack{i=1 \\ i \neq x}}^{m} ([L_i]/K_{Ri})(1 + \sum_{j=1}^{m} [L_j]/K_{j(Rx)})\}]}$$

$$(170)$$

3.5.2 Numerical solutions

When only the total concentrations of the ligands are known *a priori*, the free concentrations of all species in Scheme VI can be calculated from the equations of state for R and L_i (eqns 171 and 173). Levels of binding in turn are calculated by substituting for [R] and $[L_i]$ in eqns 145–151 or for $[L_i]$ in eqns 154–163 or 166–170. As in the case of Scheme V, it can prove convenient in subsequent analyses to have included a separate population of sites distinct from R. Such an addition is shown here as S, which is assumed for simplicity to bind L_i according to Scheme I (i.e. $[L_i][S]/[L_iS] = K_{iS}$) (eqn 172); more complex models might be indicated in particular situations, given sufficient data. The sites S could represent a distinct population labelled by a relatively unselective probe or perhaps a non-interacting, protomeric form of R:

$$-[R]_t + [R] + \sum_{i=1}^{m} ([RL_i] + [L_iR] + \sum_{j=1}^{m} [L_iRL_j]) = 0 \qquad (171)$$

$$-[S]_t + [S] + \sum_{i=1}^{m} [L_iS] = 0 \tag{172}$$

$$-[L_x]_t + [L_x] + [RL_x] + [L_xR] + \sum_{i=1}^{m} ([L_xRL_i] + [L_iRL_x]) + [L_xS] = 0. \tag{173}$$

The expansion for the calculation of [R], [S] and $[L_i]$ according to the Newton–Raphson method is shown below (eqn 174). The individual functions are defined in eqns 175–177, and the elements of the matrix of partial derivatives are defined in eqns 178–185; partial derivatives not shown are equal to zero. The quantity **x** is the vector of $2 + m$ variables (i.e. [R], [S], $[L_1]$, $\cdots$ $[L_m]$), and **a** is the current point at each iteration:

$$0 = f_R\,(\mathbf{a}) + \frac{\partial f_R}{\partial[R]}\,\delta[R] + \frac{\partial f_R}{\partial[S]}\,\delta[S] + \frac{\partial f_R}{\partial[L_1]}\,\delta[L_1] + \ldots + \frac{\partial f_R}{\partial[L_m]}\,\delta[L_m]$$

$$0 = f_S\,(\mathbf{a}) + \frac{\partial f_S}{\partial[R]}\,\delta[R] + \frac{\partial f_S}{\partial[S]}\,\delta[S] + \frac{\partial f_S}{\partial[L_1]}\,\delta[L_1] + \ldots + \frac{\partial f_S}{\partial[L_m]}\,\delta[L_m]$$

$$0 = f_{L1}\,(\mathbf{a}) + \frac{\partial f_{L1}}{\partial[R]}\,\delta[R] + \frac{\partial f_{L1}}{\partial[S]}\,\delta[S] + \frac{\partial f_{L1}}{\partial[L_1]}\,\delta[L_1] + \ldots + \frac{\partial f_{L1}}{\partial[L_m]}\,\delta[L_m] \tag{174}$$

$$\vdots \qquad\qquad \vdots \qquad\quad \vdots \qquad\qquad \vdots \qquad\qquad\qquad \vdots$$

$$0 = f_{Lm}(\mathbf{a}) + \frac{\partial f_{Lm}}{\partial[R]}\,\delta[R] + \frac{\partial f_{Lm}}{\partial[S]}\,\delta[S] + \frac{\partial f_{Lm}}{\partial[L_1]}\delta[L_1] + \ldots + \frac{\partial f_{Lm}}{\partial[L_m]}\,\delta[L_m],$$

where

$$f_R(\mathbf{x}) = -[R]_t + [R]\left[1 + \sum_{i=1}^{m}\left\{\frac{[L_i]}{K_{iR}} + \frac{[L_i]}{K_{Ri}}\left(1 + \sum_{j=1}^{m}\frac{[L_j]}{K_{j(Ri)}}\right)\right\}\right] \tag{175}$$

$$f_S(\mathbf{x}) = -[S]_t + [S]\left\{1 + \sum_{i=1}^{m}\frac{[L_i]}{K_{iS}}\right\} \tag{176}$$

$$f_{Lx}(\mathbf{x}) = -[L_x]_t + [L_x]\left\{1 + \frac{[R]}{K_{xR}} + \frac{[R]}{K_{Rx}} + \frac{[R]}{K_{Rx}}\sum_{i=1}^{m}\frac{[L_i]}{K_{i(Rx)}}\right.$$
$$\left. + \sum_{i=1}^{m}\frac{[L_i][R]}{K_{Ri}K_{x(Ri)}} + \frac{[S]}{K_{xS}}\right\} \tag{177}$$

$$\frac{\partial f_R}{\partial[R]} = 1 + \sum_{i=1}^{m}\left\{\frac{[L_i]}{K_{iR}} + \frac{[L_i]}{K_{Ri}}\left(1 + \sum_{j=1}^{m}\frac{[L_j]}{K_{j(Ri)}}\right)\right\} \tag{178}$$

$$\frac{\partial f_R}{\partial [L_x]} = [R] \left\{ \frac{1}{K_{xR}} + \frac{1}{K_{Rx}} + \frac{1}{K_{Rx}} \sum_{i=1}^{m} \frac{[L_i]}{K_{i(Rx)}} + \sum_{i=1}^{m} \frac{[L_i]}{K_{Ri}K_{x(Ri)}} \right\} \quad (179)$$

$$\frac{\partial f_S}{\partial [S]} = 1 + \sum_{i=1}^{m} \frac{[L_i]}{K_{iS}} \quad (180)$$

$$\frac{\partial f_S}{\partial [L_x]} = \frac{[S]}{K_{xS}} \quad (181)$$

$$\frac{\partial f_{Lx}}{\partial [R]} = \frac{[L_x]}{K_{xR}} + \frac{[L_x]}{K_{Rx}} + \frac{[L_x]}{K_{Rx}} \sum_{i=1}^{m} \frac{[L_i]}{K_{i(Rx)}} + \sum_{i=1}^{m} \frac{[L_i][L_x]}{K_{Ri}K_{x(Ri)}} \quad (182)$$

$$\frac{\partial f_{Lx}}{\partial [S]} = \frac{[L_x]}{K_{xS}} \quad (183)$$

$$\frac{\partial f_{Lx}}{\partial [L_x]} = 1 + \frac{[R]}{K_{xR}} + \frac{[R]}{K_{Rx}} \left\{ 1 + 2\frac{[L_x]}{K_{x(Rx)}} + \sum_{i=1}^{m} \frac{[L_i]}{K_{i(Rx)}} \right\}$$

$$+ \sum_{i=1}^{m} \frac{[L_i][R]}{K_{Ri}K_{x(Ri)}} + \frac{[S]}{K_{xS}} \quad (184)$$

$$\frac{\partial f_{Lx}}{\partial [L_y]} = \frac{[L_x][R]}{K_{Rx}K_{y(Rx)}} + \frac{[R][L_x]}{K_{Ry}K_{x(Ry)}} \quad (x \neq y). \quad (185)$$

Successful convergence to the appropriate values of [R], [S] and [L$_i$] can depend upon the initial estimates. If depletion is relatively low, a simple procedure is to assume initially that the free and total concentrations are equal for all ligands. The concentrations of all complexes can then be calculated from eqns 166–170 and an appropriate expression for [L$_i$S] (e.g. eqn 45 with $n = 1$); the corresponding values of **x** can be obtained by subtraction (eqns 171–173) and used in the first cycle of the Newton–Raphson procedure. This approach tends to fail when [L$_i$] is negative for one or more ligands; that is, when depletion is sufficiently high that the estimate of total bound L$_i$ exceeds [L$_i$]$_t$. In that event, successful convergence generally occurs if the starting value of [L$_i$] is taken as zero for all ligands.

Scheme VI with two ligands results in a 4 × 4 matrix, only two elements of which are zero. Moderate quantities of data are handled expeditiously by newer processors, particularly those equipped with a floating point accelerator or equivalent capability. CPU times may be inconveniently long with pooled data from several experiments or, for example, when serial analyses are performed to map individual parameters. For applications with particularly intensive numerical requirements, the system can be reduced from four variables to two. Eqns 175 and 176 are linear in [R] and [S] respectively; it is therefore straightforward to substitute for those quantities in the corresponding expressions for each ligand (i.e. eqn 177) and thereby obtain the function $f_{Lx}([L_1], \ldots, [L_m], [R]_t, [S]_t, \mathbf{K})$, in which $\mathbf{K}$ is the set of all dissociation constants. Manual substitution is not recommended; the final equation contains 78 terms in the simple case of Scheme VI with two ligands, and the complexity increases with the number of interacting sites on the receptor. The resulting polynomials are quartic in $[L_x]$, in contrast to the quadratic expressions shown above (eqn 177), and the increase in degree can narrow the range of starting values that lead to the appropriate root. When only two ligands are involved, particularly awkward systems can be plotted on three-dimensional co-ordinates to devise a robust strategy for selecting the initial values of $[L_x]$ (cf. *Figure 2*).

If the usefulness of the model is unclear at the outset, it may be advantageous to opt for expediency rather than efficiency in preliminary analyses. Given the speed of modern microprocessors, the fastest route to an initial answer is probably to minimize the algebra and to solve a somewhat larger system of equations; the potential for technical errors is also reduced, and the program can be used to check the results of more efficient routines developed subsequently. The solution of complicated models can also be simplified if it is known, for example, that one of the ligands undergoes no appreciable depletion (cf. eqn 68); each such assumption reduces the number of simultaneous equations by one.

3.5.3 General comments

In the above presentation, all equilibria are defined in terms of their microscopic equilibrium dissociation constants (eqns 145–151); also, the vacant receptor is potentially asymmetric with respect to L_i (i.e. $K_{iR} \neq K_{Ri}$), and all ligands can differ in affinity from each other. The question of asymmetry can be difficult to address in practice, since K_{iR} and K_{Ri} are not uniquely defined by the data typically acquired at graded concentrations of a probe or an unlabelled ligand. Practical applications of the model therefore tend to require a simplifying assumption regarding the ratio K_{iR}/K_{Ri}, which is generally taken as unity for all L_i.

It is convenient to express the affinity of the second equivalent of ligand in terms of that of the first, as illustrated in eqns 186–189 for two ligands competing for two sites:

$$K_{1(R1)} = aK_{1R} \text{ and } K_{1(1R)} = aK_{R1} \tag{186}$$

$$K_{2(R2)} = bK_{2R} \text{ and } K_{2(2R)} = bK_{R2} \tag{187}$$

$$K_{1(R2)} = cK_{1R} \text{ and } K_{2(1R)} = cK_{R2} \tag{188}$$

$$K_{2(R1)} = dK_{2R} \text{ and } K_{1(2R)} = dK_{R1}. \tag{189}$$

When K_{1R}/K_{R1} and K_{2R}/K_{R2} are taken as unity or otherwise assigned, the values of a and b can be estimated from data acquired at graded concentrations of L_1 and L_2. The mutual interactions between two different ligands are not necessarily symmetrical, but either c or d is redundant in the absence of other information; it generally is convenient to assign the same value to both parameters (i.e. $c = d$). When the probe and the unlabelled ligand are functionally identical (e.g. $K_{R1} = K_{R2}$, and $K_{1R} = K_{2R}$ in Scheme VI), the relative change in affinity effected by one equivalent in the binding of a second is the same for all possible combinations (i.e. $a = b = c = d$ in eqns 186–189). The values of c and d determine the apparent proportions of different 'states' that emerge when the inhibitory behaviour of an unlabelled ligand is described empirically in terms of the multi-site model (Scheme II).

Derivations and the resulting equations are less complex if simplifying assumptions are made at the outset, as illustrated above in the case of the multi-site model (e.g. eqns 66–68). In the long run, however, the most useful approach is to describe the model in the most general terms possible. Various assumptions can be imposed initially by appropriate switches built into the software, and the restrictions can be removed as required to incorporate further information. In the case of Scheme VI, the fitting procedure might optimize single values of K_i which would be substituted for both K_{Ri} and K_{iR} when solving eqns 166–170 or 171–173; similarly, a single value could be substituted for parameters $a–d$ (eqns 186–189) if the experiment involved the stable and tritiated forms of the same compound.

With larger oligomers, simplifying assumptions are unavoidable for most practical purposes other than simulation. The number of independent microscopic dissociation constants greatly outpaces the number of binding sites as the latter increases, particularly when two or more ligands are involved. It therefore becomes increasingly impractical to retain parameters that are likely to be redundant (e.g. K_{iR} and K_{Ri}). Even with models formulated in terms of their macroscopic dissociation constants (see Section 4.2.3), it is experimentally arduous to define the various parameters of more complex systems. Since there is often little or no structural information on the composition and nature of the putative oligomers, the binding data typically serve to define not only the parameters but also the model. An example of this dual purpose is the persistent question of the value of n when the behaviour of G protein-linked receptors is described empirically in terms of the multi-site model (Scheme II). If the model is vague, experiments are less readily optimized to yield specific parameters, and the required amount of experimental effort is correspondingly greater.

The model depicted in Scheme VI and the accompanying equations constitute a generic description of co-operativity in a bivalent receptor. The molecular basis for the interactions between sites is not explicit, and in that sense the model is more empirical than mechanistic. Although different states of affinity are implied for each site, it is important to recognize that only one state is populated in the absence of any ligand (i.e. R). The model is thus conceptually closer to that of Koshland *et al.* (9) than to that of Monod *et al.* (10), although no assumption is made *a priori* regarding the relative affinities of a ligand for the two sites when both are vacant (i.e. K_{iR} and K_{Ri}).

4. Empirical models

If the aim is simply to draw a curve through a series of experimentally acquired points, the computational problem is essentially one of interpolation or approximation. The former is typically resolved by finding the polynomial, spline or other function that passes through all of the points in an acceptable manner; the latter implies a process of smoothing in which the computed line is permitted to deviate from individual points, usually in order to eliminate fluctuations arising from experimental error. Parametric values obtained from such procedures can be accompanied by considerable uncertainty, but the parameters *per se* are often irrelevant except insofar as they determine the second and higher derivatives of the function; rather, the desired quantity is typically the value of the function at one or more points on the abscissa. Other properties of the relationship may be of interest, however, as in the example of an integral obtained by first computing the spline.

A particularly useful method of smoothing is to fit a model by adjusting the values of a manageable number of parameters. The model may be empirical in the sense that it is not presented as an explicit description of the underlying phenomenon; nevertheless, it may resemble the strictly mechanistic model, and the parameters of one will then be related qualitatively or quantitatively to those of the other. In such cases, the fitted parameters will have diagnostic or even physical relevance, depending upon the degree of similarity between the two models. Perhaps the most important advantage of such models is the quantitative summarization achieved when a set of data is expressed in terms of relatively few parameters.

4.1 The Hill equation

4.1.1 Mechanistic basis of the Hill equation

Saturable processes at equilibrium are frequently described in terms of the Hill equation, which can be expressed as in eqns 190 and 191:

$$f(x) = \frac{Y_{max}K^{n_H}}{K^{n_H} + x^{n_H}} \tag{190}$$

$$f(x) = \frac{Y_{max}x^{n_H}}{K^{n_H} + x^{n_H}} \tag{191}$$

The parameter Y_{max} represents the maximal signal, obtained when $x = 0$ (eqn 190) or as $x \to \infty$ (eqn 191); K is the value of x when $f(x) = Y_{max}/2$. The exponent n_H is the Hill coefficient and reflects the range of values of x over which $f(x)$ differs by some arbitrary value from $Y_{max}/2$ (e.g. $f(x) = 0.5 \pm 0.4$). The independent variable x is typically concentration entered in linear or, preferably, in logarithmic units (i.e. $x = [L]$, or $x = 10^{\log[L]}$).

Although presented here as an empirical expression, the Hill equation was originally put forward to describe an unlikely process in which a protein or other macromolecule binds n equivalents of a ligand (i.e. $R + nL \rightleftharpoons L_nR$, eqn 191). The scheme does not admit intermediate complexes (i.e. L_1R, L_2R, ..., $L_{n-1}R$), and the ligand therefore occupies n sites or none at all. In that event, the Hill coefficient equals the number of interacting sites (i.e. $n_H = n$). If binding is sequential, the model only fits when the system is highly co-operative with respect to each successive equivalent of the probe (e.g. $K_{1(R1)} < 10^{-4} K_{R1}$, and $K_{1(1R)} < 10^{-4} K_{1R}$ in Scheme VI); n_H equals n when the degree of positive co-operativity is sufficient for binding to be all or none in effect, albeit not in principle. Such extreme behaviour is improbable and becomes increasingly so at higher values of n.

In practice, intermediate complexes contribute appreciably to the total signal at subsaturating concentrations of the ligand; indeed, such complexes can account for virtually all of the signal when one equivalent of a ligand causes successive equivalents to bind more weakly. Total specific binding is therefore described by rational functions containing terms in intermediate powers of x (i.e. eqn 164; cf. eqn 191). The Hill coefficient can be defined as the slope of a relationship between fractional signal f_s and the concentration of the ligand:

$$n_H = \frac{d \log\{f_s/(1 - f_s)\}}{d \log [L]}. \tag{192}$$

The relationship is a straight line when binding is all or none as described above:

$$\log \frac{f_s}{1 - f_s} = n_H \log [L] + n_H \log K. \tag{193}$$

Rational functions of the form expected in practice are not linear when rearranged as $\log \{(f_s/(1 - f_s))\}$ versus $\log [L]$, and the value of n_H is therefore expected to vary with the concentration of the ligand.

When the Hill coefficient is not a constant, the fitted value of n_H from eqn 190 or 191 is an average determined by the range of values defined by the data; it sometimes is referred to as the *pseudo Hill coefficient* or *slope factor*, reflecting the likelihood of systematic discrepancies between the model and the data. Such discrepancies can vary in magnitude and may be virtually undetectable. The model nevertheless embodies a dubious mechanistic assumption, and the parameters must be interpreted with caution. These limitations notwithstanding, the Hill equation has been widely used as an empirical model. It can be helpful in identifying binding that deviates from apparent homogeneity and provides a convenient vehicle for illustrating the nature and limitations of empirical analyses.

4.1.2 Mechanistic implications of the Hill coefficient

The main advantage of the Hill equation lies in the diagnostic usefulness of the Hill coefficient. There are two conditions for the diagnosis to be mechanistically informative: first, the system must have been at thermodynamic equilibrium at the time of the measurement; second, the independent variable must represent the concentration of *unbound* ligand. If both conditions are met, the Hill coefficient is a convenient if sometimes crude index of the range of x required to define the binding profile; that in turn is a reflection of the nature of the interaction between the ligand and the receptor.

Hill coefficients indistinguishable from 1.0 occur when an increase in log ($[L]/K$) from -0.954 to $+0.954$ is accompanied by an increase in $f(x)/Y_{max}$ from 0.1 to 0.9 (eqn 191). The functional form is that of a rectangular hyperbola, which in turn is characteristic of identical and mutually independent sites. The receptors therefore behave as such with respect to the ligand at hand and under the conditions of the experiment, but little more can be inferred in the absence of further information; virtually any mechanistic model will collapse to a rectangular hyperbola with the right parameters.

Hill coefficients exceed unity when the range of log [L] required to increase fractional occupancy from 0.1 to 0.9 is less than 1.91 log unit. Such behaviour is a good indicator of positive co-operativity if artefacts have been avoided and binding is measured at thermodynamic equilibrium. The value of $\dot{n}_H$ provides a lower limit on the stoichiometry of the interaction between ligand and receptor (i.e. $n_H \leqslant n$); as noted above, n_H equals n only in the unlikely event that binding is sufficiently co-operative to be all or none. Positively co-operative systems tend to be reasonably well described by the Hill equation, the presence of intermediate complexes notwithstanding. In the example of a co-operative dimer, simulated data and best fits of eqn 191 are almost superimposable even when the fitted value of n_H is mid-way between one and two (e.g. curve c in *Figure 3*).

Hill coefficients are less than unity when the range of log [L] exceeds 1.91 log unit for an increase in fractional occupancy from 0.1 to 0.9. Such behaviour is consistent with negative co-operativity (*Figure 3*), although other phenomena can yield similar or identical results. Data from negatively co-operative systems can deviate appreciably from best fits of the Hill equation, as illustrated by the comparison shown in the inset to *Figure 3*. The model predicts two maxima in the first derivative of the semi-logarithmic binding function; in contrast, the first derivative of eqn 191 ($x = 10^{\log [L]}$) is characterized by a single maximum centred at $x = 10^{\log K}$. Systematic deviations can render the fitted values of n_H highly sensitive to the weighting

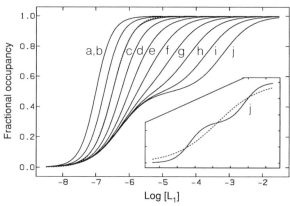

Figure 3. Binding in a co-operative system and its approximation by the Hill equation. The solid lines were calculated according to eqns 166–170 (Scheme VI) with the probe as the only ligand (i.e. $m = 1$). The value of log K_{R1} and log K_{1R} was −6.0 throughout, and the values of log a (eqn 186) for curves a–j were as follows: a, −2.0; b, −1.5; c, −1.0; d, −0.5; e, 0.0; f, +0.5; g, +1.0; h, +1.5; i, +2.0; j, +2.5. Curve j is replotted in the inset. The dotted lines represent the best fits of eqn 191 to curves c and j; Y_{max} was fixed at 1.0, and the fit was obtained by optimizing the values of K and n_H. The fitted values of n_H for curves a–j are as follows: a, 1.79; b, 1.50; c, 1.46; d, 1.25; e, 1.00; f, 0.88; g, 0.80; h, 0.65; i, 0.50; and j, 0.36.

procedure and to related factors such as whether the asymptotic values of the function are taken as fixed or variable parameters. As the magnitude of the deviations increases, the Hill coefficient becomes progressively less informative as an empirical descriptor of the data.

The Hill coefficient only distinguishes unambiguously between positive and negative co-operativity when n is two and x is the probe. When n exceeds two, successive equivalents of the probe may elicit both positive and negative co-operativity. Such complexity cannot be described by a single parameter, and a reliance on the Hill coefficient may obscure behaviour evident in the data *per se*. Ambiguity can also arise from the binding of a probe at graded concentrations of an unlabelled ligand. The Hill coefficient may be less than unity, but such a dispersion can be the net result of negative co-operativity between successive equivalents of the unlabelled ligand and positive co-operativity between the unlabelled ligand and the probe. Positive co-operativity between a probe and an unlabelled ligand can also result in a bell-shaped curve in which binding first increases and then decreases at increasing concentrations of the latter. These considerations highlight the limitations of the Hill coefficient and illustrate the constant need for visual inspection of the data and the fitted curve.

Hill coefficients less than unity are predicted by various mechanistic schemes other than negative co-operativity. An important example is that in which n classes of distinct and mutually independent sites differ in affinity for the ligand (Scheme II) (*Figure 4*). Binding is described by a rational function, and the patterns therefore resemble those obtained with negative co-operativity (see Section 4.2); the highest power is n in the denominator and either $n - 1$ or n in the numerator, depending upon whether the independent variable is the probe (eqn 48) or an unlabelled ligand (eqn 49). Terms in intermediate powers of [L] result in a functional form distinct from that of eqn 190 or 191, and the fitted curves can thus deviate appreciably from the experimental data (*Figure 4*). When L is the probe, as in the examples of *Figure 4*, the origin of the heterogeneity is clear; when L is an unlabelled ligand, the multiphasic behaviour may reflect the differential affinity of the probe, the unlabelled ligand or both.

Hill coefficients less than unity can also arise from an intrinsically homogeneous population of sites if binding is sensitive to a perturbing species X that interacts with the receptor (Schemes IV and V) (*Figure 5*). The phenomenon requires that the total concentration of X be comparable with or less than that of R. When $[X]_t$ equals $[R]_t$, the dispersion arises from the change in unbound X that occurs when the ligand promotes the formation or

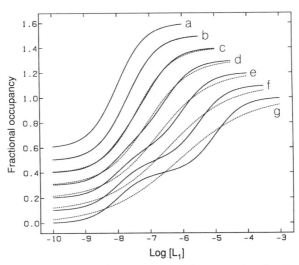

Figure 4. Binding to multiple classes of sites and its approximation by the Hill equation. The solid lines were calculated according to eqn 45 with two classes of sites (i.e., $n = 2$) and the probe as the only ligand (i.e. $m = 1$). Parametric values common to all curves were as follows: $\log K_{11} = -8.0$, $[R_1]_t/([R_1]_t + [R_2]_t) = 0.4$ and $[R_2]_t/([R_1]_t + [R_2]_t) = 0.6$. The values of $\log K_{12}$ for individual curves were as follows: a, -8.0; b, -7.5; c, -7.0; d, -6.5; e, -6.0; f, -5.5; g, -5.0. The dotted lines represent best fits of eqn 191; Y_{max} was fixed at 1.0, and the fit was obtained by optimizing the values of $\log K$ and n_H. The fitted values of n_H for cuves a–g are as follows: a, 1.00; b, 0.94; c, 0.81; d, 0.67; e, 0.55; f, 0.46, g, 0.39. Successive curves are offset on the ordinate by 0.1 unit for clarity.

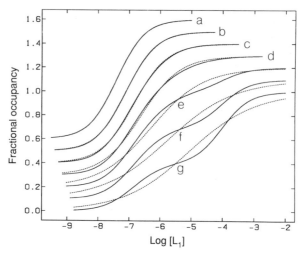

Figure 5. Binding in the presence of an allosteric modulator and approximation by the Hill equation. The solid lines were calculated according to eqns 106–107 with the probe as the only ligand to the receptor. Parametric values common to all curves were as follows: log K_{L_i} = −4.0, log $K_{L,X}$ = −8.0, log ([R]$_t$/K_X) = −1.0. The values of [G]$_t$/[R]$_t$ for individual curves were as follows: a, 4.0; b, 2.0; c, 1.3; d, 1.0; e, 0.8; f, 0.6; g, 0.4. The dotted lines represent best fits of eqn 191; Y_{max} was fixed at 1.0, and the fit was obtained by optimizing the values of log K and n_H. The fitted values of both parameters for curves a–g are as follows: a, −7.40 and 0.97; b, −7.10 and 0.91; c, −6.83 and 0.82; d, −6.60 and 0.70; e, −6.27 and 0.52; f, −5.72 and 0.41; g, −5.07 and 0.41. Successive curves are offset on the ordinate by 0.1 unit for clarity.

dissociation of the complex between R and X (e.g. $K_{L_i}/K_{L,X} = K_X/L_{XL_i}$ in Scheme V); it follows that the Hill coefficient is indistinguishable from unity if [X]$_t$ exceeds [R]$_t$ by more than about threefold. When [X]$_t$ is less than [R]$_t$, the dispersion arises from any ligand-dependent change in free X as well as from the certainty that some receptors will remain uncoupled from X under any conditions. The functional form of the model differs from eqn 190 or 191 under all conditions (cf. eqns 106–107), except when [X]$_t$ greatly exceeds [R]$_t$ and n_H is therefore unity (cf. eqns 100–101 and 103–104); best fits of eqn 190 or 191 can thus be expected to deviate systematically from the data (*Figure 5*).

The foregoing examples of Hill coefficients less than unity all concern systems at thermodynamic equilibrium. In cyclical models such as Schemes III and IV, the attainment of equilibrium requires the maintenance of microscopic reversibility (i.e. $K'_1/K''_1 = K_{11}/K_{12}$ in Scheme III, where $K''_1 = $ [L$_1$R$_1$]/[L$_1$R$_2$]; $K_X/K_{XL} = K_L/K_{LX}$ in Scheme IV); the system will otherwise attain a steady state provided that the demands of energy-dependent processes are met over the course of the experiment. A discussion of such systems is beyond the scope of this chapter, but it is worth noting that the steady state equations defining the concentration of each species typically yield rational functions in which total binding is at least quadratic in [L] (cf. eqn 48). Even a model such as the simplest variant of Scheme III (i.e. $m = 1$, $n = 1$) can exhibit time-independent binding patterns characterized by Hill coefficients greater or less than unity, depending upon the values of individual rate constants (11).

4.1.3 Artefactual effects on the Hill coefficient

In the various examples described above, the Hill coefficient is a reflection of behaviour intrinsic to the model. It is assumed that the system is at thermodynamic equilibrium and that the receptor only binds a negligible fraction of the ligand present. A violation of either assumption can lead to Hill coefficients greater than otherwise would be obtained. The effect is illustrated by the examples in *Figure 6*; in each case, binding at equilibrium is described by a rectangular hyperbola with respect to the concentration of unbound ligand (i.e. $n_H = 1$).

The data in *Figure 6(a)* were computed according to eqn 194 (12), which is the integrated rate equation for the reversible interaction between a probe L and a receptor R as shown below; the model corresponds to Scheme II or Scheme III with one ligand and one form or state of the receptor ($m = 1, n = 1$):

$$L + R \overset{k_{+1}}{\underset{k_{-1}}{\rightleftharpoons}} LR$$

$$[LR] = \frac{[L]_t[R]_t}{\frac{1}{2}([L]_t + [R]_t + K_1) + \beta\coth(\beta k_{+1} t)}, \tag{194}$$

where $K_1 = \dfrac{k_{-1}}{k_{+1}}$

$$\beta^2 = \frac{1}{4}([L]_t + [R]_t + K_1)^2 - [L]_t[R]_t.$$

The parameter k_{+1} is the second-order rate constant for the association of L with R; k_{-1} is the first-order rate constant for dissociation of the LR complex and enters into the equation via the equilibrium dissociation constant K_1. Both the ligand and the receptor enter as total concentration (i.e. $[L]_t = [L] + [LR]$, and $[R]_t = [R] + [LR]$), but the values of $[R]_t$ and K_1 used in the simulations are such that $[LR] \ll [L]_t$ at any time t (i.e. $[L] \approx [L]_t$). Upon attainment of equilibrium, the expected binding profile is characterized by a Hill coefficient of unity and a value of K (eqn 191) equal to K_1 (eqn 194) (curve a in *Figure 6(a)* and *Table 2*). Measurements at progressively earlier times result in a steepening of the curve and an increase in the concentration of L required to achieve half-maximal occupancy (EC_{50}, *Table 2*).

The data in *Figure 6(b)* were computed according to eqn 63 for the binding of L to a single class of sites (i.e. Scheme II, $m = 1$ and $n = 1$). At concentrations of receptor well below the equilibrium dissociation constant,

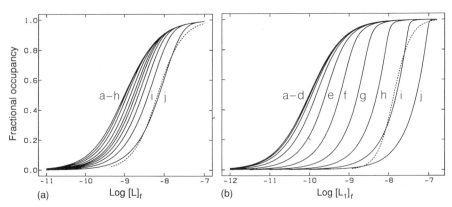

Figure 6. Effects of pre-equilibrium and depletion on the binding of a probe. The solid lines were simulated according to eqn 194 (a) or 63 (b), as described in the footnote to *Table 2*. Curve a in each frame represents binding at equilibrium and sufficiently low concentrations of receptor to be without effect on the concentration of unbound probe. Curves b–j represent binding at times prior to equilibration (a) and at depleting concentrations of receptor (b). The dashed lines represent best fits of eqn 191 to the simulated data, as described in the footnote to *Table 2* (a, curve j; b, curve i).

the binding profile is characterized by a Hill coefficient of unity and a value of K (eqn 191) equal to K_{11} (eqn 63) (curve a in *Table 2* and *Figure 6(b)*). Higher concentrations result in a steepening of the curve and an increase in EC_{50}. These effects reflect the appreciable fraction of total probe that binds to the receptor at equilibrium when the concentration of the latter approximates or exceeds the equilibrium dissociation constant. At very high concentrations (i.e. $[R_1]_t > 10K_{11}$), the receptor binds virtually all of the probe until the latter is present in equimolar amounts or greater; under such conditions, binding increases linearly with $[L_1]_t$, and the value of EC_{50} is defined by $[R_1]_t$ rather than K_{11} (curves f–j, *Table 2*).

Binding profiles acquired prior to the attainment of equilibrium or at depleting concentrations of receptor are not well described by the Hill equation, although the Hill coefficient will tend to exceed unity as illustrated by the results summarized in *Table 2*. The discrepancies are illustrated for two of the examples plotted in *Figure 6*, where the dashed lines represent best fits of eqn 191 to the simulated data (curve j, *Figure 6(a)*; curve i, *Figure 6 (b)*); the lack of agreement is particularly striking in *Figure 6 (b)*, where fractional occupancy is a linear function of $[L_1]_t$ for over 80% of the curve. Substantial differences in functional form between that described by the data and that of the expression being fitted render the parametric values highly sensitive to the weighting procedure and to the completeness of the data.

In the examples of *Figure 6(b)*, identical and mutually independent sites appear to exhibit positive co-operativity with respect to the binding of a single ligand. More subtle manifestations of depletion can arise when two or more

Table 2. Artefactual effects on the binding of a probe[a]

Curve	Pre-equilibrium (*Figure 6(a)*)				Depletion of unbound probe (*Figure 6(b)*)					
	Eqn 194		Eqn 191		Eqn 63				Eqn 191	
	Time t (min)	$-\log EC_{50}$	$-\log K$	n_H	$[R_1]_t$ (nM)	$-\log EC_{50}$	EC_{50} (nM)	$EC_{50}/[R]_t$	$-\log K$	n_H
a	600	9.000	9.00	1.00	0.0001	10.000	0.100	1000	10.00	1.00
b	180	8.978	8.97	1.04	0.0100	9.979	0.105	10.5	9.98	1.02
c	120	8.931	8.93	1.09	0.0316	9.936	0.116	3.66	9.94	1.06
d	80	8.851	8.86	1.16	0.100	9.824	0.150	1.50	9.83	1.15
e	60	8.776	8.79	1.21	0.316	9.588	0.258	0.816	9.61	1.34
f	50	8.723	8.74	1.24	1.00	9.222	0.600	0.600	9.26	1.61
g	40	8.653	8.68	1.27	3.16	8.775	1.68	0.532	8.82	1.85
h	30	8.554	8.58	1.31	10.0	8.293	5.10	0.510	8.35	2.00
i	20	8.405	8.44	1.35	31.6	7.799	15.9	0.503	7.86	2.08
j	10	8.132	8.17	1.39	100	7.300	50.1	0.501	7.36	2.06

[a] Data were simulated to illustrate the consequences of failing to attain equilibrium (eqn 194) and of depletion of the unbound probe through formation of the ligand–receptor complex (eqn 63). The values of t in eqn 194 are listed in the table, and other parameters were as follows: $K_{+1} = 10\,\mu M^{-1}\,min^{-1}$, log $K_1 = -9.0$ and $[R]_t = 1.0\,pM$. The values of $[R_1]_t$ in eqn 63 are listed in the table, and K_{11} was 0.1 nM (log $K_{11} = -10.0$). The values of EC_{50} refer to the simulated data and indicate the total concentration of the probe corresponding to 50% occupancy of the sites. The values of log K and n_H are from best fits of eqn 191 to the simulated data; Y_{max} was fixed at the asymptotic value of the function as $[L] \rightarrow \infty$, and no weighting was applied. The data are illustrated in *Figure 6*.

ligands are present, as occurs when unlabelled ligands are characterized via their inhibitory effect on the binding of a probe. One example involves G protein-linked receptors, which can often be described empirically in terms of two classes of sites (Scheme II, $n = 2$) differing in affinity for agonists L_2 but not antagonists L_1. The latter are exploited as probes in many studies, and their affinities can be comparable with the total concentration of receptors; accordingly, the unbound probe is usually depleted to some extent. At low concentrations of an agonist, that quantity of probe otherwise bound to the sites of higher affinity for the agonist is distributed between the sites of lower affinity and the pool of free ligand. Binding patterns obtained for the unlabelled ligand are thus steepened to an extent that depends upon the depletion of the probe, irrespective of whether or not the unlabelled ligand is itself depleted.

The artefact is illustrated in *Figure 7*, where the data have been calculated according to eqn 68 ($n = 2$). Curves a–h reflect increasing depletion of the probe, achieved by decreasing the value of K_{11}. Curves b and g from *Figure 7* are redrawn as the dashed lines in *Figure 8(a)* and (*b*) respectively. The surfaces in *Figure 8* have been calculated according to eqn 45 and illustrate the dependence of total binding on the free concentrations of both ligands. A comparison of the dashed lines in *Figure 8* reveals the changes in [L_1] that are

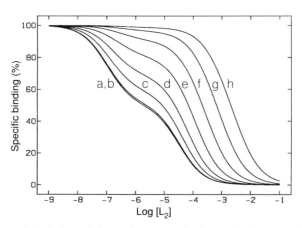

Figure 7. Effect of depletion of the probe on the inhibitory behaviour of an unlabelled ligand in the multi-site model. The data were calculated according to eqn 68 (Scheme II) and normalized to the asymptotic value of [L_1]$_b$ taken as 100. There are two classes of sites (i.e. [R_3]$_t$ = 0), and parametric values common to all of the curves were as follows: log K_{21} = -7.0, log K_{22} = -4.5, [R_1]$_t$ = 250 pM, [R_2]$_t$ = 250 pM and [L_1]$_t$ = 0.1 nM. Values of log K_{11} are as follows for curves a–h in order of increasing depletion of L_1: a, -7.0; b, -8.0; c, -9.0; d, -9.5; e, -10.0; f, -10.5; g, -11.0; h, -11.5. Occupancy of the sites by L_1 when [L_2] equals zero is 0.099%, 0.994%, 6.38%, 11.7%, 16.1%, 18.6%, 19.5% and 19.8% respectively; depletion of [L_1] when [L_2] equals zero is 0.497%, 4.72%, 31.9%, 58.3%, 80.7%, 92.8%, 97.6% and 99.2% respectively. Curves b and g are shown in *Figure 8*.

disregarded when the three-dimensional system is mapped onto two-dimensional co-ordinates as in *Figure 7*. Essentially the same phenomenon accounts for the dispersion encountered with Scheme IV and equal concentrations of R and X (curve d, *Figure 5*), although in that case there need be no depletion of the probe.

The artefacts described above relate exclusively to specific binding and, in particular, to the shape or steepness of the binding profile. Non-specific binding generally reflects the concentration of unbound probe (eqn 9) and is thus sensitive to any reduction brought about through binding to the receptor

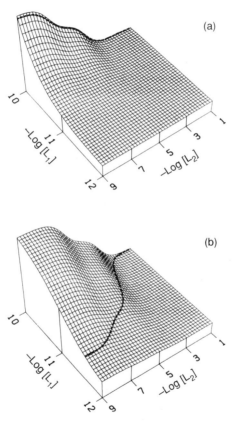

Figure 8. Effect of depletion of the probe on the path traced by an unlabelled ligand in the multi-site model. The surfaces were calculated according to eqn 45 with two ligands ($m = 2$) and two classes of sites ($n = 2$). Parametric values were as listed in the legend to *Figure 7*((a), curve b, log $K_{11} = -8.0$; (b), curve g, log $K_{11} = -11.0$). The ordinate in each frame represents total specific binding of the probe $[L_1]_b$. Maximal binding plotted in the figure represents 0.985% of $[R]_t$ in (a) and 90.9% of $[R]_t$ in (b) (log $[L_1] = -10.0$, log $[L_2] = -9.0$); the scales therefore differ by about 100-fold. The dashed lines correspond to curves b and g in *Figure 7* and indicate the level of binding when the concentrations of free and bound probe sum to 0.1 nM.

(eqn 14); when depletion is appreciable, naïve analytical procedures will underestimate the total number or concentration of sites (eqn 22). The propensity of lipophilic probes to locate in the membrane exacerbates the effects of depletion when the ligand binds to receptors accessible from the aqueous phase; also, the lower effective concentration may increase the time required to reach equilibrium under the conditions of the experiment. Finally, non-specific binding may itself be overestimated or underestimated, perhaps owing to a false assumption regarding the level of binding in the presence of an unlabelled ligand; such errors are generally reflected in the Hill coefficient. An awkward property of artefacts is that one often leads to another, and the cumulative effect of several may render the data virtually uninterpretable. It is helpful to distinguish between artefacts arising from the reaction *per se* and those introduced subsequently through ill advised manipulations of the data; the latter can generally be avoided if the equations are derived to reflect the form in which the data are acquired.

The scope for artefacts is limited only by the perversity of biological systems and our own ingenuity. The list includes irreversible processes such as proteolysis of the receptor and hydrolysis or decomposition of the ligand, technical slips such as an error in the specific radioactivity, transport of the ligand, internalization of the receptor, receptors accessible to one ligand but not another, racemates, chemically impure probes, and more. The only protection is vigilance.

4.1.4 Practical implementation of the Hill equation

i. The probe as independent variable

In binding studies with a radiolabelled probe, the signal usually consists of a saturable and a non-saturable contribution. The latter is absent from eqn 191, and the simplest practical formula incorporating the Hill equation is therefore eqn 195; B_{obsd} represents total binding of the probe L, and $B_{max,sp}$ represents maximal saturable binding (cf. Y_{max} in eqn 191):

$$B_{obsd} = \frac{B_{max,sp} [L]^{n_H}}{K^{n_H} + [L]^{n_H}} + B_1 [L] + B_{[L]=0} \tag{195}$$

Non-saturable binding is shown as a linear function of [L] with slope B_1 and intercept $B_{[L]=0}$; the latter can presumably be taken as zero if the background signal from the counting machine has been subtracted previously.

In a well behaved and well characterized system, there is equivalence between the terms 'saturable' and 'specific' on the one hand and 'non-saturable' and 'non-specific' on the other; moreover, the former is a non-linear function of [L], and the latter is linear. Such terms can be ambiguous, however, as in the case when n_H equals unity and K exceeds the highest concentration of L. Eqn 195 then reduces to the expression $B_{obsd} = (B_{max,sp}/$

$K + B_1)$ [L] $+ B_{[L]=0}$, in which both the saturable and the non-saturable contributions are linear in [L].

In principle, all of the parameters in eqn 195 can be estimated by means of the radioligand alone. In practice, specific binding is a minor fraction of the total signal at values of [L] sufficient to define B_1, and B_1 is undefined when [L] is sufficiently small to permit the characterization of specific binding. The various parameters can be highly correlated during the fitting procedure, and parametric values may be accompanied by an unacceptable degree of uncertainty; also; the signal required to define B_1 may dominate chi-square and thereby subordinate that associated with $B_{max, sp}$, K and n_H. To avoid such problems, one generally obtains an independent estimate of B_1 by measuring binding under conditions when the quantity $B_{max, sp}$ $[L]^{n_H}/(K^{n_H} + [L]^{n_H})$ is effectively zero; that is typically achieved by including sufficient unlabelled ligand to preclude specific binding of the probe.

Experiments of this nature leave the investigator with two sets of data, which represent two extreme paths across the surface of B_{obsd} as defined by [L] and the concentration of the unlabelled ligand. Specific binding is the difference in B_{obsd} at each value of [L], and a convenient way to extract the various parameters is to fit eqn 195 simultaneously to all of the data. If all measurements were made under identical conditions apart from the concentrations of the two ligands, both sets of data can share single values of all parameters except $B_{max, sp}$. The latter can be assigned by means of switches included in the software; at each point, the fitting routine can select either a finite and probably optimized value, for data acquired in the absence of unlabelled ligand, or zero, for data acquired in the presence of unlabelled ligand.

The presence or absence of unlabelled ligand L_2 can be flagged by including the concentration *per se* in the data file. In the example shown below, y and x represent the dependent and independent variables respectively; of the latter, $x_{i,1}$ is the probe, and $x_{i,2}$ is the unlabelled ligand. Measurements in the absence and presence of the unlabelled ligand $(x_{1,2}, x_{2,2})$ result in $m + n$ values of y at different concentrations of the probe.

$$
\begin{array}{lll}
y_1 & , x_{1,1} & , x_{1,2} \\
y_2 & , x_{2,1} & , x_{1,2} \\
y_3 & , x_{3,1} & , x_{1,2} \\
\vdots & \vdots & \vdots \\
y_m & , x_{m,1} & , x_{1,2} \\
y_{m+1} & , x_{m+1,1} & , x_{2,2} \\
y_{m+2} & , x_{m+2,1} & , x_{2,2} \\
\vdots & \vdots & \vdots \\
y_{m+n} & , y_{m+n,1} & , x_{2,2}.
\end{array}
$$

359

Each value of $x_{i,2}$ can be tested, and the value of $B_{max,sp}$ is taken as zero whenever $[L_2]$ is sufficiently large; $B_{max,sp}$ is otherwise taken as the current value provided by the fitting routine. If concentrations are presented on a logarithmic scale (i.e. $x = 10^{\log[L]}$), the value of log $[L_2]$ can be entered as a very small number when L_2 is absent (e.g. $x = -15$). The numerical values of $x_{i,2}$ are irrelevant when B_{obsd} is calculated according to eqn 195 and other expressions in which the probe is the only independent variable; accordingly, other means of flagging the presence of the unlabelled ligand are equally effective. One advantage of the above arrangement is that the same files can be used in subsequent analyses with models, particularly mechanistic models, in which both ligands appear as explicit variables.

A switching procedure is unnecessary if the empirical model itself accommodates both ligands, as illustrated in eqn 196:

$$B_{obsd} = B_{max,sp} \left\{ \frac{[L_1]^{n_{H(1)}}}{K_1^{n_{H(1)}} + [L_1]^{n_{H(1)}}} \right\} \left\{ \frac{K_2^{n_{H(2)}}}{K_2^{n_{H(2)}} + [L_2]^{n_{H(2)}}} \right\} + B_1[L_1] + B_{[L_1]=0}.$$
(196)

For measurements at graded concentrations of L_1, either alone or together with a saturating concentration of L_2, neither K_2 nor $n_{H(2)}$ is defined by the data. It is therefore necessary to assign arbitrary values to both parameters, bearing in mind that the quantity $K_2^{n_{H(2)}}/(K_2^{n_{H(2)}} + [L]_2^{n_{H(2)}})$ must be virtually zero at the concentration of L_2 used in the experiment. It is up to the investigator to confirm that this is indeed so.

When the concentration of L_2 is either saturating or zero, the choice between eqn 195 with an explicit switch and eqn 196 is largely a matter of taste. Both procedures allow all parameters to be estimated simultaneously from the original data; $B_{[L]=0}$ can usually be fixed at zero if the background signal from the counting machine has been subtracted previously. Simultaneous optimization of all parameters, including B_1, is generally to be preferred when the model is expected to describe the data. As illustrated in *Figures 2–5*, however, specific binding can be expected to deviate systematically from the Hill equation, and the deviations are likely to be appreciable at values of n_H less than unity. It follows that only the non-specific component of binding may be represented accurately by eqn 195 (i.e. when $B_{max,sp} = 0$). This disparity can be avoided by estimating B_1 and $B_{[L]=0}$ from the estimates of non-specific binding taken separately; those parameters then can be fixed accordingly when eqn 195 is fitted to the data acquired in the absence of the unlabelled ligand. In most situations, inherent discrepancies between the Hill equation and the data represent a minor fraction of chi-square, and the parametric values are essentially the same irrespective of whether B_1 is estimated separately or together with the other parameters. Practical limitations on the concentration of the probe generally preclude the acquisition of data over a range sufficient for the most prominent discrepancies to be visible.

A somewhat different approach is first to subtract the appropriate estimate of non-specific binding from B_{obsd} measured in the absence of L_2, and then to analyse the difference in terms of eqn 191. Estimates of non-specific binding ($[L_2]^{n_{H(2)}} \gg K_2^{n_{H(2)}}$) and of specific plus non-specific binding ($[L_2] = 0$) are typically made in parallel at each concentration of the radioligand; the former can then be taken either as the measured value or as the smoothed value obtained by fitting a straight line to the relevant data. Only the smoothed values are available if measurements with and without the unlabelled ligand were made at different concentrations of the probe.

Fitting eqn 195 or 196 to the data taken as measured (i.e. B_{obsd}) is to be preferred over fitting eqn 191 to the difference between total and non-specific binding. The latter procedure underestimates the experimental error at higher concentrations of the radioligand as described below. If such an approach must be taken, it is generally preferable to subtract the smoothed rather than the measured estimates of non-specific binding; individual errors are not then transferred wholesale to the corresponding estimates of saturable binding submitted to the fitting procedure.

An awkward problem arises when supposedly non-specific binding is not linear with respect to the concentration of L_1. A resolution can be fudged by a point-by-point subtraction of measured values or by using a more suitable function to smooth the data acquired in the presence of L_2 (e.g. $B_2[L_1]^2 + B_1[L_1] + B_{[L]=0}$ instead of $B_1[L_1] + B_{[L]=0}$ in eqns 195 and 196). That may be overly simplistic, however, even for an empirical analysis; any curvature in the concentration-dependence of what is defined as non-specific binding probably warrants further investigation.

ii. The probe as independent variable with depletion

As illustrated in *Figure 6(b)*, the mechanistic implications of n_H are obscured when the receptor binds an appreciable fraction of the probe. If depletion is unavoidable, the free concentration must be either measured or calculated; the latter is less tedious but ought to be confirmed at least once by direct measurement. An explicit calculation requires that apparent specific binding B'_{sp} first be estimated from measurements of total and non-specific binding at the same total concentration of the probe (eqn 197, cf. eqn 20); if non-specific binding is indeed linear in $[L]_t$, B_1 can be obtained from the best straight line through the data acquired at saturating concentrations of an unlabelled ligand. A correction for the difference in non-specific binding between the two measurements yields specific binding B_{sp} (eqn 198, cf. eqn 23), which in turn is used to obtain the free concentration of the probe (eqn 199). Estimates of the Hill coefficient and other parameters then can be obtained from a best fit of eqn 191:

$$B'_{sp} = B_{obsd} - B_1 - B_{[L]_t=0} \tag{197}$$

$$B_{sp} = B'_{sp}/(1 - B_1) \tag{198}$$

$$[L] = [L]_t - B_{sp}. \tag{199}$$

Calculations to obtain B_{sp} and $[L]$ underestimate the error on the former in the absence of an appropriate weighting function (see Section 5.3) or other corrections (e.g. eqns 200–202):

$$(A \pm a) + (B \pm b) = (A + B) \pm (a^2 + b^2)^{1/2} \tag{200}$$

$$(A \pm a)(B \pm b) \simeq AB \pm (A^2b^2 + B^2a^2)^{1/2} \tag{201}$$

$$\frac{(A \pm a)}{(B \pm b)} \simeq \frac{A}{B} \pm \frac{A}{B}\left\{\frac{a^2}{A^2} + \frac{b^2}{B^2}\right\}^{1/2} \tag{202}$$

Also, experimental variations in B_{sp} are copied to $[L]$, and an error is therefore introduced into the supposedly independent variable. These problems are avoided if the data are analysed according to eqn 203, in which the dependent variable is total binding taken as measured, and the independent variable is total concentration of the probe (cf. eqns 14 and 18):

$$B_{obsd} = \{B_{sp} + B_1([L]_t - B_{sp})\} \cdot SA \cdot 2.22 \times 10^{12} + B_{[L]_t=0}. \tag{203}$$

Specific binding can be calculated from eqns 204 and 205 by means of the Newton–Raphson method:

$$f(B_{sp}) = -B_{sp} + B_{max,sp}\frac{([L]_t - B_{sp})^{n_H}}{K^{n_H} + ([L]_t - B_{sp})^{n_H}} \tag{204}$$

$$\frac{df(B_{sp})}{dB_{sp}} = -1 + B_{max,sp}\frac{n_HK^{n_H}([L]_t - B_{sp})^{n_H-1}}{\{K^{n_H} + ([L]_t - B_{sp})^{n_H}\}^2}. \tag{205}$$

This function is deceptively simple, and it can be tricky to find the root if Newton–Raphson is not combined with bracketing and bisection (see Section 3.3.3).

iii. The unlabelled ligand as independent variable

Data acquired at a single concentration of the probe and graded concentrations of an unlabelled ligand can be analysed in terms of eqn 206, in which $B_{[L_2]=0}$ and $B_{[L_2]\to\infty}$ represent the asymptotic values of the function in the limits of $[L_2]$:

$$B_{obsd} = B_{[L_2]\to\infty} + (B_{[L_2]=0} - B_{[L_2]\to\infty})\frac{K_2^{n_{H(2)}}}{K_2^{n_{H(2)}} + [L_2]^{n_{H(2)}}}. \tag{206}$$

The expression contains one independent variable, but the parameters are related to those of eqn 196 as follows:

$$B_{[L_2]=0} = \frac{B_{max,sp}[L_1]^{n_{H(1)}}}{K_1^{n_{H(1)}} + [L_1]^{n_{H(1)}}} + B_1[L_1] + B_{[L_1]=0} \tag{207}$$

$$B_{[L2]\to\infty} \geq B_1[L_1] + B_{[L_1]=0}. \tag{208}$$

It follows that the data can also be analysed in terms of eqn 196 if the value of $[L_1]$ is made available to the fitting procedure. Since eqn 206 is usually fitted by optimizing the values of $B_{[L_2]=0}$ and $B_{[L_2]\to\infty}$ in addition to those of K_2 and $n_{H(2)}$, analyses with eqn 196 will require fixed values of either $B_{max,sp}$ or $K_1^{n_{H(1)}}$ and of either B_1 or $B_{[L_1]=0}$; only one parameter from each pair is defined at a single concentration of the radioligand.

The uncertainty over $B_{[L_2]\to\infty}$ in eqn 208 illustrates the limitations of empirical models. If binding is mutually exclusive, sufficient concentrations of an unlabelled ligand are expected to reduce total binding of the probe to non-specific levels (i.e. $B_{[L_2]\to\infty} = B_1[L_1] + B_{[L_1]=0}$); $B_{[L_2]\to\infty}$ will otherwise exceed non-specific levels to the extent that the probe interacts with whatever complex is formed between the receptor and the unlabelled ligand. In this regard, it is worth recalling that eqn 206 can be rearranged to obtain eqn 209:

$$B_{obsd} = B_{[L2]=0} + (B_{[L2]\to\infty} - B_{[L2]=0})\frac{[L_2]^{n_{H(2)}}}{K_2^{n_{H(2)}} + [L_2]^{n_{H(2)}}}. \tag{209}$$

If $B_{[L_2]\to\infty}$ exceeds $B_{[L_2]=0}$, B_{obsd} increases with increasing $[L_2]$, thereby implying a heterotropic and positively co-operative effect of L_2 on the binding of L_1. Such a pattern would suggest that the parameters of eqn 196 ought to be redefined if that expression were to be used in favour of eqn 206 or 209.

Although two ligands may be present in the assay, attempts to accommodate both in an empirical model are questionable. Naïve variations on the Hill equation are probably wrong, except for extreme situations. Eqn 196 implies, for example, that $K_2^{n_{H(2)}}$ is independent of $[L_1]$ and that $K_1^{n_{H(1)}}$ is independent of $[L_2]$. Neither assumption is likely to be true; moreover, whatever relationships exist will be complicated by the differences in functional form between the Hill equation and any mechanistic scheme of physical relevance. The concentration of one ligand must therefore be the same among data that are to share the same parameters, except in the special case of non-specific binding defined by high concentrations of L_2. This caveat applies to measurements not only within the same experiment but also among different experiments if the results are eventually to be pooled or averaged. The converse of the above is that the values of empirical parameters are only relevant to the conditions under which they were determined.

Exceptions to this practice are sometimes made for G protein-linked receptors when the sites appear homogeneous to a radiolabelled antagonist ($n_{H(1)}$ = 1) and heterogeneous to agonists ($n_{H(2)} < 1$). If the parameter K_2 in eqn 196 or 206 is substituted according to eqn 210, the inhibitory potency of the agonist in the limit as $[L_1] \to 0$ (i.e. K_2') can be obtained directly from the fitting procedure. Such an inference is only valid if binding is strictly competitive, or at least appears so under the relevant conditions; also, the intrinsic value of $n_{H(1)}$ must equal unity at all concentrations of L_2. While these assumptions seem reasonable, they are seldom confirmed:

$$K_2 = K_2' \left\{ 1 + \frac{[L_1]}{K_{L_1}} \right\}. \tag{210}$$

iv. Multiple contributions to the specific signal

The saturable component of eqn 195 can be expanded to include multiple contributions as shown in eqn 211:

$$B_{\text{obsd}} = \sum_{j=1}^{n} \frac{B_{\text{max,sp}(j)} [L_1]^{n_{\text{H}(1j)}}}{K_{1j}^{n_{\text{H}(1j)}} + [L_1]^{n_{\text{H}(1j)}}} + B_1 [L_1] + B_{[L_1]=0}. \tag{211}$$

The quantity $B_{\text{max, sp}(j)}$ represents maximal specific binding of the probe to the sites of type j, where $j = 1, 2, \ldots, n$; K_{1j} is the concentration of L_1 that yields half-maximal occupancy at the sites of type j, and $n_{\text{H}(1j)}$ is the corresponding Hill coefficient. When specific binding is a summation of individual contributions, it is sometimes convenient to represent the maximal contribution of each as a fraction F_j of the total specific signal ($B_{\text{max, sp}}$) as illustrated in eqn 212. The total number of parameters in eqns 211 and 212 is the same, since $\sum_{j=1}^{n} F_j$ must equal unity; n classes of sites thus yield $n = 1$ values of F_j:

$$B_{\text{obsd}} = B_{\text{max,sp}} \sum_{j=1}^{n} \frac{F_j [L_1]^{n_{\text{H}(1j)}}}{K_{1j}^{n_{\text{H}(1j)}} + [L_1]^{n_{\text{H}(1j)}}} + B_1 [L_1] + B_{[L_1]=0}. \tag{212}$$

where

$$F_j = \frac{B_{\text{max,sp}(j)}}{B_{\text{max,sp}}}$$

$$B_{\text{max,sp}} = \sum_{j=1}^{n} B_{\text{max,sp}(j)}.$$

The relationship between total binding of the probe L_1 and the concentration of an unlabelled ligand L_2 can be described by eqn 213 (cf. eqn 206). The parameter F_j' differs from F_j in eqn 212, in that the former represents that fraction of inhibitable binding defined by $K_{2j}^{n_{\text{H}(2j)}}$ and thus depends upon the concentration of the probe (cf. eqn 47):

$$B_{\text{obsd}} = B_{[L_2] \to \infty} + (B_{[L_2]=0} - B_{[L_2] \to \infty}) \sum_{j=1}^{n} \frac{F_j' K_{2j}^{n_{\text{H}(2j)}}}{K_{2j}^{n_{\text{H}(2j)}} + [L_2]^{n_{\text{H}(2j)}}}. \tag{213}$$

Equations 211–213 are of limited usefulness for most applications. Unless the different processes are well separated on the abscissa (i.e. log $K_{L_{2j}} \ll$ log $K_{L_{2(j+1)}}$), or the values of $n_{\text{H}(j)}$ exceed unity, the various parameters are likely to be highly correlated and hence poorly defined. Such lack of definition can result in numerical failure during the fitting procedure. Eqns 211 and 212 are especially to be avoided, particularly when the values of $n_{\text{H}(1j)}$ are less than unity. Even in relatively favourable situations, the value of $B_{\text{max, sp}}$ can be elusive owing to the tendency of non-specific binding to preclude useful measurements at saturating concentrations of the probe. Lower Hill co-efficients require a greater range of $[L_1]$ to define the binding curve and therefore demand higher values of $B_{max,sp}/K_{L_1}^{n_{H(1j)}}$ relative to B_1; the problem is often difficult with only a single contribution to the signal (i.e. eqn 195 or 211 when $n = 1$), and it is likely to preclude any useful calculation when there are multiple contributions (i.e. eqn 211, $n > 1$). Eqn 213 is less prone to catas-trophic failure, since the limits at $[L_2] = 0$ and $[L_2] \to \infty$ can usually be defined by data. It generally behaves as one expects for the inhibition of one ligand by another, particularly at each end of the abscissa on semi-logarithmic co-ordinates. One therefore avoids some of the problems encountered with splines or other functions. The parameters may be poorly defined, but eqn 213 can serve admirably as a french curve.

4.2 Rational functions

4.2.1 Sums of hyperbolic terms

When n exceeds unity in eqns 211–213, parametric values tend to be poorly defined owing in part to correlations among parameters during the optimiza-tion. Such functions are considerably more useful when the values of $n_{H(ij)}$ can be taken as constants, and an important case is that in which all are equal to unity. Binding at graded concentrations of a probe L_1 is then described by eqn 214 or 215 (cf. eqn 211 or 212). The distinction between total binding with and without an unlabelled ligand can be made as described above for eqn 195 or 196:

$$B_{obsd} = \sum_{j=1}^{n} \frac{B_{max,sp(j)} [L_1]}{K_{1j} + [L_1]} + B_1 [L_1] + B_{[L_1]=0} \tag{214}$$

$$B_{obsd} = B_{max,sp} \sum_{j=1}^{n} \frac{F_j[L_1]}{K_{1j} + [L_1]} + B_1 [L_1] + B_{[L_1]=0}, \tag{215}$$

where $\quad F_j = \dfrac{B_{max,sp(j)}}{B_{max,sp}}$

$\quad B_{max,sp} = \sum_{j=1}^{n} B_{max,sp(j)}.$

Although less trouble-prone than eqns 211 and 212, eqns 214 and 215 can both yield spurious results owing to the nature of data acquired at graded concentrations of the probe. A commonly observed pattern results when the Hill coefficient is clearly less than unity (eqn 195) but $B_{max,sp}$ is undefined owing to practical limitations imposed by the level of non-specific binding. Appropriate statistical tests may indicate that eqn 214 yields an appreciably better fit with two contributions to the specific signal rather than one (i.e. $n = 2$). Also, the 'sites' ostensibly of higher affinity may be reasonably defined by the data (i.e. $[L_1] > 10K_{11}$), and uncorrelated estimates can therefore be obtained for K_{11} and $B_{max,sp(1)}$. If the sites of lower affinity are poorly defined owing to practical limits on $[L_1]$ (i.e. $[L_1] < K_{12}$), that contribution to the specific signal may be indistinguishable from a straight line (i.e. $B_{max,sp(2)} = [L_1]/K_{12}$); in such an event, a unique value can only be obtained for the ratio $B_{max,sp(2)}/K_{12}$. Attempts to estimate both parameters typically involve an initial estimate of K_{12} somewhere within the range of values of $[L_1]$. Since the model then predicts more curvature than exists in the data, the fitting process increases K_{12} until the model more closely approximates a straight line; this tends to occur with values of K_{12} comparable with or somewhat higher than the highest concentration of L_1 used in the experiment. The 'apparent affinity' and hence the corresponding capacity are thus determined by the last point or two on the binding curve.

The situation described above illustrates the common problem of parameters that are not defined by experimental data. If no further experiments are to be performed, the best that one can do is to determine the lower limit on K_{12} by mapping its effect on chi-square. If the mechanistic model is available, it may be possible to obtain a unique minimum for the parameters corresponding to K_{12} by pooling data acquired at multiple concentrations of both the probe and an unlabelled but otherwise identical analogue. If the situation demands an unequivocal estimate of $B_{max,sp(2)}$ (eqn 214), however, measurements are required at much higher concentrations of the probe; the quantity $B_{max,sp(2)}/(B_{max,sp(2)} + B_1[L_1])$ must therefore be increased, either by increasing $B_{max,sp(2)}$ or by decreasing B_1.

Non-specific binding determines not only the upper limit on useful concentrations of the probe, as described above, but also the best ratio of signal to noise achievable at any concentration. The latter is of particular concern in the characterization of unlabelled ligands via their effect on the binding of the probe. The relative contribution of specific to total binding is maximal when the former is linear with respect to L_1 (i.e. $[L_1] \ll K_{1j}$), and eqn 214 can be simplified as shown in eqn 216:

$$B_{obsd} = \left\{ \sum_{j=1}^{n} \frac{B_{max,sp(j)}}{K_{1j}} + B_1 \right\} [L_1] + B_{[L_1]=0}. \qquad (216)$$

Once that condition is met, any further improvement requires a decrease in K_{1j}, a selective increase in $B_{max,sp}$, or a selective decrease in B_1. The specific radioactivity of the probe is of interest only insofar as it determines the sensitivity of the assay and hence the lowest value of $[L_1]$ at which measurements are possible; proportionally equal increases in $B_{max,sp}$ and B_1 similarly allow binding to be measured at lower values of $[L_1]$. Neither can affect the ratio of signal to noise under conditions when specific binding is linearly dependent upon the probe.

Binding at a single concentration of the probe L_1 and graded concentrations of an unlabelled ligand L_2 can be described by eqn 217 (cf. eqn 213). The value of B_{obsd} when $[L_2]$ equals zero is defined empirically in terms of eqn 215 (eqn 218), but only a lower limit can be placed on the asymptotic value as $[L_2] \to \infty$ (eqn 219):

$$B_{obsd} = B_{[L_2] \to \infty} + (B_{[L_2]=0} - B_{[L_2] \to \infty}) \sum_{j=1}^{n} \frac{F_j' K_{2j}}{K_{2j} + [L_2]}, \tag{217}$$

where

$$B_{[L_2]=0} = B_{max,sp} \sum_{j=1}^{n} \frac{F_j [L_1]}{K_{1j} + [L_1]} + B_1 [L_1] + B_{[L_1]=0} \tag{218}$$

$$B_{[L_2] \to \infty} \geq B_1 [L_1] + B_{[L_1]=0} \tag{219}$$

$$\sum_{j=1}^{n} F_j' = 1.$$

Although formally indistinguishable from various mechanistic models, eqns 215 and 217 are empirical in the sense that no physical significance has been assigned to the various parameters; accordingly, the relationships between $[L_1]$, F_j and K_{1j} on the one hand (eqn 215) and $[L_2]$, F_j' and K_{2j} on the other (eqn 217) are undefined. Also, a process that is distinct with respect to $[L_1]$ may be unobservable at graded values of $[L_2]$ or vice versa; the empirically determined value of n may therefore differ with eqns 215 and 217.

When the mechanism of binding is unknown or irrelevant, eqns 215 and 217 serve primarily to smooth and perhaps to summarize the data. In such applications the fitted parameters may be of less interest than the concentration of L_i that corresponds to a specific signal of 50% (i.e. $[L_{i,50}]$). The required parameter is defined by eqns 220 and 221 respectively, both of which represent a polynomial of degree n in $[L_{i,50}]$:

$$\sum_{j=1}^{n} \frac{F_j [L_{1,50}]}{K_{1j} + [L_{1,50}]} = 0.5 \tag{220}$$

$$1 - \sum_{j=1}^{n} \frac{F_j' K_{2j}}{K_{2j} + [L_{2,50}]} = 0.5. \tag{221}$$

Eqn 222 is the cubic expression obtained when n equals three; the appropriate root can be evaluated by the Newton–Raphson procedure. By substituting $K_{2j}^{n_{H(2j)}}$ for K_{2j}, the same expression can be used to evaluate $[L_{2,50}]$ for best fits of eqn 213:

$$\varphi_0 + \varphi_1 [L_{i,50}] + \varphi_2 [L_{i,50}]^2 + \varphi_3 [L_{i,50}]^3 = 0, \tag{222}$$

where
$$\varphi_0 = K_1K_2K_3$$

$$\varphi_1 = K_1K_2(1 - 2F_3) + K_1K_3(1 - 2F_2) + K_2K_3(1 - 2F_1)$$

$$\varphi_2 = K_1(1 - 2F_2 - 2F_3) + K_2(1 - 2F_1 - 2F_3) + K_3(1 - 2F_1 - 2F_2)$$

$$\varphi_3 = -1$$

$$K_j = K_{1j} \quad \text{(eqn 220)} \quad \text{or} \quad K_j = K_{2j} \quad \text{(eqn 221)}$$

$$F_j = F_j \quad \text{(eqn 220)} \quad \text{or} \quad F_j = F_j' \quad \text{(eqn 221)}.$$

If F_j' is positive for all j, the first derivative of eqn 217 is either positive or negative at all values of $[L_2]$. In co-operative systems, however, $f'(x)$ can traverse zero one or several times when the stoichiometry is two or more with respect to L_i (e.g. Scheme VI, $a = 1$, $b > 1$, $c > 0.3$, $d > 2$ in eqns 186–189). The presence of maxima and minima in B_{obsd} can be described by eqn 217 with values of F_j' less than zero and greater than unity; alternatively, such behaviour can be accommodated by functional relationships of the form $([L]/([L] + K_1))/(K_2/([L] + K_2))$. An example of the latter is shown in eqn 223, where each term is itself a sum of multiple contributions characterized by distinct values of K. Ligand-dependent changes in B_{obsd} are determined by the asymptotic values of the function and by the intrinsic amplitude of the peak or trough; the latter can be defined in absolute terms (a) or in relative terms (b), whichever is convenient. The behaviour of eqn 223 is illustrated in *Figure 9*:

$$B_{obsd} = B_{[L_i]=0} + a\left\{\sum_{j=1}^{m} \frac{F_{1j}[L_i]}{K_{1j} + [L_i]}\right\}\left\{\sum_{j=1}^{n} \frac{F_{2j}(K_{2j} + b[L_i])}{K_{2j} + [L_i]}\right\}, \tag{223}$$

where $ab = B_{[L_i] \to \infty} - B_{[L_i]=0}$.

4.2.2 Ratios of polynomials

Eqns 214 and 215 are rational functions of the form shown in eqn 224, where φ_0' is $B_{[L_1]=0}$. If the background signal from the counting machine is subtracted prior to the analysis, total binding can be described by eqn 225 (i.e. $B_{[L_1]=0} = 0$). The expression for specific binding alone is eqn 226 (i.e. $B_1[L_1] + B_{[L_1]=0} = 0$). The definitions of φ_j' and φ_j in eqns 224–226 differ accordingly:

$$B_{obsd} = \frac{\sum_{j=0}^{n+1} \varphi_j'[L_1]^j}{1 + \sum_{j=1}^{n} \varphi_j[L_1]^j} \tag{224}$$

$$B_{obsd} = \frac{\sum_{j=1}^{n+1} \varphi_j'[L_1]^j}{1 + \sum_{j=1}^{n} \varphi_j[L_1]^j} \tag{225}$$

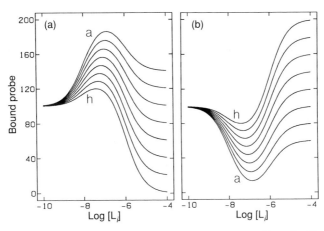

Figure 9. Empirical modelling of co-operativity between two ligands. The lines were calculated according to eqn 223 ($m = 1$, $n = 1$); parametric values common throughout are as follows: $\log K_{11} = -8.0$, $\log K_{21} = -6.0$ and $B_{[L_i]=0} = 100$. The amplitude associated with K_1 is positive in (a) and negative in (b), and individual values of the parameter a for curves a–h are as follows: a, ± 100; b, ± 90; c, ± 80; d, ± 70; e, ± 60; f, ± 50; g, ± 40; h, ± 30. The value of $B_{(L_i)\to\infty}$ for curves a–h is incremented by -20 from 140 to 0 in (a) and by $+20$ from 60 to 200 in (b).

$$B_{sp} = \frac{\sum_{j=1}^{n} \varphi'_j [L_1]^j}{1 + \sum_{j=1}^{n} \varphi_j [L_1]^j}, \tag{226}$$

where $\quad \varphi'_{n+1} = \dfrac{B_{[L_1]\to\infty}}{\prod\limits_{j=1}^{n} K_{1j}} \qquad$ (in eqns 224 and 225)

$\qquad \varphi'_n = \dfrac{B_{max,sp}}{\prod\limits_{j=1}^{n} K_{1j}} \qquad$ (in eqn 226)

$\qquad \varphi_n = \dfrac{1}{\prod\limits_{j=1}^{n} K_{1j}} \qquad$ (in eqns 224–226).

Owing to the presence of a non-saturable component, the highest power in the numerator of eqns 224 and 225 is one more than that in the denominator. Since specific binding is saturable, the polynomials of eqn 226 are of the same degree in both numerator and denominator. Eqn 226 is indistinguishable from the corresponding functions derived for the multi-site model (eqn 48) and, when φ'_j equals $j\varphi_j$, for co-operative models (eqn 164).

Eqns 217 and 223 are rational functions of the form shown in eqn 227. Total binding is asymptotic at high concentrations of the unlabelled ligand, but the relative values of $B_{[L_2]=0}$ and $B_{[L_2]\to\infty}$ determine whether the net change is positive or negative. In the extreme situation when L_2 reduces total binding to non-specific levels (i.e. $B_{[L_2]\to\infty} = B_1[L_1] + B_{[L_1]=0}$ in eqn 219), the specific component can be described by eqn 228:

$$B_{obsd} = \frac{\sum_{j=0}^{n} \psi_j'[L_2]^j}{1 + \sum_{j=1}^{n} \psi_j[L_2]^j},$$

(227)

where $\quad \psi_0' = B_{[L_2]=0} \quad$ (in eqns 217 and 223)

$$\psi_n' = \frac{B_{[L_2]\to\infty}}{\prod_{j=1}^{n} K_{2j}} \quad \text{(in eqn 217)} \quad \text{or} \quad \frac{B_{[L_1]\to\infty}}{\prod_{j=1}^{m} K_{1j} \prod_{j=1}^{n} K_{2j}} \quad \text{(in eqn 223)}$$

$$\psi_n = \frac{1}{\prod_{j=1}^{n} K_{2j}} \quad \text{(in eqn 217)} \quad \text{or} \quad \frac{1}{\prod_{j=1}^{m} K_{1j} \prod_{j=1}^{n} K_{2j}} \quad \text{(in eqn 223)}$$

$$B_{sp} = \frac{\sum_{j=1}^{n-1} \psi_j'[L_2]^j}{1 + \sum_{j=1}^{n} \psi_j[L_2]^j}.$$

(228)

A comparison of eqn 228 with eqns 49 and 165 indicates that the expression is formally indistinguishable from that derived for the effect of an unlabelled ligand in either the multi-site model or the co-operative model. Values of ψ_n'/ψ_n (eqn 227) that exceed $B_1[L_1] + B_{[L_1]=0}$ imply that binding is not mutually exclusive at saturating concentrations of the unlabelled ligand, as in the example of Scheme V (cf. eqns 105 and 217); similarly, any value of B_{obsd} that exceeds ψ_0' implies positive co-operativity.

4.2.3 General comments

All of the mechanistic models described in Section 3 are rational functions and, except as noted below, are generically equivalent to eqns 214, 215, 217 or 223. In many situations, the empirical parameters are related to those of Schemes II–VI in a straightforward manner; the relationship can readily be determined by comparing the empirical formula with the corresponding expression derived from the model. There is a direct correspondence between the parameters of eqns 214–223 and those of the multi-site model. A frequently encountered example is the relationship between eqn 217 and eqn 47, particularly with regard to the value of K_{2j} in eqn 27 (see, for example, ref. 13). Similar comparisons can be made with co-operative models (e.g. eqns

217 and 105), but the complexity of the relationships can limit the insight to be gained from empirically determined parameters.

A sum of hyperbolic terms can always be used to describe negative co-operativity; the multi-site model and co-operative models are thus indistinguishable when there is only one independent variable. When the independent variable is the probe (eqn 214 or 215), there is a simple relationship between K_{1j} and the successive dissociation constants of co-operative models, at least for small values of n; also, the smallest value of F_j places a lower limit on the stoichiometry of binding (i.e. $n \geq 1/F_j$). With two or more ligands (eqns 217 and 223), the definition of K_{1j} becomes more complex. Moreover, F_j is determined in part by the stoichiometry and in part by co-operative effects among the different ligands; accordingly, there is no restriction on ψ'_j and ψ_j in eqn 228, in contrast to the requirement that φ'_j equal $j\varphi_j$ in eqn 226 (cf. eqns 164 and 165).

For most practical purposes, the description of positive co-operativity in terms of eqns 215–223 is limited to interactions between successive equivalents of labelled and unlabelled ligands. Positive co-operativity between successive equivalents of the same ligand leads to imaginary numbers (i.e. $a + b\sqrt{-1}$); the fitting programme must therefore be coded to handle complex variables if a sum of hyperbolic terms is to describe data that reveal a Hill coefficient greater than unity. The restriction on positive co-operativity can be illustrated by expressing eqn 226 in terms of eqn 215 on the one hand and a co-operative dimer on the other. It is convenient to assume that the dimer is symmetrical with respect to the probe (i.e. $K_{1R} = K_{R1}$ in Scheme VI) and that the probe is the only ligand present. The appropriate expansion of eqn 215 includes only the saturable portion (i.e. $B_{sp} = B_{obsd} - B_1[L_1] - B_{[L_1]=0}$), which comprises two hyperbolic terms of equal amplitude (i.e. $n = 2$, $F_j = 0.5$); the restriction on F_j is necessary to obtain an Adair equation (i.e. $\varphi'_j = j\varphi_j$ in eqn 226), as required for equivalence with co-operative models (cf. eqn 164). Under these conditions, eqn 215 is also identical with the special case of the multi-site model in which all classes are of equal size (i.e. eqn 45, $m = 1$, $[R_j]_t = [R]_t/n$). The relevant form of eqn 215 is shown as eqn 229:

$$B_{sp} = \frac{B_{max,sp}}{2} \frac{\{1/K_{11} + 1/K_{12}\}\,[L_1] + (2/K_{11}K_{12})[L_1]^2}{1 + \{1/K_{11} + 1/K_{12}\}\,[L_1] + (1/K_{11}K_{12})[L_1]^2}. \quad (229)$$

The analogous expression for a symmetrical dimer can be obtained from eqns 154–157 and is shown as eqn 230 ($K_1 = K_{R1} \equiv K_{1R}$, $K_2 = K_{1(R1)}$):

$$[L_1]_b = [R]_t \frac{(2/K_1)[L_1] + (2/K_1K_2)[L_1]^2}{1 + (2/K_1)[L_1] + (1/K_1K_2)[L_1]^2}. \quad (230)$$

The coefficients of eqns 229 and 230 can be equated to obtain relationships between the apparent affinities K_{1j} of eqn 215 and the microscopic dissociation constants K_j of the dimer (eqns 231 and 232):

$$K_{11} = K_2 - (K_2^2 - K_1 K_2)^{1/2} \tag{231}$$

$$K_{12} = K_2 + (K_2^2 - K_1 K_2)^{1/2}. \tag{232}$$

The empirical parameters K_{11} and K_{12} are real numbers when K_2 equals or exceeds K_1 (i.e. $K_2^2 - K_1 K_2 \geq 0$), but they are complex numbers when K_1 exceeds K_2 (i.e. $K_2^2 - K_1 K_2 < 0$). To avoid the latter, systems that reveal positive homotropic co-operativity can be analysed empirically by including a Hill coefficient in the relevant term of eqns 215–223, or, somewhat more appropriately, by fitting one of eqns 224–228. When the probe is the only ligand present, the coefficients of eqn 226 ought to be constrained such that φ_j' equals $j\varphi_j$. The relationship between a sum of hyperbolic terms and the Adair equation has been described in detail by Fletcher and Spector (14) and by Klotz (15, 16).

For the sequential binding of a single ligand in a co-operative system, the coefficients of the Adair equation (i.e. eqn 164) can readily be expressed in terms of the macroscopic dissociation constants K_i' at each level of stoichiometry (eqn 233):

$$\varphi_j = \frac{1}{\displaystyle\prod_{i=1}^{j} K_i'}. \tag{233}$$

The microscopic dissociation constant K_i for the ith ligand can be calculated according to eqn 234 on the assumption that all vacant sites are equivalent:

$$K_i = K_i' \frac{n + 1 - i}{i}. \tag{234}$$

The relationship between macroscopic and microscopic binding constants has been described at length by Klotz and Hunston (17) and Edsall and Gutfreund (18). A description of a co-operative system in terms of its macroscopic binding constants for two ligands has been presented by De Lean *et al.* (19).

Mechanistic models are only described by rational functions if one can disregard the equations of state for all interacting species except the receptor. It follows that explicit relationships between the parameters of eqns 214–223 and Schemes II–VI or similar proposals are invalid if binding to the receptor depletes the free concentration of one or more ligands. An exception to this qualification arises when there is an excess of receptors relative to a particular ligand, virtually all of which remains bound under the conditions of the experiment. If the sequestered ligand occludes the site recognized by the probe, the total number of available sites will be reduced accordingly. A somewhat different pattern can emerge in the case of Scheme IV or V, where the interactions between a set of ligands L_i and a constituent X are strictly allosteric. If there is a molar excess of R over X, and if X is fully coupled to R at all concentrations of L_i, the system will be indistinguishable from the multi-site model.

The preference for explanations over descriptions fosters a tendency to interpret empirical parameters in mechanistic terms; this practice is encouraged by the form of eqns 214–223, which assigns recognizable units to the various parameters. Any aid to intuition is welcome; also, the way in which parameters are defined determines the constraints that can be applied in simultaneous analyses. With complex systems, however, empirical parameters may represent little more than mathematically refined puzzlement; it may sometimes be as helpful simply to fit one of eqns 224–228 until a mechanistic model is available. The degree of the polynomial in the denominator places a lower limit on the number of binding sites for the ligand.

5. Curve fitting

5.1 Non-linear regression

A mathematically formulated model can be useful in and of itself, and there is a cottage industry given to the exploration of mechanistic schemes via simulation. Such excursions can foster new insight and may help to keep one's intuition in touch with physical–chemical principles; they can be instrumental in the conception and design of new experiments. Sooner or later, however, the model and the data must be compared, preferably in a quantitative and unbiased manner. Since all measurements are subject to error, such comparisons involve finding the 'best fit' of the model to the data. The measure of agreement is often taken as the sum of squared residuals, where each residual is the difference between the measured and theoretical values of the dependent variable. The smaller the sum of squares, the better the fit. This section contains a brief description of least-squares fitting and the information that it can provide. The reader is referred to previous reviews for additional material of an introductory nature (20, 21) and to statistical texts for further details.

In practice, the measure of agreement is the *weighted* sum of squares SS as defined in eqn 235:

$$SS(\mathbf{a}) = \sum_{i=1}^{n} \frac{1}{\sigma_i^2} \{y_i - f(\mathbf{x}_i, \mathbf{a})\}^2. \tag{235}$$

The minimum in SS is achieved by adjusting a set of m parameters whose current values are shown as the vector $\mathbf{a}$; the ith set of independent variables from a total of n observations is shown as $\mathbf{x}_i$. The measured value of the dependent variable is y_i, and $f(\mathbf{x}_i, \mathbf{a})$ is the corresponding value computed according to the model; σ_i^2 is the current estimate of the variance on y_i, and the quantity $y_i - f(\mathbf{x}_i, \mathbf{a})$ is the residual z. As described below, the weight is taken as the reciprocal of the estimated variance. Implicit in the model represented by $f(\mathbf{x}_i, \mathbf{a})$ are any constants (e.g. $f(\mathbf{x}_i, \mathbf{a}, \mathbf{c})$); independent variables and constants are assumed to be without error.

In models of saturable processes, the dependent variable is generally non-linear in at least one parameter; in Scheme II, for example, specific binding of the radioligand is linear in $[R_j]_t$ but non-linear in K_{ij} (eqn 45). When $f(\mathbf{x}_i, \mathbf{a})$ is non-linear in $\mathbf{a}$, the minimization of $SS(\mathbf{a})$ requires a procedure for non-linear regression; linearity or the lack of it in $\mathbf{x}$ is irrelevant to whether or not linear regression can be used to optimize $\mathbf{a}$. Simple models can be rearranged, or 'reparameterized', to obtain linearity in a set of parameters expressed in terms of $\mathbf{a}$ as originally defined; the process also involves redefining x and y, and in the best known examples the outcome is a straight line. Linear transformations may be of some diagnostic value, but generally ought to be avoided for purposes of optimization. The application of linear regression to transformed data is statistically awkward, and generations of authors have alerted us to the pitfalls of the Scatchard plot and the dreaded Lineweaver–Burk plot. With many models of practical relevance, the transform merely exchanges one curve for another; in such cases, the statistical indignities have been inflicted without achieving linearity in x.

The distinction among parameters, independent variables and constants is somewhat arbitrary. Generally speaking, parameters are the quantities to be optimized and thus associated with some uncertainty; independent variables and constants are both assumed to be without error, and measurements are taken at multiple values of the former. Potential parameters that are fixed throughout the fitting procedure or independent variables that remain unchanged throughout the experiment are in effect constants for the purpose of the analysis. Binding studies typically involve at least two ligands (e.g. $[L_1]$ and $[L_2]$ in eqns 45–47 or eqns 113–116), only one of which is varied over the course of an experiment. The practical implementation of relatively complex models can be expected to require different concentrations of multiple ligands, and the data are likely to be acquired in a series of experiments involving changes in one ligand or another. While only one ligand may be a variable in any one experiment, all ligands are in effect variables when the data are analysed simultaneously. Receptors and related proteins are traditionally handled as parameters (e.g. $[R]_t$ or $[X]_t$ in eqns 113–116); with reconstitution or controlled expression, however, any component of the system may be exploited as an independent variable. It can be helpful to consider such possibilities when coding the programmes that are to perform the analyses.

The method suggested by Marquardt (22) has become the standard for the non-linear optimization of parameters in binding studies and related situations. It is rapid and well suited to the data and models typically encountered; also, the computation involves a matrix of partial derivatives that yields not only the parametric values but also information on their statistical properties. A brief description of Marquardt's procedure is given below, and a general discussion of multidimensional minimization can be found in ref. 2. Alternative procedures are not considered here except for a word on the simplex method, which can be used in situations that are incompatible or are likely to prove troublesome with Marquardt's algorithm. Derivatives are not required, and the function need not be continuous; also, the method is less prone to failure when the investigator cannot acquire the data best suited to the parameters of interest. Statistical details are not available automatically; if the function is differentiable, the required matrix can be computed at any stage of the minimization, including the end.

The function SS(**a**) is minimal when the *gradient* with respect to all **a** is zero (eqn 235): that is, at the point in m-dimensional parameter space where an increase or decrease in the value of any parameter is accompanied by an increase in SS. The optimal values of **a** are obtained in an iterative process that begins with an initial guess for each parameter. Depending upon the degree to which all parameters are defined by the data, the starting values can be critical to whether the process converges happily to the true minimum, finds a local minimum that does not represent the best fit, or departs for some physically irrelevant region with often dire numerical consequences. At each iteration, the current parameters are adjusted by increments δa_k obtained by solving a system of linear equations defined by the partial derivatives of SS with respect to **a** (eqn 236). The process is convergent upon the minimum and must be terminated when there is no appreciable change in successive estimates of either SS or all a_k:

$$0 = \frac{\partial SS(\mathbf{a})}{\partial a_1} + \lambda' \frac{\partial^2 SS}{\partial a_1 \partial a_1} \delta a_1 + \frac{\partial^2 SS}{\partial a_1 \partial a_2} \delta a_2 + \cdots + \frac{\partial^2 SS}{\partial a_1 \partial a_m} \delta a_m$$

$$0 = \frac{\partial SS(\mathbf{a})}{\partial a_2} + \frac{\partial^2 SS}{\partial a_2 \partial a_1} \delta a_1 + \lambda' \frac{\partial^2 SS}{\partial a_2 \partial a_2} \delta a_2 + \cdots + \frac{\partial^2 SS}{\partial a_2 \partial a_m} \delta a_m \quad (236)$$

$$\vdots$$

$$0 = \frac{\partial SS(\mathbf{a})}{\partial a_m} + \frac{\partial^2 SS}{\partial a_m \partial a_1} \delta a_1 + \frac{\partial^2 SS}{\partial a_m \partial a_2} \delta a_2 + \cdots + \lambda' \frac{\partial^2 SS}{\partial a_m \partial a_m} \delta a_m$$

where $\dfrac{\partial SS(\mathbf{a})}{\partial a_k} = 2\sum_{i=1}^{n} \dfrac{\{y_i - f(\mathbf{x}_i, \mathbf{a})\}}{\sigma_i^2} \dfrac{\partial f(\mathbf{x}_i, \mathbf{a})}{\partial a_k}$

$$\frac{\partial^2 SS}{\partial a_k \partial a_l} \approx 2\sum_{i=1}^{n} \frac{1}{\sigma_i^2} \left\{ \frac{\partial f(\mathbf{x}_i, \mathbf{a})}{\partial a_k} \frac{\partial f(\mathbf{x}_i, \mathbf{a})}{\partial a_l} \right\}$$

$$\lambda' = 1 + \lambda.$$

When λ equals zero, eqn 236 can be expressed more succinctly in matrix notation as shown in eqn 237; $\mathbf{H}$ is the square matrix of partial derivatives, and Δ is the vector of increments in a_k.

$$0 = \mathbf{f}'(\mathbf{a}) + \mathbf{H}\Delta. \tag{237}$$

Marquardt's contribution relates to the changing nature of the procedure whereby one obtains successive estimates of δa_k, which involves a continuous interpolation from the method of *steepest descent* at the outset to the *Gauss–Newton* method near the solution. The former involves repeatedly selecting the steepest gradient at each step down the side of the depression in SS($\mathbf{a}$), and the latter is analogous to the Newton–Raphson procedure (cf. eqns 70 and 236). The blend is controlled by the parameter λ, which is large when $\mathbf{a}$ is far from the minimum and approaches zero as $\mathbf{a}$ becomes optimal. Large values of λ cause the diagonal elements of eqn 236 to dominate, and the estimation of δa_k is predominantly by the method of steepest descent; when λ approaches zero, the estimation of δa_k is essentially by the Gauss–Newton method. The algorithm has been described in detail by Marquardt (22) and by Press *et al.* (2).

As described in the previous section, the Newton–Raphson procedure involves a square matrix of partial *first derivatives* of the relevant functions (a *Jacobian matrix*) (e.g. eqn 70); in contrast, the Gauss–Newton procedure involves a square matrix ($m \times m$) of partial *second derivatives* of SS($\mathbf{a}$) (a *Hessian matrix*) (i.e. $\mathbf{H}$ in eqn 237). The difference arises from the nature of multidimensional root-finding on the one hand and minimization on the other. The former seeks the values of $\mathbf{x}$ at which several functions all equal zero, while the latter seeks the values of $\mathbf{a}$ at which the gradient or first derivative of a single function equals zero (e.g. eqn 236). Minimization of SS($\mathbf{a}$) thus represents an application of the one-dimensional Newton method in m-dimensional space, where m is the number of parameters.

The partial derivatives of $f(\mathbf{x}_i, \mathbf{a})$ can be calculated from explicit expressions, since the function itself is known. This is the most accurate approach and sometimes is the fastest in terms of CPU time; it is therefore recommended for a well established model in a programme designed for a specific set of conditions. The relevant expressions are tedious to derive, however, and analytic derivatives constitute a considerable inconvenience to the investigator whose model or models are under constant revision; also, it is not trivial to code in a manner that accommodates all possible restraints on the parameters in simultaneous analyses of data from multiple experiments. Such problems are aggravated with models that lack the symmetry amenable to clever solutions. For most situations, the practical alternative is to evaluate the derivatives numerically. One small subroutine can handle any model, and the numerical inaccuracies are negligibly small, at least for the functions generally encountered in the study of ligand–receptor interactions. The need to compute the model for incremented values of a_k may increase the computing time, but the real times involved are a minor concern with the microprocessors now available to every laboratory.

The major question with numerical derivatives is the size of the relative increment ε used to define the finite difference in each parameter (e.g. eqn 238 or 239):

$$\frac{\partial f(\mathbf{x}_i, \mathbf{a})}{\partial a_k} = \frac{f(\mathbf{x}_i, \mathbf{a})_{a_k} - f(\mathbf{x}_i, \mathbf{a})_{a_k + \Delta a_k}}{\Delta a_k} \quad \text{where } \Delta a_k = \varepsilon a_k \qquad (238)$$

$$\frac{\partial f(\mathbf{x}_i, \mathbf{a})}{\partial a_k} = \frac{f(\mathbf{x}_i, \mathbf{a})_{a_k - \Delta a_k} - f(\mathbf{x}_i, \mathbf{a})_{a_k + \Delta a_k}}{2\Delta a_k} \quad \text{where } \Delta a_k = \varepsilon a_k/2 \qquad (239)$$

The increment is important, as it can affect the final values of the parameters of the model $\hat{\mathbf{a}}$ and of the statistical parameters calculated from the matrix of partial derivatives (see below). Too large a value of ε is inconsistent with the notion of ∂a and will tend to damp the signal. Too small a value will increase unnecessarily the number of iterations required to find the minimum of SS; exceptionally small values will lead to errors and possibly outright failure through loss of significance. While it is convenient to decide on ε and then to use the same value throughout, it is important to realize that the acceptable upper limit can depend upon the parameter being incremented and upon the values of other parmeters and $\mathbf{x}_i$. In the author's experience, the value of ε can usually be between 0.0001 and 0.01 for variables coded in single precision with a 32-bit word. This issue is explored further in some texts on numerical analysis (23, 24).

5.2 Statistical parameters

5.2.1 General comments

The covariance matrix $\mathbf{C}$ corresponding to the fitted parameters is the inverse of the Hessian matrix at the minimum in SS (eqn 240):

$$\mathbf{C} = \mathbf{H}^{-1}. \qquad (240)$$

It is thus straightforward to estimate the uncertainty associated with the fitted value of each parameter, but the validity of these estimates rests upon four assumptions: first, that the model fits the data; second, that errors on the measurement are normally distributed; third, that the error on one point is independent of the error on another; and fourth, that the model is linear in all parameters or at least behaves as such. The goodness of fit can be tested as described below, and all bets are off should it prove to be unacceptable; there is little point in discussing the statistical niceties or the parameters of a scheme that does not work.

The distribution of experimental error is seldom examined in practice. Perhaps the most harmful deviations from a normal distribution occur with the occasional fumbles that lead to outliers, which are often readily identified; a reaction tube may receive two aliquots of the radioligand rather than one, or the pellet may be lost from the microcentrifuge tube during the superficial wash that typically follows centrifugation. More subtle deviations may also occur, however, particularly with regard to the usual requirement for separation of bound and unbound probe; such procedures are more likely to reduce the signal than to decrease it. It is therefore important to recall that statistical parameters calculated from the covariance matrix become less meaningful as the distribution becomes non-normal. The assumption of linearity with respect to the parameters and the model is the so-called *linear hypothesis*. The extent to which it is true will determine the degree to which computed estimates of statistical variability reflect the real values. True chi-square boundaries cannot be calculated explicitly for parameters evaluated by non-linear regression, but they can be estimated from Monte Carlo simulations.

5.2.2 Standard errors and confidence intervals

The approximate standard error s_k on each parameter can be calculated from the diagonal elements of the covariance matrix C_{kk} according to eqn 241; the difference between the number of measurements n and the number of parameters m is the degrees of freedom:

$$s_k \approx \pm \left\{ C_{kk} \sum_{i=1}^{n} \frac{\{y_i - f(\mathbf{x}_i, \mathbf{\hat{a}})\}^2}{\sigma_i^2(n - m)} \right\}^{\frac{1}{2}}. \tag{241}$$

Approximate confidence intervals for each parameter can be estimated from s_k and the appropriate value of the t-statistic. A subroutine can be included in the fitting programme to provide values of t for any level of confidence and any number of degrees of freedom.

Standard errors and confidence intervals calculated in this manner are only approximations of the true values and ought not to be taken literally in further calculations. Nevertheless, they are a reflection of the range over which the parameter is without appreciable effect on the sum of squares; that range is an important indicator of the degree of uncertainty associated with the fitted value of the parameter. A small interval indicates that the parameter is well defined by the data and lies at the bottom of a steep depression in the relationship between SS and a_k; a larger interval indicates that the depression is relatively broad, and the parameter is therefore being estimated from inappropriate data.

Highly scattered data can result in large errors on most or all parameters, even when the experiment has been well designed with regard to the distribution and resolution of $\mathbf{x}_i$ relative to $\mathbf{a}$. The problem here is essentially one of signal to noise. Even at the minimum, the sum of squares tends to be large owing to the inevitability of large differences between $f(\mathbf{x}_i, \mathbf{\hat{a}})$ and most y_i (eqn 235); a significant increase in SS may therefore require comparatively large perturbations about $\mathbf{\hat{a}}$.

When there is relatively little scatter, large errors typically signal an attempt to estimate parameters from data acquired at inappropriate values of the independent variable; if the model is overly complex, a parameter may be undefined at any value of the independent variable used in the experiment. The problem can be illustrated by the example of a probe and a single class of sites (e.g. Scheme II, eqn 45, $m = 1$, $n = 1$). Well defined estimates of capacity require that $[L_1]$ exceed K_{11} by a factor of ten or more; as the highest value of $[L_1]$ becomes smaller, the value of $[R_1]_t$ is obtained by progressively longer extrapolation. Well defined estimates of affinity require data at concentrations of the probe sufficient to define $[R_1]_t$ *and* at concentrations comparable with K_{11}. When binding is measured only at high concentrations of $[L_1]$ (i.e. $[L_1] > K_{11}$) the sum of squares will be sensitive to changes in $[R_1]_t$ but relatively insensitive to K_{11}; at low concentrations of $[L_1]$, the depression in SS will be shallow with respect to both parameters. This example also illustrates the tendency of parameters associated with large errors to be correlated with other parameters, and such interrelationships can be identified by means of a correlation matrix as described below.

Parametric errors computed from the covariance matrix do not necessarily reflect goodness of fit. A sharp minimum in ss(**a**) is no guarantee that the data are distributed randomly about the fitted curve; conversely, a random distribution does not necessarily imply a sharp minimum.

Standard errors computed from the formal covariance matrix can also be used to estimate confidence intervals about the fitted curve. The practice is subject to the caveats noted above, but it can be helpful when the investigator is attempting to identify outliers that might be omitted from subsequent analyses. Confidence intervals calculated according to eqn 241 relate to the effect on SS of perturbations in each parameter taken separately; in contrast, confidence intervals on $f(\mathbf{x}_i, \hat{\mathbf{a}})$ reflect the cumulative effect of the uncertainty associated with all parameters. The appropriate standard error on the function at each $\mathbf{x}_i$ is calculated according to eqn 242, and $s_{f(x)}$ can be multiplied by the appropriate t-statistic to obtain the desired confidence interval:

$$s_{f(x_i)} \approx \left\{ \sum_{k=1}^{m} \sum_{l=1}^{m} \frac{\partial f(\mathbf{x}_i, \hat{\mathbf{a}})}{\partial a_k} \frac{\partial f(\mathbf{x}_i, \hat{\mathbf{a}})}{\partial a_l} C_{kl} \right\}^{1/2} \left\{ \sum_{j=1}^{n} \frac{\{y_i - f(\mathbf{x}_j, \hat{\mathbf{a}})\}}{\sigma_j^2 (n - m)} \right\}^{1/2}. \quad (242)$$

5.2.3 Correlation matrix

Correlations arise when two or more parameters compete for the same data. Highly correlated values are meaningless, and the effect can be monitored by the degree to which a change in one parameter is absorbed by all others. When there are only two parameters, any change not transferred from one to the other must be taken up in the sum of squares; multiple parameters increase the likelihood that a change in one can be transferred to another, and the distribution among parameters will depend upon the relative independence of each. It is therefore necessary to consider all possible correlations, and m, parameters result in $m(m - 1)/2$ combinations. The correlation coefficient between any two parameters r_{kl} is given by the scaled, off-diagonal elements of the co-variance matrix (eqn 243). The parameters are mutually independent when r_{kl} is zero and can compensate exactly when r_{kl} is ± 1.0:

$$r_{kl} = \frac{C_{kl}}{(C_{kk}C_{ll})^{1/2}}. \tag{243}$$

With a single class of sites (e.g. eqn 45, $m = 1$, $n = 1$), for example, neither K_{11} nor $[R_1]_t$ is well defined by the data when the highest concentration of the probe $[L_1]_{max}$ occupies only a minor fraction of the sites (i.e. $[L_1]_{max} < K_{11}$); in the limit as $[L_1]_{max} + K_{11}$ approaches K_{11} (i.e. $[L_1]_{max} \to 0$), the individual parameters merge into the ratio $[R_1]_t/K_{11}$ as eqn 45 collapses into the expression $[L_1R_1] = [L_1][R_1]_t/K_{11}$. The correlation coefficient for $[R_1]_t$ and K_{11} is thus near $+1.0$ at very low concentrations of the probe; the value decreases toward zero as $[L_1]_{max}$ approaches and exceeds K_{11}. It follows that the correlation between K_{11} and $[R_1]_t$ can be reduced simply by extending the range of $[L_1]$ in an otherwise identical experiment; from a practical point of view, the eventual severity of the problem is determined by the highest level of occupancy (i.e. $[L_1R_1]/[R_1]_t$) that is measurable given the concentration-dependence of non-specific binding. As models become more complex, one tends to encounter parameters that are defined at appropriate values of one independent variable but are undefined at any value of another. Perhaps the simplest example occurs when two ligands compete for a uniform population of sites (e.g. eqn 45, $m = 2$, $n = 1$). There are two dissociation constants (K_{11} and K_{21} in eqn 45), and estimates of both require data at graded concentrations of both ligands; when one ligand is varied at a single concentration of the other, K_{11} and K_{21} are indistinguishable (i.e. $r_{kl} = 1$) and must be represented as a single parameter if no other information is available (e.g. $IC_{50} = K_{21}(1 + [L_1]/K_{11})$.

5.3 Weighting

5.3.1 General comments

Weighting schemes exist to deal with experimental error; the smaller the error, the less relevant is the issue of weighting for any model that fits. If noiseless data are simulated and then fitted according to the same model, any weighting scheme will yield a fitted curve superimposable with the data; similarly, the parametric values will be identical with those used in the simulation, at least within the precision of the processor. Since it may not be clear how best to weight, it is prudent to achieve a level of error at which any reasonable weighting scheme leads to the same conclusions. The importance of the weighting scheme is also dependent upon the degree to which all parameters are defined by the data. Good resolution throughout leaves relatively less room to manoeuvre; the more a parameter is determined by interpolation or extrapolation, the latter being especially dangerous, the more a weighting scheme can affect the results.

When the weight w_i is the reciprocal of the variance (eqn 244, cf. eqn 235), the problem becomes one of estimating the variance on y_i for each x_i:

$$w_i = \frac{1}{\sigma_i^2}. \tag{244}$$

If the weighting scheme is doing its job, the weighted residuals corresponding to the best fit will be comparable for all of the data included in the analysis. Routine inspection of weighted residuals is thus recommended, and either a plot or the values themselves ought to be readily available from any fitting programme worth its salt. The selection of weighting schemes is often a process of trial and error; in the final analysis, an investigator will want the one that works.

Various weighting schemes and their limitations are considered below. Since weighting factors adjust the relative contribution of each residual to the sum of squares, it is relative rather than the absolute values of w_i that are of concern in most cases.

5.3.2 Constant error

The variance is assumed to be constant for all y_i, and w_i is typically set at 1.0. This is seldom appropriate for the types of experiments under consideration here.

5.3.3 Poisson error

Random processes with a constant probability of occurrence can be described by the binomial distribution, which approaches a limiting case known as the Poisson distribution when the probability of occurrence is small and the number of events is large. The mean and variance of a Poisson distribution are both equal to the number of events per interval, and w_i can therefore be taken as $1/y_i$. The decay frequencies of atomic nuclei and photo-excited atoms exhibit Poisson distributions, which therefore contribute to the distribution of the measured signal in experiments utilizing radiolabelled or fluorescent ligands. Experimental error is not determined exclusively by the number of disintegrations per minute, however, and data from binding studies are more appropriately described by a Gaussian distribution under most circumstances. The Poisson distribution approaches the Gaussian distribution when the frequency of events is large, and weighting according to $1/y_i$ may be adequate if not strictly correct in some situations.

5.3.4 Proportional error

The sample standard deviation on estimates of binding is often directly proportional to y_i, at least for values of y_i comfortably greater than zero;

similarly, the sample variance is directly proportional to y_i^2, and w_i can be taken as $1/y_i^2$. Proportional weighting can be imprecise, since individual estimates of y_i are usually based on very few degrees of freedom. That uncertainty is of particular concern with highly scattered data, and an alternative is to weight according to a smoothed value of the dependent variable (i.e. $w_i = 1/\hat{y}_i^2$). The most convenient smoothing function is the model itself (i.e. $\hat{y}_i = f(\mathbf{x}, \hat{\mathbf{a}})$), and $\hat{y}_i$ can be constantly updated by substituting the current values of $\mathbf{a}$ at each cycle of the fitting procedure.

These procedures cannot be used for very small values of y_i or $\hat{y}_i$, since the corresponding residuals will be grossly overweighted and can dominate the fitting process. The problem often does not arise with data acquired at a fixed concentration of the probe and graded concentrations of an unlabelled ligand; when the data are taken as measured, non-specific binding tends to ensure that no value of y_i is dangerously small. This built-in safeguard is lost when an estimate of non-specific binding is subtracted from y_i prior to the analysis; a particularly awkward situation occurs when one or more of the corrected y_i equal zero. With data acquired at graded concentrations of the probe, the problem is unavoidable if total binding at the lowest concentrations is virtually zero. In most situations, values of y_i near zero underestimate the error by violating the assumption of direct proportionality between σ_i^2 and y_i^2. With the naïve subtraction described above, y_i^2 is rendered artefactually small without a corresponding change in σ_i^2; when total binding is near zero, σ_i^2 is dominated by background noise and is therefore independent of y_i.

5.3.5 Measured error

When each point is the mean of several replicates, measured values of σ_i^2 are available and can be used to calculate w_i. This approach works when the sample variance reasonably approximates the population variance. As with proportional weighting, however, individual estimates of σ_i^2 are typically based on a small number of observations and are therefore imprecise. Poor estimates can lead to an inappropriate weight in a number of anomalous but familiar situations. It is not unusual, for example, to encounter outliers with relatively low error; similarly, points with relatively high error can often be found on or near the fitted curve.

5.3.6 Error functions

Imprecision in measured estimates of the variance tends to be a fact of life, since there are practical limitations on the number of observations at each $\mathbf{x}_i$. The consequences of individual variability in y_i and σ_i^2 can be reduced in severity if a smoothed estimate of the variance $\hat{\sigma}_i^2$ is obtained by fitting an empirical function to the relevant variables. Polynomials are particularly useful for this purpose (e.g. eqn 245, and refs 25, 26), as they are linear in all parameters and can therefore be handled by analytical rather than iterative methods:

$$f(Y) = \varphi_0 + \varphi_1 Y_i + \varphi_2 Y_i^2. \tag{245}$$

Prior to the first cycle of Marquardt's procedure, general linear regression is used to fit eqn 245 with $f(Y)$ and Y is taken as the measured values of σ_i^2 and y_i respectively; the fitted coefficients φ_j and y_i are then substituted back to obtain smoothed estimates of the variance $\hat{\sigma}_i^2$, and these in turn are used to calculate the weights (i.e. $w_i = 1/\hat{\sigma}_i^2$). Values of w_i estimated from σ_i^2 and y_i can also be used in all subsequent iterations. Alternatively, the weights can be refined after each cycle by fitting eqn 245 with $f(Y)$ taken as σ_i^2, as before, but with Y taken as the computed value of $\hat{y}_i$ for the current values of $\mathbf{a}$ (i.e. $\hat{y}_i = f(\mathbf{x}_i, \mathbf{a})$); the new values of φ_j are then used together with $\hat{y}_i$ to calculate $\hat{\sigma}_i^2$ and the revised values of w_i. It is a good practice to plot the fit of $f(y)$ to $\{\sigma_i^2\}$ at the minimum in SS, at least from time to time; the parameters of the error function (e.g. $\hat{\varphi}_j$ in eqn 245) and the relevant data therefore ought to be available from the fitting programme upon request.

5.3.7 Robust weighting

The influence of outliers can be avoided altogether by their removal prior to the first analysis or, preferably, prior to a second analysis if they lie outside a specified confidence interval around the fitted curve from the first. An alternative approach is that of robust weighting, in which the weight is reduced to zero as the residual becomes large relative to the norm. One example of the technique is the method described by Mosteller and Tukey (27).

5.4 Goodness of fit

5.4.1 General comments

Eventually, there must be a decision as to whether or not the fit is acceptable. The model may be wrong, or it may never have had a chance if the fitting process took off in the wrong direction. Goodness of fit relates in part to the distribution of the data about the fitted curve, which ought to be random, and in part to the variance of residuals. The assessment often tends to be subjective, statistical tests notwithstanding, and it behoves the investigator always to inspect a plot of the fitted curve superimposed on the data; accordingly, fitting programmes ought to produce such plots with the same agility as they produce the parametric values. With plots of $f(x)$ versus x, however, the visual impression of the fit can be misleading in regions of the curve where $df(x)/dx$ is relatively large; also, the distribution can be difficult to assess if the residuals are small relative to $f(x)$, particularly when the diameter of the symbols is relatively large. One's perception is much enhanced by holding the page at a slight angle to the line of vision and sighting along the fitted curve. An alternative presentation that highlights the distribution is a plot of the residuals versus x or $\hat{y}$, with the ordinate scaled to the value of the largest (>0) or smallest (<0) residual. A striking illustration of the former has been presented by Birdsall *et al.* (28); examples also can be found in the monograph by Bates and Watts (24).

Statistical tests for goodness of fit are an important complement to visual inspection and allow the model to be accepted or rejected at a known level of confidence. The first procedure listed below assesses the relative merit of two fits; in the others, the assessment is based on tests of internal consistency. The latter are helpful when the model has been fed all of the parameters required by the hypothesis or justified by the data. Two such procedures considered here assess the relationship between each residual and its neighbour along the fitted curve (Sections 5.4.4 and 5.4.5). It is therefore important that the data be evaluated in order of increasing or decreasing x_i. A forgiving programme will recognize any points that are out of order and sort the data before performing the calculation. Ordering of the data poses special problems when data from separate experiments share common parameters; in such situations, it is not uncommon to find that individual sets of data lie predominantly above or below the fitted curves, while the overall distribution is random. When data from separate experiments are analysed independently, a global assessment of the distribution can be performed if the residuals are available from the fitting programme; the last value from one set of data must not be compared with the first value from another.

When evaluating the results of simultaneous analyses, it is important that different sets of data make comparable contributions to the weighted sum of squares. Experiments with highly scattered or abnormally distributed data render the others irrelevant to the extent that they dominate the fit or the statistical parameters in question. Statistical details ought therefore to be available not only for the data taken together but also for the data from individual experiments taken separately.

5.4.2 Extra sum of squares

The weighted sum of squares for a given pool of data is inversely dependent upon the number of parameters in the model. Additional parameters thus tend to improve the fit, but successive additions yield diminishing returns until the change is not significant. The upper limit on the number of parameters is the number of observations; when the two are equal, the result is no longer a fit but a solution achieved by iterative rather than analytic means. In practice, additional parameters ought to be unprofitable at values of m well below that at which the degrees of freedom is zero.

When comparing hierarchical models, the probability that additional parameters are without effect on the sum of squares is defined by an F-distribution (29). The F-statistic is calculated according to eqn 246, where SS_x

is as defined in eqn 235 and df_x is the degrees of freedom. It is assumed that $1/\sigma_i^2$ is reasonably approximated by w_i:

$$F = \frac{(SS_1 - SS_2)/(df_1 - df_2)}{SS_2/df_2}. \qquad (246)$$

The two values of SS_x and $df_x(x = 1$ or $2)$ are obtained from two analyses differing only in the number of parameters (i.e. $df_x = n - m_x$, $df_2 < df_1$). Levels of significance corresponding to the calculated value of F can be estimated by comparison with tabulated values or, more conveniently, can be computed from the probability function of the F-distribution; the number of degrees of freedom is $df_1 - df_2$ and df_2 in the numerator and denominator respectively. When df_1 equals df_2, the sums of squares are not independent.

When evaluating the F-statistic, the relevant pool of data may represent one experiment or several. Similarly, the variance from several sets of data may represent a single analysis or separate analyses of the data grouped in one way or another. Comparisons based on the multi-site model (Scheme II, eqn 45) and related empirical models frequently involve a particular pool of data and analyses in terms of n and $n + 1$ classes of sites; similarly, two pools of data can be analysed with a single and separate values of particular parameters (e.g. K_{ij} in eqn 45). The latter comparison is an alternative to a t-test when each pool represents a number of replicated experiments that could have been analysed separately to yield average values ($\pm$ SEM) of the parameter under consideration. Examples of these procedures can be found in Wong *et al.* (30).

When the results of multiple experiments are involved, simultaneous analyses ought to include only those data that share one or more of the parameters being optimized. It is technically possible to optimize several independent sets of parameters, where each set is exclusive to a distinct sub-pool of the data, but the practice serves only to reduce the relative sensitivity of SS to one set of parameters or another. The effect is to decrease the ratio of signal to noise, which can result in increased uncertainty over the parametric values and convergence to a false or local minimum.

One generally expects the sum of squares to be less with fewer degrees of freedom; in practice, however, additional parameters sometimes result in an increase (i.e. $SS_2 > SS_1$, $df_2 < df_1$). When the model is essentially the same, as in the case of two rational functions, the anomaly typically reflects parameters that are poorly defined and highly correlated; the extra parameter has destabilized the successive estimation of one or more δa_k (eqn 236), and the process has converged to a local minimum. With mathematically distinct models, a function that is mechanistically correct may describe the data with fewer parameters than are required with a function that is mechanistically wrong. Such models are not hierarchical, however, and the ratio does not have an F-distribution.

5.4.3 Variance ratio

For a good fit, the distribution of residuals is expected to be consistent with the experimental error. If the weight has been taken as the reciprocal of the estimated variance on y_i (i.e. $w_i = 1/\sigma_i^2$ or $w_i = 1/\hat{\sigma}_i^2$), the residuals and the sample error are in the same units. A value near unity is therefore expected for the ratio SS/df, in which SS is the sum of weighted residuals as in eqn 246. The probability that SS/df significantly exceeds unity is defined by a chi-square distribution with df degrees of freedom. A large value suggests that the scatter of the data about the fitted curve exceeds that expected on the basis of the sample variance (20).

5.4.4 Run test

The sign of each residual identifies its position as either above (i.e. $y_i - f(x_i, \hat{a}) > 0$) or below (i.e. $y_i - f(x_i, \hat{a}) > 0$) the fitted curve. A run is defined as a string of successive residuals, all of the same sign; the number of runs $\hat{r}$ can therefore vary from 1 to the number of residuals n. For a good fit, the sign of the residual is expected to change randomly with increasing or decreasing x_i; for a particular number of residuals, randomness is associated with an optimal number of runs provided that each residual represents an independent event. An inappropriate model generally results in too few runs (e.g. $+++++----$ $-+++++$; $n = 14$, $\hat{r} = 3$); that is, $\hat{r}$ is small for the given n, and the residuals are clustered as the fitted curve weaves back and forth through the data. A large $\hat{r}$ indicates regular change (e.g. $+-+-+-+-+++-+-$; $n = 14$, $\hat{r} = 12$), but the fit may not be considered poor according to criteria based on the sum of squares. The problem may reflect systematic errors related to experimental design. The randomness of the sequence can be tested by means of the z-statistic, which describes a standard normal distribution (31).

5.4.5 Correlation of neighbouring residuals

An approach that considers both sign and magnitude is to evaluate the correlation between each successive residual z_i and its nearest neighbour z_{i+1} (32). The weighted residuals are written into two arrays as illustrated below, and the correlation coefficient r is obtained by linear regression:

$$
\begin{array}{ccc}
z_1 & , & z_2 \\
z_2 & , & z_3 \\
\vdots & & \vdots \\
z_i & , & z_{i+1} \\
\vdots & & \vdots \\
z_{n-1} & , & z_n.
\end{array}
$$

Fisher's *t*-statistic can be calculated from eqn 247 and used in turn to compute the corresponding level of significance:

$$t_F = r \left\{ \frac{n-3}{1-r^2} \right\}^{1/2}.$$ (247)

Inappropriate models generally result in positively correlated residuals. Negatively correlated residuals arise in situations that would yield too many runs in a run test. The correlation or lack of one can be illustrated in a plot of z_i versus z_{i+1}, sometimes referred to as a *lag plot*; examples can be found in Bates and Watts (24) and Wong *et al.* (30).

5.5 Bounded parameters

The parameters of mechanistic models are often defined by upper or lower bounds. Values beyond the prescribed limits may be ruled out by other data or be uninterpretable in physical terms; the latter can be especially irksome, as they tend to result in numerical failure. It is tempting to intervene in the fitting process when one is about to be presented with an unpalatable value.

Zero is the lower limit for many parameters entered on an arithmetic scale. Values less than zero can be avoided by redefining a_k as log a_k whenever the increment produced by Marquardt (i.e. δa_k in eqn 236) would result in the parameter becoming negative. Such reparameterization affects both the value *per se* and the corresponding estimate of uncertainty computed from the diagonal element of the covariance matrix. If the result is required in the original units, both the value and the error must be converted back once the minimum has been found. For quantities of the form $X \pm x$ in which x is relatively small, the expansion in terms of a function f can be approximated by the first two terms of Taylor's series (eqn 248):

$$f(X + x) \approx f(X) + \frac{df(X)}{dX} x.$$ (248)

Equations 249 and 250 can thus be used to convert errors from linear units x_X to logarithmic units $x_{\log X}$ and vice versa:

$$x_{\log X} = \frac{x_X \log e}{X}$$ (249)

$$x_X = x_{\log X} X \ln_{10}.$$ (250)

A questionable but more general approach to bounded parameters is to increase λ (eqn 236) and adjust the angle of descent whenever the parameter is about to go out of bounds. In this way, a value just inside the prescribed limit will eventually be found. Estimates of the standard error become meaningless upon such intervention; also, the successive adjustments can destabilize other parameters and result in convergence to a local minimum.

Generally speaking, parameters ought to be processed in the form most relevant to the physical quantity that they represent. Affinities are related exponentially to the free energy of the interaction (i.e. $\Delta G = -RT \ln K$); similarly, the activity of an interacting species is related exponentially to its chemical potential (i.e. $\mu - \mu^\circ = RT \ln a$). Parameters that represent affinity and concentration therefore ought to be optimized throughout in logarithmic form, although the models themselves eventually require the linear form. It has been shown experimentally that estimates of affinity are distributed geometrically rather than arithmetically (33).

Arbitrary limits on parameters are in effect the imposition of a bias and are best avoided. If the process is converging to supposedly inappropriate values, it is usually preferable to identify the cause and then to proceed accordingly; the experiment may have been poorly designed, or there may be a problem with the model. In some instances, unexpected values may be a source of new insight into the system under study. Limits are useful in practice, however, if only insofar as they avoid numerical problems that bring the operation to an inelegant halt. A spurious answer is probably better than no answer at all, provided that the investigator is aware of the intervention. If a limit must be enforced, the best approach generally is to repeat the analysis with the offending parameter reclassified as a constant and assigned the desired value. Parameters near zero pose special problems related not only to the possibility of a change in sign but also to loss of significance. If the value is sufficiently small to require special treatment, in effect it is known and can be fixed accordingly.

A final word is in order regarding other people's black boxes. Numerical methods often require traps and interventions in order to guarantee an answer; some are permissible and even advisable, while others may compromise the validity of the process. Prepackaged software cannot and indeed ought not to be avoided in many situations; there is no point in reinventing the wheel, and one's home-built contraption may be decidedly inferior to a professional product. Nevertheless, it is important to be aware of the procedures involved and to be alerted when problems arise.

Acknowledgements

The author is grateful to Dr John Batchelor, Peter Chidiac, Gary Eppel, Ricky Fung, Dr Peter Pennefather and William Sinkins for many helpful discussions before or during the preparation of this chapter. Dr Laszlo Endrenyi and Dr Andreas Schwab are thankfully acknowledged for their comments regarding the section on statistical procedures. Special thanks are also due to the staff of the Division of Computing Services, University of Toronto; Terry Jones, Dr Lee Oattes, Dr John Roth and Dr Andrzej Pindor are particularly acknowledged for their assistance with the numerical procedures and with computer-related problems in general. Mrs Roselie Damian is gratefully acknowledged for her secretarial expertise and heroic efforts with the word processor. This investigation has been supported by grants from the Medical Research Council of Canada and the Heart and Stroke Foundation of Ontario.

References

1. Seeman, P., Ulpian, C., Wreggett, K. A., and Wells, J. W. (1984). *J. Neurochem.*, **43**, 221–35.
2. Press, W. H., Flannery, B. P., Teukolsky, S. A., and Vetterling, W. T. (1986). *Numerical Recipes: The Art of Scientific Computing.* Cambridge University Press. (This edition contains subroutines coded in FORTRAN; editions in Pascal and C are also available.)
3. Acton, F. S. (1970). *Numerical Methods that Work.* Harper and Row, New York.
4. Feldman, H., Rodbard, D., and Levine, D. (1972). *Anal. Biochem.*, **45**, 530–56.
5. Pauling, L., Pressman, D., and Grossberg, A. L. (1944). *J. Am. Chem. Soc.*, **66**, 784–92.
6. Sips, R. (1950). *J. Chem. Phys.*, **18**, 1024–26.
7. Bruni, C., Germani, A., Koch, G., and Strom, R. (1976). *J. Theor. Biol.*, **61**, 143–70.
8. Tobler, H. J. and Engel, G. (1983). *Naunyn-Schmiedebergs Arch. Pharmakol.*, **322**, 183–92.
9. Koshland, D. E., Nemethy, G., and Filmer, D. (1966). *Biochemistry*, **5**, 365–85.
10. Monod, J., Wyman, J., and Changeux, J.-P. (1965). *J. Mol. Biol.*, **12**, 88–118.
11. Boeynaems, J. M. and Dumont, J. E. (1980). *Outlines of Receptor Theory.* Elsevier–North-Holland, Amsterdam.
12. Moelwyn-Hughes, E. A. (1971). *The Chemical Statics and Kinetics of Solutions.* Academic Press, London.
13. Cheng, Y. and Prusoff, W. (1973). *Biochem. Pharmacol.*, **22**, 3099–108.
14. Fletcher, J. E. and Spector, A. A. (1977). *Mol. Pharmacol.*, **13**, 387–99.
15. Klotz, I. M. (1983). *Trends Pharmacol. Sci.*, **4**, 253–5.
16. Klotz, I. M. (1989). In *Protein Function: a Practical Approach* (ed. T. E. Creighton), pp. 25–54. IRL Press, Oxford.
17. Klotz, I. M. and Hunston, D. L. (1975). *J. Biol. Chem.*, **250**, 3001–9.
18. Edsall, J. T. and Gutfreund, H. (1983). *Biothermodynamics: The Study of Biochemical Processes at Equilibrium.* Wiley, Chichester.
19. De Lean, A., Munson, P. J., and Rodbard, D. (1979). Multi-subsite receptors for multivalent ligands: *Mol. Pharmacol.*, **15**, 60–70.
20. Landaw, E. M. and DiStefano, J. J. (1984). *Am. J. Physiol.*, **246**, R665–R677.
21. Motulsky, H. J. and Ransnas, L. A. (1987). *FASEB J.*, **1**, 365–74.
22. Marquardt, D. W. (1963). *J. Soc. Indust. Appl. Math.*, **2**, 431–41.
23. Dennis, J. E. and Schnabel, R. B. (1983). *Numerical Methods for Unconstrained Optimization and Nonlinear Equations.* Prentice Hall, N.J.
24. Bates, D. M. and Watts, D. G. (1988). *Non-Linear Regression Analysis and Its Applications.* Wiley, New York.

25. Rodbard, D. (1974). *Clin. Chem.*, **20**, 1255–70.
26. Rodbard, D., Lenox, R. H., Wray, H. L., and Ramseth, D. (1976). *Clin. Chem.*, **22**, 350–8.
27. Mosteller, F. and Tukey, J. W. (1977). *Data Analysis and Regression.* Addison-Wesley, New York.
28. Birdsall, N. J. M., Hulme, E. C., and Burgen, A. S. V. (1980). *Proc. R. Soc. London*, **207**, 1–12.
29. Bard, Y. (1974). *Non-Linear Parameter Estimation.* Academic Press, New York.
30. Wong, H.-M. S., Sole, M. J., and Wells, J. W. (1986). *Biochemistry*, **25**, 6995–7008.
31. Sachs, L. (1984). *Applied Statistics: A Handbook of Techniques* (2nd edn). Springer, New York.
32. Reich, J. G., Wangerman, G., Falck, M., and Rohde, K. (1972). *Eur. J. Biochem.*, **26**, 368–79.
33. De Lean, A., Hancock, A. A., and Lefkowitz, R. J. (1982). *Mol. Pharmacol.*, **21**, 5–16.

Further reading

The following list is necessarily selective. The monographs by Acton (1981), Boeynaems and Dumont (1980), and Press *et al.* (1986) are particularly recommended as general references.

General topics

Boeynaems, J. M. and Dumont, J. E. (1980). *Outlines of Receptor Theory.* Elsevier–North-Holland, Amsterdam.
Edsall, J. T. and Gutfreund, H. (1983). *Biothermodynamics: The Study of Biochemical Processes at Equilibrium.* Wiley, Chichester.
Hollenberg, M. D. (1985). Receptor models and the action of neurotransmitters and hormones: some new perspectives. In *Neurotransmitter Receptor Binding* (2nd edn), (ed. H. Yamamura, S. J. Enna, and M. J. Kuhar), pp. 1–39. Raven Press, New York.
Kenakin, T. P. (1987). *Pharmacologic Analysis of Drug–Receptor Interaction.* Raven Press, New York.
Klotz, I. M. (1989). Ligand–protein binding affinities. In *Protein Function: a Practical Approach* (ed. T. E. Creighton), pp. 25–54. IRL Press, Oxford.
Levitzki, A. (1984). *Receptors: A Quantitative Approach.* Benjamin/Cummings, Menlo Park, CA.
Moelwyn-Hughes, E. A. (1971). *The Chemical Statics and Kinetics of Solutions.* Academic Press, London.
Segel, I. H. (1975). *Enzyme Kinetics: Behavior and Analysis of Rapid Equilibrium and Steady State.* Wiley, New York.

Multi-site model

Bruni, C., Germani, A., Koch, G., and Strom, R. (1976). Derivation of antibody distribution from experimental binding data. *J. Theor. Biol.*, **61**, 143–70.

Feldman, H. A. (1972). Mathematical theory of complex ligand-binding systems at equilibrium: some methods for parameter fitting. *Anal. Biochem.*, **48**, 317–38.

Feldman, H., Rodbard, D., and Levine, D. (1972). Mathematical theory of cross-reactive radiommunoassay and ligand-binding systems at equilibrium. *Anal. Biochem.*, **45**, 530–56.

Munson, P. J. (1983) LIGAND: a computerized analysis of ligand binding data. *Methods Enzymol.*, **92**, 543–76.

Munson, P. J. and Rodbard, D. (1980). LIGAND: a versatile computerized approach for characterization of ligand-binding systems. *Anal. Biochem.*, **107**, 220–39.

Pauling, L., Pressman, D., and Grossberg, A. L. (1944). The serological properties of simple substances. VII. A quantitative theory of the inhibition by haptens of the precipitation of heterogeneous antisera with antigens, and comparison with experimental results for polyhaptenic simple substances and for azoproteins. *J. Am. Chem. Soc.*, **66**, 784–92.

Sips, R. (1950). On the structure of a catalyst surface. *J. Chem. Phys.*, **18**, 1024–26.

Tobler, H. J. and Engel, G. (1983). Affinity spectra: a novel way for the evaluation of equilibrium binding experiments. *Naunyn-Schmiedebergs Arch. Pharmakol.*, **322**, 183–92.

Mobile or floating receptor model

Abramson, S. N., McGonigle, P., and Molinoff, P. B. (1987). Evaluation of models for analysis of radioligand binding data. *Mol. Pharmacol.*, **31**, 103–11.

Boeynaems, J. M. and Dumont, J. E. (1977). The two-step model of ligand–receptor interaction. *Mol. Cell. Endocrinol.*, **7**, 33–47.

De Haen, C. (1976). The non-stoichiometric floating receptor model for hormone sensitive adenylyl cyclase. *J. Theor. Biol.*, **58**, 383–400.

De Lean, A., Stadel, J. M., and Lefkowitz, R. J. (1980). A ternary complex model explains the agonist-specific binding properties of the adenylate cyclase-couple β-adrenergic receptor. *J. Biol. Chem.*, **255**, 7108–17.

Ehlert, F. J. and Rathbun, B. E. (1990). Signaling through the muscarinic receptor-adenylate cyclase system of the heart is buffered against GTP over a range of concentrations. *Mol. Pharmacol.*, **38**, 148–58.

Jacobs, S. and Cuatrecasas, P. (1976). The mobile receptor hypothesis and 'cooperativity' of hormone binding: application to insulin. *Biochim. Biophys. Acta*, **433**, 482–95.

Lee, T. W. T., Sole, M. J., and Wells, J. W. (1986). Assessment of a ternary model for the binding of agonists to neurohumoral receptors. *Biochemistry*, **25**, 7009–20.

Leung, E., Jacobson, K. A., and Green, R. D. (1990). Analysis of agonist–antagonist interactions at A_1 adenosine receptors. *Mol. Pharmacol.*, **38**, 72–83.

Minton, A. P. and Sokolovsky, M. (1990). A model for the interaction of muscarinic receptors, agonists, and two distinct effector substances. *Biochemistry*, **29**, 1586–93.

General co-operativity

Adair, G. S. (1925). The hemoglobin system. *J. Biol. Chem.*, **63**, 493–545 (see especially pp. 529–45).

De Lean, A., Munson, P. J., and Rodbard, D. (1979). Multi-subsite receptors for multivalent ligands: applications to drugs, hormones, and neurotransmitters. *Mol. Pharmacol.*, **15**, 60–70.

Eigen, M. (1967). Kinetics of reaction control and information transfer in enzymes and nucleic acids. In *Fast Reactions and Primary Processes in Chemical Kinetics* (ed. S. Claesson), pp. 333–69. Almquist and Wiksell, Stockholm.

Koshland, D. E. (1970). The molecular basis for enzyme regulation. In *The Enzymes*, Vol. 1, *Structure and Control* (ed. P. D. Boyer), pp. 341–96. Academic Press, New York.

Koshland, D. E., Nemethy, G., and Filmer, D. (1966). Comparison of experimental binding data and theoretical models in proteins containing subunits. *Biochemistry*, **5**, 365–85.

Mattera, R., Pitts, B. J. R., Entman, M. L., and Birnbaumer, L. (1985). Guanine nucleotide regulation of a mammalian myocardial muscarinic receptor system. Evidence for homo– and heterotropic cooperativity in ligand binding analyzed by computer-assisted curve fitting, *J. Biol. Chem.*, **360**, 7410–21.

Monod, J., Wyman, J., and Changeux, J.-P. (1965). On the nature of allosteric transitions: a plausible model. *J. Mol. Biol.*, **12**, 88–118.

Neet, K. E. (1980). Cooperativity in enzyme function: equilibrium and kinetic aspects. *Methods Enzymol.*, **64**, 138–92.

Co-operativity and the Hill equation

Cornish-Bowden, A. and Koshland, D. E. (1975). Diagnostic uses of the Hill (Logit and Nernst) plots. *J. Mol. Biol.*, **95**, 201–12.

Hill, A. V. (1913). The combinations of haemoglobin with oxygen and with carbon monoxide. *Biochem. J.*, **7**, 471–80.

Wyman, J. (1964). Linked functions and reciprocal effects in hemoglobin: a second look. *Adv. Protein Chem.*, **19**, 223–86.

Interpretation of data

Cheng, Y. and Prusoff, W. (1973). Relationship between the inhibition constant (K_I) and the concentration of inhibitor which causes 50 per cent inhibition (I_{50}) of an enzymatic reaction. *Biochem. Pharmacol.*, **22**, 3099–108.

DeBlasi, A. and Motulsky, H. J. (1987). Alternative molecular interpretations of binding curves: compelling competition? *Trends Pharmacol. Sci.*, **8**, 421–3.

Fletcher, J. E. and Spector, A. A. (1977). Alternative models for the analysis of drug–protein binding. *Mol. Pharmacol.*, **13**, 387–99.

Klotz, I. M. (1983). Ligand–receptor interactions: what we can and cannot learn from binding measurements, *Trends Pharmacol. Sci.*, **4**, 253–5.

Klotz, I. M. and Hunston, D. L. (1984). Mathematical models for ligand–receptor binding. Real sites, ghost sites. *J. Biol. Chem.*, **259**, 10060–2.

Klotz, I. M. and Hunston, D. L. (1975). Protein interactions with small molecules. Relationships between stoichiometric binding constants, site binding constants, and empirical binding parameters. *J. Biol. Chem.*, **250**, 3001–9.

Minton, A. P. (1979). On the interpretation of binding isotherms in complex biological systems. Apparent homogeneity of some heterogeneous systems. *Biochim. Biophys. Acta*, **558**, 179–86.

Prinz, H. (1983). On the interpretation of equilibrium binding studies. *J. Receptor Res.*, **3**, 239–48.

Tomlinson, G. (1988). Inhibition of radioligand binding to receptors: a competitive business. *Trends Pharmacol. Sci.*, **9**, 159–62.

Wong, J. T.-F. and Endrenyi, L. (1971). Interpretation of nonhyperbolic behavior in enzymic systems. I. Differentiation of model mechanisms. *Can. J. Biochem.*, **49**, 568–80.

Artefacts

Burgisser, E., Hancock, A. A., Lefkowitz, R. J., and De Lean, A. (1981). Anomalous equilibrium binding properties of high-affinity racemic radioligands. *Mol. Pharmacol.*, **19**, 205–16.

Chang, K.-J., Jacobs, S., and Cuatrecasas, P. (1975). Quantitative aspects of hormone–receptor interactions of high affinity. Effect of receptor concentration and measurement of dissociation constants of labeled and unlabeled hormones. *Biochim. Biophys. Acta*, **406**, 294–303.

Goldstein, A. and Barrett, R. W. (1987). Ligand dissociation constants from competition binding assays: errors associated with ligand depletion. *Mol. Pharmacol.*, **31**, 603–9.

Jacobs, S., Chang, K.-J., and Cuatrecasas, O. (1975). Estimation of hormone receptor affinity by competitive displacement of labeled ligand: effect of concentration of receptor and of labeled ligand. *Biochem. Biophys. Res. Commun.*, **66**, 687–92.

Kermode, J. C. (1989). The curvilinear Scatchard plot: experimental artifact or receptor heterogeneity? *Biochem. Pharmacol.*, **38**, 2053–60.

Munson, P. J. (1983). Experimental artifacts and the analysis of ligand binding data: results of a computer simulation. *J. Receptor Res.*, **3**, 249–59.

Seeman, P., Ulpian, C., Wreggett, K. A., and Wells, J. W. (1984). Dopamine receptor parameters detected by [^{3}H]spiperone depend on tissue concentration: analysis and examples. *J. Neurochem.*, **43**, 221–35.

Swillens, S. and Dumont, J. (1985). A pitfall in the interpretation of data on ligand–protein interaction. *Biochem, J.*, **149**, 779–82.

Wells, J. W., Birdsall, N. J. M., Burgen, A. S. V., and Hulme, E. C. (1980). Competitive binding studies with multiple sites: effects arising from depletion of the free radioligand. *Biochim. Biophys. Acta*, **632**, 464–9.

Linear transformations and their limitations

Atkins, G. L. and Nimmo, I. A. (1975). A comparison of seven methods for fitting the Michaelis–Menten equation. *Biochem. J.*, **149**, 775–7.

Burgisser, E. (1984). Radioligand–receptor binding studies: what's wrong with the Scatchard analysis? *Trends Pharmacol. Sci.*, **5**, 142–4.

Feldman, H. A. (1983). Statistical limits in Scatchard analysis. *J. Biol. Chem.*, **258**, 12865–7.

Klotz, I. M. (1982). Numbers of receptor sites from Scatchard graphs: facts and fantasies. *Science*, **217**, 1247–9.

Klotz, I. M. and Hunston, D. L. (1984). Properties of graphical representations of multiple classes of binding sites. *J. Biol. Chem.*, **259**, 10060–2.

Lineweaver, H. and Burk, D. (1934). The determination of enzyme dissociation constants. *J. Am. Chem. Soc.*, **56**, 658–66.

Scatchard, G. (1949). The attractions of proteins for small molecules and ions. *Ann. N.Y. Acad. Sci.*, **51**, 600–72.

Numerical methods and statistical procedures

Acton, F. S. (1970). *Numerical Methods that Work.* Harper and Row, New York.

Bard, Y. (1974). *Non-Linear Parameter Estimation.* Academic Press, New York.

Bates, D. M. and Watts, D. G. (1988). *Non-Linear Regression Analysis and Its Applications.* Wiley, New York.

Birdsall, N. J. M., Hulme, E. C., and Burgen, A. S. V. (1980). The character of the muscarinic receptors in different regions of the rat brain. *Proc. R. Soc. London,* **207**, 1–12.

Box, G. E. P., Hunter, W. G., and Hunter, J. S. (1978). *Statistics for Experimenters: An Introduction to Design, Data Analysis, and Model Building.* Wiley, New York.

Carroll, R. J. and Ruppert, D. (1988). *Transformation and Weighting in Refression.* Chapman and Hall, New York.

Crabbe, M. J. C. (1985). Computers in biochemical analysis. *Methods Biochem. Anal.*, **31**, 417–74.

De Lean, A., Hancock, A. A., and Lefkowitz, R. J. (1982). Validation and statistical analysis of a computer modeling method for quantitative analysis of radioligand binding data for mixtures of pharmacological receptor subtypes. *Mol. Pharmacol.*, **21**, 5–16.

Dennis, J. E. and Schnabel, R. B. (1983). *Numerical Methods for Unconstrained Optimization and Nonlinear Equations.* Prentice Hall. N.J.

Endrenyi, (ed.) (1981). *Kinetic Data Analysis: Design and Analysis of Enzyme and Pharmacokinetic Experiments.* Plenum Press, New York.

Hamming, R. W. (1973). *Numerical Methods for Scientists and Engineers* (2nd edn). McGraw-Hill, New York.

Johnson, L. W. and Riess, R. D. (1982). *Numerical Analysis.* Addison-Wesley, Reading, MA.

Landaw, E. M. and DiStefano, J. J. (1984). Multiexponential, multicompartmental, and noncompartmental modeling. II. Data analysis and statistical considerations. *Am. J. Physiol.*, **246**, R665–R677.

Leatherbarrow, R. J. (1990). Using linear and non-linear regression to fit biochemical data. *Trends Biochem. Sci.*, **15**, 455–8.

Marquardt, D. W. (1963). An algorithm for least-squares estimation of nonlinear parameters. *J. Soc. Indust. Appl. Math.*, **2**, 431–41.

Mosteller, F. and Tukey, J. W. (1977). *Data Analysis and Regression.* Addison-Wesley, New York.

Motulsky, H. J. and Ransnas, L. A. (1987). Fitting curves to data using nonlinear regression: a practical and nonmathematical review. *FASEB J.*, **1**, 365–74.

Press, W. H., Flannery, B. P., Teukolsky, S. A., and Vetterling, W. T. (1986). *Numerical Recipes: The Art of Scientific Computing.* Cambridge University Press. (This edition contains subroutines coded in FORTRAN; editions in Pascal and C are also available.)

Ratkowsky, D. A. (1990). *Handbook of Nonlinear Regression Models*. Marcel Dekker, New York.

Reich, J. G., Wangerman, G., Falck, M., and Rohde, K. (1972). A general strategy for parameter estimation from isosteric and allosteric-kinetic data and binding measurements. *Eur. J. Biochem.*, **26**, 368–79.

Rodbard, D. (1974). Statistical quality control and routine data processing for radioimmunoassays and immunoradiometric assays. *Clin. Chem.*, **20**, 1255–70.

Rodbard, D., Lenox, R. H., Wray, H. L., and Ramseth, D. (1976). Statistical characterization of the random errors in the radioimmunoassay dose-response variable. *Clin. Chem.*, **22**, 350–8.

Sachs, L. (1984). *Applied Statistics: A Handbook of Techniques* (2nd edn). Springer, New York.

Seber, G. A. F. and Wild, C. J. (1989). *Nonlinear Regression*. J. Wiley, New York.

Vetterling, W. T., Teukolsky, S. A., Press, W. H., and Flannery, B. P. (1985). *Numerical Recipes Example Book (FORTRAN)*. Cambridge University Press.

Wong, H.-M. S., Sole, M. J., and Wells, J. W. (1986). Assessment of mechanistic proposals for the binding of agonists to cardiac muscarinic receptors. *Biochemistry*, **25**, 6995–7008.

Appendices

A1

Radioligands for the study of receptor–ligand interactions: Amersham International plc.

K. McFARTHING

Table 1. Endothelins

Code	Product	Specific activity	Pack sizes	Comments
IM.223	(3-[^{125}I]*Iodotyrosyl*) Endothelin-1 Freeze-dried solid	~74 TBq/mmol, ~2000 Ci/mmol	185 kBq, 5 µCi	Used to investigate endothelin receptors, and in radioimmunoassays.
IM.226	(3-[^{125}I]*Iodotyrosyl*) Endothelin-2 Freeze-dried solid	~74 TBq/mmol, ~2000 Ci/mmol	185 kBq, 5 µCi 925 kBq, 25 µCi	An iso-form of Endothelin that was found to give 50% greater maximal vasoconstriction in pig coronary arteries than ET-1 (1).
IM.228	(3-[^{125}I]*Iodotyrosyl*[6]) Endothelin-3 Freeze-dried solid	~74 TBq/mmol, ~2000 Ci/mmol	185 kBq, 5 µCi 926 kBq, 25 µCi	Also known as rat-Endothelin, shown to be a potent constrictor of rat blood vessels *in vitro*, and an effective pressor agent *in vivo* (2).

IM.231	(3-[^{125}I]Iodotyrosyl^{13}) Big endothelin 1-38, human Freeze-dried solid	~74 TBq/mmol, ~2000 Ci/mmol	185 kBq, 5 µCi 925 kBq, 25 µCi	Believed to be the circulating precursor of ET-1, (3), and found present at levels 2.5 times lower than ET-1 (4).
IM.235	(3-[^{125}I]Iodotyrosyl) Sarafotoxin S6b Freeze-dried solid	~74 Tbq/mmol, ~2000 Ci/mmol	185 kBq, 5 µCi 925 kBq, 25 µCi	Shows a high degree of sequence homology to the endothelins, and has been shown to share a common binding site in brain and cardiac tissues with endothelin (5).
IM.239	(3-[^{125}I]Iodotyrosyl) Vasoactive intestinal contractor Freeze-dried solid	~74 Tbq/mmol, ~2000 Ci/mmol	185 kBq, 5 µCi 925 kBq, 25 µCi	Also known as mouse β-endothelin Studies on its biological activity show stronger contractile activity on guinea-pig and mouse ileum (6), however ET-1 is a more potent vasoconstrictor in porcine coronary artery

References

1. *The first William Harvey workshop on Endothelin*, December 6th, 1988, London.
2. Yanagisawa, M. et al. (1988). *Proc. Natl. Acad. Sci. USA*, **85**, 6964.
3. Yanagisawa, M. et al. (1988). *Nature*, **332**, 411.
4. Kimura, G. Tsukuba University, Japan. (Personal communication.)
5. Ambar, I., Kloog, Y., Schwartz, I., Hazum, E., and Sokolovsky, M. (1989). *BBRC*, **158(1)**, 195.
6. Ishida, N., Tsukjioka, K., Tomoi, M., Saida, K., and Mitsui, Y. (1989). *FEBS. Lett*, **247(2)**, 337.

Table 2. γ-Aminobutyric acid/benzodiazepine ligands

Code	Ligand	Specific activity	Pack sizes	Comments
TRK.527	4-amino-n-[2,3-^{3}H] butyric acid([^{3}H]GABA) Aqueous solution containing 2% ethanol, sterilized	1.85–3.0 TBq/mmol, 50–80 Ci/mmol	9.25 MBq/250 μCi 37 MBq/1 mCi	Labels GABA$_A$ receptors in presence of L-baclofen; GABA$_B$ receptors in presence of isoguvacine and Ca^{2+}
TRK.586	[Methylamine-^{3}H]Muscimol 0.02 M acetic acid/ethanol (1:1) solution	185–740 GBq/mmol, 5–20 Ci/mmol	9.25 MBq/250 μCi 37 MBq/1 mCi	GABA$_A$ agonist; can be used for photoaffinity labelling
TRK.849	[^{3}H]TBOB Toluene/triethylamine (99:1)	0.74–2.2 TBq/mmol, 20–60 Ci/mmol	9.25 MBq/250 μCi 37 MBq/1 mCi	GABA$_A$ antagonist
TRK.572	[N-methyl-^{3}H]Diazepam Ethanol solution	2.6–3.1 TBq/mmol, 70–85 Ci/mmol	9.25 MBq, 250 μCi	Benzodiazepine agonist
TRK.590	[N-methyl-^{3}H]Flunitrazepam Ethanol solution	2.6–3.1 TBq/mmol, 70–85 Ci/mmol	9.25 MBq/250 μCi 37 MBq/1 mCi	Benzodiazepine agonist; can be used for photoaffinity labelling
TRK.212	7-[^{125}I]Iodoclonazepam Water/acetonitrile (1:1) solution	~74 TBq/mmol, ~2000 Ci/mmol	925 kBq/25 μCi 1.85 MBq/50 μCi	Benzodiazepine agonist

Table 3. Histamine ligands

Code	Ligand	Specific activity	Pack sizes	Comments
TRK.631	[2,5-³H]Histamine dihydrochloride 0.01 M HCl/ethanol (1:1) solution	0.9–1.85 TBq/mmol, 25–50 Ci/mmol	9.25 MBq, 250 μCi 37 MBq, 1 mCi	H₁ antagonist
TRK.608	[pyridinyl-5-³H]Pyrilamine Ethanol solution containing 0.25% ascorbic acid	0.74–1.1 TBq/mmol, 20–30 Ci/mmol	9.25 MBq, 250 μCi 37 MBq, 1 mCi	
TRK.615	[N-methyl-³H]Cimetidine Ethanol solution	0.37–1.1 TBq/mmol, 10–30 Ci/mmol	9.25 MBq, 250 μCi 37 MBq, 1 mCi	H₂ antagonist

Table 4. Serotonin/5-hydroxytryptamine ligands

Code	Ligand	Specific activity	Pack sizes	Comments
TRK.223	5-hydroxy[G-^{3}H]tryptamine, creatinine sulphate Aqueous solution containing 2% ethanol, sterilized	370–740 GBq/mmol, 10–20 Ci/mmol	37 MBq, 1 mCi 185 MBq, 5 mCi	5-HT uptake blocker
TRK.553	[^{3}H]Imipramine Ethanol solution	0.74–1.1 TBq/mmol, 20–30 Ci/mmol	9.25 MBq, 250 Ci	5-HT uptake blocker
TRK.850	8-hydroxy[^{3}H]DPAT Ethanol solution	2.6–4.1 TBq/mmol, 70–110 Ci/mmol	9.25 MBq, 250 μCi 37 MBq, 1 mCi	5-HT$_{1A}$ agonist
TRK.845	[N-6-methyl-^{3}H]Mesulergine Ethanol solution	2.6–3.1 TBq/mmol, 70–85 Ci/mmol	9.25 MBq, 250 μCi	5-HT$_{1C}$ agonist
IM.252	7-amino-8-[^{125}I]Iodoketanserin	~74 TBq/mmol, ~2000 Ci/mmol	1.85 MBq, 50 μCi 3.7 MBq, 100 μCi 9.25 MBq, 250 μCi	5-HT$_2$ antagonist
TRK.570	[phenyl-4-^{3}H]Spiperone Ethanol solution	0.55–1.1 TBq/mmol, 15–30 Ci/mmol	9.25 MBq, 250 μCi 37 MBq, 1 mCi	5-HT$_2$ antagonist
TRK.818	[^{3}H]Spiperone Toluene/ethanol (1:1) solution	2.2–3.7 TBq/mmol, 60–100 Ci/mmol	9.25 MBq, 250 μCi 37 MBq, 1 mCi	5-HT$_2$ antagonist
Melatonin				
IM.215	2-[^{125}I]Iodomelatonin Water/isopropanol (75:25) solution	~74 TBq/mmol, ~2000 Ci/mmol	925 kBq, 25 μCi 1.85 MBq, 50 μCi 3.7 MBq, 100 μCi	

Table 5. Cholinergic ligands

Code	Ligand	Specific activity	Pack sizes	Comments
TRK.593	[Methyl-^{3}H]Choline chloride Ethanol solution	2.8–3.1 TBq/mmol, 75–85 Ci/mmol	9.25 MBq, 250 µCi 37 MBq, 1 mCi 185 MBq, 5 mCi	For studies of high affinity choline uptake.
TRK.179	[Methyl-^{3}H]Choline chloride Ethanol solution	~550 GBq/mmol, ~15 Ci/mmol	37 MBq, 1 mCi 185 MBq, 5 mCi	
TRK.701	Acetyl [methyl-^{3}H]Choline chloride Ethanol/water (1:1) solution	2.6–3.1 TBq/mmol, 70–85 Ci/mmol	9.25 MBq/250 µCi 37 MBq/1 mCi	Can be used to label high affinity agonist binding sites of both mAChRs and nAChRs in the presence of AChE inhibitors.
TRA.277	[^{3}H]Acetylcholine chloride Ethanol/water (1:1) solution	18.5–75 GBq/mmol, 0.5–2.0 Ci/mmol	37 MBq, 1 mCi	
TRK.604	L-Quinuclidinyl [phenyl-4-^{3}H] benzilate Ethanol solution	1.1–2.2 TBq/mmol, 30–60 Ci/mmol	9.25 MBq, 250 µCi 37 MBq, 1 mCi	Very high-affinity non-selective tertiary amine muscarinic antagonist ($K_d \sim 20$ pM).
TRK.666	L-[N-methyl-^{3}H]Scopolamine methyl chloride Ethanol solution	2.2–3.1 TBq/mmol, 60–85 Ci/mmol	9.25 MBq, 250 µCi 37 MBq, 1 mCi	Very high-affinity non-selective quaternary amine muscarinic antagonist ($K_d \sim 100$ pM).
TRK.841	(−)-[N-methyl-^{3}H]Nicotine Ethanol solution containing 0.01 M HCl	2.2–3.1 TBq/mmol, 60–85 Ci/mmol	37 MBq, 1 mCi	Labels some categories of neuronal nicotinic receptors.
IM.209	(3-[^{125}I]Iodotyrosyl^{54})α-bungarotoxin	~74 TBq/mmol, ~2000 Ci/mmol	1.85 MBq, 50 µCi 9.25 MBq, 250 µCi	Very high affinity for nicotinic AChRs of neuromuscular junction, and electric organs.
IM.109	(3-[^{125}I]Iodotyrosyl)α-bungarotoxin	> 7.4 TBq/mmol, > 200 Ci/mmol	3.7 MBq, 100 µCi 18.5 MBq, 500 µCi 37 MBq, 1 mCi	Very high affinity for nicotinic AChRs of neuromuscular junction, and electric organs.
TRK.603	α-Bungarotoxin, N-[propionyl-^{3}H] propionylated	1.5–2.6 TBq/mmol, 40–70 Ci/mmol	370 kBq, 10 µCi 1.85 MBq, 50 µCi	Very high affinity for nicotinic AChRs of neuromuscular junction, and electric organs.

Table 6. Adenosine ligands

Code	Ligands	Specific activity	Pack sizes	Comments
TRK.423	[2-^{3}H]Adenosine Aqueous solution, sterilized	740–925 GBq/mmol, 20–25 Ci/mmol	37 MBq, 1 mCi 185 MBq, 5 mCi	
TRK.609	[2,5′,8-^{3}H]Adenosine Ethanol/water (1:1) solution	1.5–2.2 TBq/mmol, 40–60 Ci/mmol	37 MBq, 1 mCi	
TRK.783	(−)-N^6-R-[G-^{3}H] Phenylisopropyladenosine Water/ethanol (7:3) solution	1.3–2.2 Tbq/mmol, 35–60 Ci/mmol	9.25 mBq, 250 μCi 37 mBq, 1 mCi	A$_1$ agonist
TRK.897	[^{3}H]DPCPX Ethanol solution	3.33–4.07 BTq/mmol, 90–110 Ci/mmol	9.25 MBq, 250 μCi 37 MBq, 1 mCi	A$_1$ antagonist
TRK.689	5′-N-Ethylcarboxamido [8(n)-^{3}H]adenosine Ethanol/water (1:1) solution	0.74–1.5 TBq/mmol, 20–40 Ci/mmol	9.25 MBq, 250 μCi 37 MBq, 1 mCi	A$_2$ agonist

Table 7. Amino acid neurotransmitters

Code	Ligand	Specific activity	Pack sizes	Comments
TRK.606	D-[2,3-³H]Aspartic acid Aqueous solution containing 2% ethanol, sterilized	0.37–1.1 TBq/mmol, 10–30 Ci/mmol	9.25 MBq, 250 µCi 37 MBq, 1 mCi	
TRK.574	L-[2,3-³H]Aspartic acid Aqueous solution containing 2% ethanol, sterilized	0.55–1.5 TBq/mmol, 15–40 Ci/mmol	37 MBq, 1 mCi 185 MBq, 5 mCi	
TRK.445	L-[G-³H]Glutamic acid Aqueous solution containing 2% ethanol, sterilized	0.74–1.5 TBq/mmol, 20–40 Ci/mmol	9.25 MBq, 250 µCi 37 MBq, 1 mCi 185 MBq, 5 mCi	NMDA receptor agonist; also labels AMPA receptor
TRK.71	[2-³H]Glycine Aqueous solution, sterilized	0.37–1.1 TBq/mmol, 10–30 Ci/mmol	9.25 MBq, 250 µCi 37 MBq, 1 mCi 185 MBq, 5 mCi	
TRK.577	[2,4,11-³H]Strychnine sulphate Ethanol/water (4:1) solution	0.55–1.1 TBq/mmol, 15–30 Ci/mmol	9.25 MBq, 250 µCi 37 MBq, 1 mCi	Glycine antagonist; can be used for photoaffinity labelling.
TRK.566	[G-³H]Kainic acid Aqueous solution containing 2% ethanol, sterilized	74–370 GBq/mmol, 2–10 Ci/mmol	9.25 MBq, 250 µCi	Kainate receptor agonist
TRK.573	[1,2-³H]Taurine Aqueous solution containing 2% ethanol, sterilized	185–740 GBq/mmol, 5–20 Ci/mmol	9.25 MBq, 250 µCi	

Table 8. Opioid ligands

Code	Ligand	Specific activity	Pack sizes	Comments
IM.155	(3-[^{125}I]Iodotyrosyl) Enkephalin (5-L-leucine) Freeze-dried solid	~74 BTq/mmol, ~2000 Ci/mmol	370 kBq, 10 μCi 1.85 MBq, 50 μCi	Used in radioimmunoassay.
IM.156	(3-[^{125}I]Iodotyrosyl) Enkephalin (5-L-methionine) Freeze-dried solid	~74 TBq/mmol, ~2000 Ci/mmol	370 kBq, 10 μCi 1.85 MBq, 50 μCi	Used in radioimmunoassay.
TRK.730	[15,16(n)-^{3}H]Diprenorphine Ethanol solution	0.9–1.85 TBq/mmol 25–50 Ci/mmol	9.25 MBq, 250 μCi 37 MBq, 1 mCi	Universal ligand used in autoradiography and binding experiments. Apparent K_i ~0.2 nM at μ site, ~0.18 nM at δ site, ~0.5 nM at κ site, ~0.47 nM at ε site in rat brain membranes.
TRK.476	(15,16-^{3}H]Etorphine Ethanol solution	1.1–2.2 TBq/mmol, 30–60 Ci/mmol	9.25 MBq, 250 μCi 37 MBq, 1 mCi	Used to investigate opioid binding sites. Apparent K_i ~0.15 nM at μ site, ~0.3 nM at δ site, ~0.48 nM at κ site, ~4.7 nM at ε site in rat brain membranes.

Code	Compound	Specific activity	Amounts	Application
TRK.896	[³H] U-69593 Ethanol solution containing 0.01 M hydrochloric acid	1.48–2.22 TBq/mmol, 40–60 Ci/mmol	9.25 MBq, 250 μCi 37 MBq, 1 mCi	Kappa (κ) agonist
TRK.819	D-pen², D-pen⁵ [tyrosyl-3,5-³H] Enkephalin (DPDPE) solution in 0.05 M acetic acid/acetonitrile (3:1)	0.9–1.8 TBq/mmol 25–50 Ci/mmol	1.85 MBq, 50 μCi 9.25 MBq, 250 μCi	Used in binding studies of the δ type receptor. More selective than DADLE and DSLET with $K_d \sim 7.2$ nM in rat brain and $K_d \sim 1.61$ nM in guinea-pig brain.
TRK.681	D-ala, ²N-methyl-phe⁴glyol⁵ [tyrosyl-3,5-³H] Enkephalin (DAGO) 0.07 M Triethylammonium phosphate buffer (pH4.2) acetonitrile (22:3) solution	1.1–2.2 TBq/mmol, 30–60 Ci/mmol	9.25 MBq, 250 μCi 37 MBq, 1 mCi	Used in binding studies due to its low non-specific binding and enzymatic stability to aminopeptidases and peptidyl dipeptidases *in vivo*. Very selective for μ binding sites with $K_i \sim 3.6$ nM for μ site. ~ 700 nM for δ site.
IM.162	(3-[¹²⁵I]Iodotyrosyl²⁷) β-Endorphin Freeze-dried solid	~ 74 TBq/mmol ~ 2000 Ci/mmol	370 kBq, 10 μCi 1.85 MBq, 50 μCi	Used to investigate β-endorphin binding sites. β-Endorphin has the highest affinity for benzomorphan (ε) sites, but is also a potent agonist for μ and δ sites.
TRK.619	[N-allyl-2,3,³H]Naloxone Ethanol solution	1.5–2.2 TBq/mmol, 40–60 Ci/mmol	9.25 MBq, 250 μCi	Opioid antagonist. Used to assay opioid binding sites. Selective for μ site with $K_i \sim 1$ nM.

Table 9. Catecholamine ligands

Code	Ligand	Specific activity	Pack sizes	Comments
Adrenergic ligands				
TRK.174	DL-[7-³H]Noradrenaline Aqueous solution	37–40 GBq/mmol, 10–20 Ci/mmol	9.25 MBq, 250 µCi 37 MBq, 1 mCi	Used to study the mechanism of action and metabolism of noradrenaline. [³H]noradrenaline requires the inclusion of an antioxidant, a monoamine oxidase inhibitor and pyrocatechol to prevent non-specific catechol interactions. K_d for racemate ~ 3–10 nM in brain.
TRK.584	L[7,8-³H]Noradrenaline Dilute buffer pH 4.5/ethanol (9:1) solution	1.1–1.85 TBq/mmol, 30–50 Ci/mmol	9.25 MBq, 250 µCi 37 MBq, 1 mCi	
TRA.584	L-[7,8-³H]Noradrenaline Dilute buffer pH 4.5/ethanol (9:1) solution	300–500 GBq/mmol, 8–15 Ci/mmol	9.25 MBq, 250 µCi 37 MBq, 1 mCi 185 MBq, 5 mCi	
α-Adrenergic				
TRK.555	[9,10(n)-³H]9,10-Dihydroergocryptine Ethanol solution	0.55–1.1 TBq/mmol, 15–30 Ci/mmol	9.25 MBq, 250 µCi 37 MBq, 1 mCi	Used in displacement experiments to determine α_1 and α_2 populations, and total α-receptor populations. Non-selective ligand with some affinity for dopamine and 5-HT receptors. Apparent K_d ~ 1.9 nM in rat cerebral cortex membranes.
TRK.579	[³]WB4101 Ethanol solution ANTAGONIST	0.74–1.48 TBq/mmol, 20–40 Ci/mmol	9.25 MBq, 250 µCi	Used to identify α-adrenergic binding sites in brain. K_d ~ 0.48 nM in rat brain membranes, also has some selectivity for 5-HT$_1$ receptors.

α_1-Adrenergic

Code	Compound	Specific activity	Quantity	Description
TRK.647	[furanyl-5-^{3}H]Prazosin Ethanol/0.01 M HCl (1:1) solution ANTAGONIST	0.4–1.1 TBq/mmol, 10–30 Ci/mmol	9.25 MBq, 250 µCi	Used to identify receptor binding sites due to its high affinity and high selectivity. $K_d \sim 0.1$–0.5 nM in brain.
TRK.843	[7-methoxy-^{3}H]Prazosin Ethanol/0.01 M HCl (1:1) solution. ANTAGONIST	2.4–3.2 TBq/mmol, 65–85 Ci/mmol	9.25 MBq, 250 µCi	Higher specific activity than TRK647.

α_2-Adrenergic

Code	Compound	Specific activity	Quantity	Description
TRK.685	[o-methyl-^{3}H] Rauwolscine Ethanol solution ANTAGONIST	2.6–3.1 TBq/mmol, 70–85 Ci/mmol	9.25 MBq, 250 µCi	Used to investigate α_2 binding sites. It is selective for high and low affinity states of the receptor with $K_d \sim 1$–4 nM in rat cortex. Rauwolscine is more selective than yohimbine for the receptor.
TRK.684	[o-methyl-^{3}H]Yohimbine Ethanol solution ANTAGONIST	2.6–3.1 TBq/mmol, 70–85 Ci/mmol	9.25 MBq, 250 µCi 37 MBq, 1 mCi	Used to investigate α_2, receptors as it is selective for high and low affinity states with $K_d \sim 10$–13 nM in rat cortex. Diastereoisomer of rauwolscine.
TRK.799	[^{2}H]RX.781094 (Idazoxan) Ethanol solution containing 0.05 M HCl ANTAGONIST	1.5–2.2 TBq/mmol, 40–60 Ci/mmol	9.25 MBq, 250 µCi 37 MBq, 1 mCi	Used to characterize α_2 receptors, for example in rat cortical membranes. It selectively labels equally the high (α_2 (H)) and low (α_2 (L)) affinity states. $K_d \sim 3.9$ nM in rat cortical membranes

Table 9. *Contd.*

Code	Ligand	Specific activity	Pack sizes	Comments
TRK.621	[phenyl-4-³H]Clonidine hydrochloride Ethanol/0.01 M HCl (1:1) solution AGONIST	0.74–1.1 TBq/mmol, 20–30 Ci/mmol	9.25 MBq, 250 µCi	Used in binding studies, it labels the α_2 (L) state at high concentrations and is a partial α_2 (H) agonist with $K_d \sim 5.8$ nM in rat brain membranes. Labels same binding sites as [³H] adrenaline and is a more stable ligand.
TRK.914	[-³H]RX821002, Ethanol solution	1.1–2.2 TBq/mmol, 30–60 Ci/mmol	9.25 MBq, 250 µCi	α_2-antagonist

β-Adrenergic

Code	Ligand	Specific activity	Pack sizes	Comments
TRK.551	L-[propyl-2,3-³H] Dihydroalprenolol Ethanol solution ANTAGONIST	1.1–2.2 TBq/mmol, 30–60 Ci/mmol	9.25 MBq, 250 µCi 37 MBq, 1 mCi	Most widely used radioligand for characterization and quantification of β-adrenergic receptors with $K_d \sim 0.5$–2 nM.
TRK.649	L-[4,6-propyl-³H] Dihydroalprenolol Ethanol/water (9:1) solution ANTAGONIST	2.6–4.1 TBq/mmol, 70–110 Ci/mmol	9.25 MBq, 250 µCi 37 MBq, 1 mCi	Used in binding assays and antidepressant drug screening, the higher specific activity leading to higher sensitivity.
TRK.495	DL[4-³H]Propranolol hydrochloride Ethanol solution ANTAGONIST	0.55–1.1 TBq/mmol, 15–30 Ci/mmol	9.25 MBq, 250 µCi 185 MBq, 5 mCi	Used for β-adrenergic receptor binding studies, affinity is relatively low and exhibits non-specific binding. $K_d \sim 2.5$ nM in turkey erythrocyte ghosts.

	Specific activity	Amounts	Description
IM.142 (−)-3-[¹²⁵I]Iodocyanopindolol Methanol solution ANTAGONIST	~74 TBq/mmol, ~2000 Ci/mmol	3.7 MBq, 100 μCi 18.5 MBq, 500 μCi 37 MBq, 1 mCi	Used in binding studies with the advantages of high specific activity, high affinity, and of a pure stereoisomer such that only half the radioactivity is required to achieve the same amount of binding as the racemic mixture $K_d \sim 10$–20 pM.
IM.149 (±)-3-[¹²⁵I]Iodocyanopindolol diazirine Methanol solution ANTAGONIST	74 TBq/mmol, 2000 Ci/mmol	3.7 MBq, 100 μCi 18.5 MBq, 500 μCi	This photoaffinity probe is used in binding studies, for example in turkey erythrocyte membranes ($K_d \sim 60$ pM) and guinea-pig lung membranes where it binds with high affinity. Photolysis can be carried out under very mild conditions.
TRK.751 (±)-[³H]CGP12177 Ethanol solution ANTAGONIST	1.1–1.85 TBq/mmol, 30–50 Ci/mmol	9.25 MBq, 250 μCi 37 MBq, 1 mCi	Used in binding studies where it exhibits low non-specific binding. It is a highly hydrophilic antagonist with high affinity, $K_d \sim 0.37$ nM in living heart cells
TRK.835 (−)-[³H]CGP-12177 Ethanol solution	1.11–2.22 TBq/mmol, 30–60 Ci/mmol	9.25 MBq, 250 μCi 37 MBq, 1 mCi	Used to determine rate constants of unlabelled antagonists. Twice as potent as the racemic mixture and is advantageous for studies in tissues and intact cells possessing high numbers of non-specific binding sites.

Table 10. Dopamine ligands

Code	Ligand	Specific activity	Pack sizes	Comments
TRK.582	[7,8-^{3}H]Dopamine 0.02 M Acetic acid/ethanol (1:1) solution AGONIST	1.5–2.2 TBq/mmol, 40–60 Ci/mmol	9.25 MBq, 250 μCi 37 MBq, 1 mCi	Used in the assay of dopaminergic receptors, and to study uptake and release of dopamine.
TRK.284	[2,5,6-^{3}H]Dopamine hydrochloride 0.02 M Acetic acid/ethanol (1:1) solution, sterilized AGONIST	185–550 GBq/mmol, 5–15 Ci/mmol	9.25 MBq, 250 μCi 37 MBq, 1 mCi	
D_2 Ligands				
IM.203	[^{125}I]Iodosulpiride D_2 ANTAGONIST	~74 TBq/mmol, ~2000 Ci/mmol	925 kBq, 25 μCi 1.85 MBq, 50 μCi	Used in binding studies and tissue autoradiography. High specific activity of ^{125}I and high affinity of [^{125}I]iodosulpiride for the D_2 receptors, K_d 1.6 nM ± 0.3 nM: rat striatal membranes.
TRK.570	[phenyl -4-^{3}H]Spiperone Ethanol solution D_2 ANTAGONIST	0.55–1.1 TBq/mmol, 15–30 Ci/mmol	9.25 MBq, 250 μCi 37 MBq, 1 mCi	Used in binding studies. Spiperone has equal binding affinity for 5-HT$_2$ receptors, but this is overcome by saturating the receptors with specific non-radioactive 5-HT$_2$ ligands. [^{3}H]Spiperone does not wash off readily, leading to good reproducibility. It can be used with very small quantities of tissue due to its high affinity for the D_2 receptor. K_d ~0.6–2.3 nM in rat brain.
TRK.818	[^{3}H]Spiperone Ethanol toluene (1:1) solution D_2 ANTAGONIST	2.2–3.7 TBq/mmol, 60–100 Ci/mmol	9.25 MBq, 250 μCi 37 MBq, 1 mCi	

Code	Product	Specific activity	Amounts	Application
TRK.881	[N-propyl-2,3-^{3}H]-Sandoz 205-501, Ethanol, D$_2$ ANTAGONIST	3.0–4.4 TBq/mmol, 80–120 Ci/mmol	9.25 MBq, 250 µCi 37 MBq, 1 mCi	
TRK.879	[^{3}H]N-0437 D$_2$ AGONIST	1.85–2.96 TBq/mmol, 50–80 Ci/mmol	9.25 MBq, 250 µCi 37 MBq, 1 mCi	

D$_1$ Ligands

Code	Product	Specific activity	Amounts	Application
TRK.848	[^{3}H]SKF 38393 Methanol/water/acetic acid (1:1:0.05) solution containing ascorbic acid (8 mg/ml) D$_1$ AGONIST	0.92–1.85 TBq/mmol, 25–50 Ci/mmol	9.25 MBq, 250 µCi	Used as a selective D$_1$ agonist in autoradiographic studies to localize D$_1$ receptors, for example in rat brain.
TRK.876	[N-methyl-^{3}H]SCH 23390 D$_1$ AGONIST	2.2–3.3 TBq/mmol, 60–90 Ci/mmol	3.7 MBq, 100 µCi 9.25 MBq, 250 µCi 37 MBq, 1 mCi	
TRK.848	[^{3}H]SKF-38393 D$_1$ AGONIST	0.92–1.85 TBq/mmol, 25–50 Ci/mmol	9.25 MBq, 100 µCi 37 MBq, 1 mCi	

Dopamine-related products

Code	Product	Specific activity	Amounts	Application
TRK.873	[N-methyl-^{3}H]MPTP ([N-methyl-^{3}H]1-Methyl-4-phenyl-1,2,3,6-tetrahydropyridine) (Ethanol)	2.22–3.33 TBq/mmol, 60–90 Ci/mmol	9.25 MBq, 250 µCi 37 MBq, 1 mCi	Used in binding and autoradiographic studies in connection with Parkinson's disease.
TRK.874	[N-methyl-^{3}H]MPP+ ([N-methyl-^{3}H]1-Methyl-4-phenyl-pyridinium acetate)	1.85–3.3 TBq/mmol, 50–90 Ci/mmol	9.25 MBq, 250 µCi 37 MBq, 1 mCi	For use in binding experiments in the study of Parkinsonism. MPP+ is a metabolite of MPTP and is considered to be the neurotoxic component.

Table 11. Growth factors

Code	Ligand	Specific activity	Pack sizes	Comments
IM.219	(3-[125I]Iodotyrosyl) Erythropoietin, high specific activity (human, recombinant)	111–148 TBq/mmol 3000–4000 Ci/mmol	74 kBq, 2 µCi 370 kBq, 10 µCi	For use in radioimmunoassay using several different antisera to erythropoietin and receptor studies showing specific binding to murine fetal liver cells with low non-specific binding
IM.178	(3-[125I]Iodotyrosyl) Erythropoietin (human, recombinant)	11–33 TBq/mmol 300–900 Ci/mmol	74 kBq, 2 µCi 370 kBq, 10 µCi	For use in radioimmunoassay (exhibits low non-specific binding) and receptor binding assays. Purified by HPLC.
IM.194	(3-[125I]Iodotyrosyl) Transferrin, human	25.9–29.6 TBq/mmol 700–800 Ci/mmol	370 kBq, 10 µCi 1.85 MBq, 50 µCi	For use in localizing and quantifying the transferrin receptor and to measure transferrin in conjunction with RPN 1941 antiserum
IM.206	([125I]Iodotyrosyl) Tumour necrosis factor-α ([125I]TNF-α)(human, recombinant)	15–30 TBq/mmol 400–800 Ci/mmol at date of first availability	185 kBq, 5 µCi 925 kBq, 25 µCi	Tumour necrosis factor-α (TNF-α), also known as cachectin, is a 157 amino acid polypeptide cytokine synthesized by activated macrophages. Its major functions include suppression of lipoprotein lipase (an essential enzyme involved in fat synthesis), necrosis and regression of some solid tumours *in vivo* and direct cytolytic or cytostatic activity on certain transformed cells *in vitro*.
IM.172	(3-[125I]Iodotyrosyl) Insulin-like growth factor-1, natural sequence ([125I]IGF-1)(recombinant)	~74 TBq/mmol ~2000 Ci/mmol	195 kBq, 5 µCi 925 kBq, 25 µCi	For use in radioimmunoassay and radioreceptor assays showing low non-specific binding and high binding to excess antiserum
IM.1721	Insulin-like growth factor-1, natural sequence (Somatomedin C) reagent pack for RIA		1 Pack	

Code	Product	Activity	Specific activity	Description
IM.166	(3-[125I]Iodotyrosyl A14) Insulin (human)	370 kBq, 10 µCi 1.85 MBq, 50 µCi	~74 TBq/mmol, ~2000 Ci/mmol	Insulin contains tyrosine residues at positions 14 and 19 in the A chain and positions 16 and 26 in the B chain. Thus on iodination, 4 possible monoiodinated forms of the hormone can be produced and these are generally described as the (A14), (A19), (B16) and (B26) isomers. [125I]-(A14)-monoiodinated insulin has been shown to behave similarly to native insulin both in RIA and in receptor studies with isolated adipocytes, IM.9-lymphocytes cultured fibroblasts and isolated hepatocytes. The ligand has been fully tested in our laboratories against guinea-pig antiserum to insulin, code RPN 1661, giving low blank binding and high binding to excess antiserum.
IM.167	(3-[125I]Iodotyrosyl B26) Insulin (human)			
IM.207	(3-[125I]Iodotyrosyl) Murine 2.5S nerve growth factor ([125I]NGF)	370 kBq, 10 µCi 1.85 MBq, 50 µCi	~74 TBq/mmol, ~2000 Ci/mmol at date of first availability ~55.5 TBq/mmol, ~1500 Ci/mmol at activity reference date	For use in radioimmunoassay and receptor binding. [125I]NGF has been shown to bind specifically to rat phaeochromocytoma (PC-12) cells
IM.208	([125I]Iodotyrosyl) Fibroblast growth factor basic (bovine, recombinant)	370 kBq, 10 µCi 1.85 MBq, 50 µCi	~520 kBq/mmol, ~1400 Ci/mmol at date of first availability	Basic fibroblast growth factor has the widest range of effects of any growth factor described to date, stimulating the proliferation (and sometimes the differentiation) of many cells of mesodermal and neuroectodermal origin including vascular endothelial cells, lens epithelial cells, neurons and bone cells, and may have an important *in vivo* role in

415

Table 11. *Contd.*

Code	Ligand	Specific activity	Pack sizes	Comments
				many different processes including wound healing, angiogenesis and mesodermal induction and neuronal development in the embryo.
IM.124	[125I]Epidermal growth factor (mouse) ([125I]EGF)	~3.7 MBq/µg, ~100 µCi/µg	370 kBq, 10 µCi 1.85 MBq, 50 µCi	Used in receptor binding studies, for example in cell culture and to study receptor mediated endocytosis.
IM.202	(3-[125I]Iodotyrosyl) Interferon-gamma ([125I]IFN-γ) (human, recombinant)	~37 TBq/mmol, ~1000 Ci/mmol at date of first availability ~25 TBq/mmol, ~675 Ci/mmol at activity reference date (exact value given with each despatch)	370 kBq, 10 µCi 1.85 MBq, 50 µCi	For use in studying IFN-γ receptors. IFN-γ exhibits two functional domains—one includes fibroblasts to resist viral infection, and the other induces macrophages to become cytotoxic. This implies that the IFN-γ receptors on macrophages and fibroblasts differ. [125I]IFN can be used for both fields of study as it has been shown to bind both monocytic (h.U937) and fibroblast (h.MRC-5) cell lines in a specific and saturable manner.
IM.205	(3-[125I]Iodotyrosyl) Interleukin-1α ([125I]IL-1α) (human recombinant)	~74 TBq/mmol ~2000 Ci/mmol	185 kBq, 5 µCi 925 kBq, 25 µCi	For use in radioimmunoassay and receptor binding. [125I]IL-1α has been shown to bind specifically to 3T3 and YT cells.
IM.222	[125I]Interleukin-1β ([125I]IL-1β) (human, recombinant)	>37 TBq/mmol >1000 Ci/mmol at date of first availability	185 kBq, 5 µCi 925 kBq, 25 µCi	Interleukin-1β is a 153 amino acid polypeptide hormone which, along with interleukin-1α (IL-1α), acts as a soluble mediator of inflammatory responses and has interleukin-1 (IL-1) biological activity.

Code	Product	Specific activity	Radioactivity	Description
IM.197	(3-[125I]Iodotyrosyl) Interleukin-2 [met⁰, ala¹²⁵] (human, recombinant) ([125I]IL-2)	>22.2 TBq/mmol, >600 Ci/mmol	185 kBq, 5 µCi; 925 kBq, 25 µCi	IL-1 is a thymocyte co-mitogenic factor that has a range of biological activities including enhancement of B-lymphocyte proliferation and maturation, stimulation of thymocyte proliferation and fibroblast proliferation. For use in receptor binding studies and radioimmunoassay.
IM.213	[125I]Platelet-derived growth factor (c-sis) (PDGF) (recombinant)	>37 TBq/mmol, >1000 Ci/mmol at date of first availability; >26 TBq/mmol, >700 Ci/mmol at activity reference date (exact value given with each despatch)	185 kBq, 5 µCi; 925 kBq, 25 µCi	For use in receptor binding studies [125I]PDGF (c-sis) has been shown to bind to PDGF receptors on human foreskin fibroblasts and 3T3 cells. Binding occurs in a specific, saturable manner with low NSB and kinetics similar to natural PDGF.
IM.196	(3-[125I]Iodotyrosyl) Epidermal growth factor ([125I]EGF), (human, recombinant)	>33 TBq/mmol, >900 Ci/mmol at date of first availability; >27.7 TBq/mmol, >750 Ci/mmol at activity reference date	185 kBq, 5 µCi; 925 kBq, 25 µCi	For use in radioimmunoassay and receptor binding. [125I]-h-EGF has been shown to bind specifically to EGF receptors on both mouse 3T3 cells and human A431 cells.

Table 12. Blood pressure regulators

Code	Ligand	Specific activity	Pack sizes	Comments
IM.187	(3[^{125}I]*Iodotyrosyl*^{28}) Human α-atrial natriuretic peptide (α-hANP) Freeze-dried solid	~74 TBq/mmol, ~2000 Ci/mmol	185 kBq, 5 μCi 925 kBq, 25 μCi	For use in radioimmunoassay, binding and autoradiographic studies. Contains <5% ANP sulphoxide. IM.187 has been found to interact with ANP receptors in human brain, kidney, and adrenals.
IM.186	(3-[^{125}I]*Iodotyrosyl*^{28}) Rat atrial natriuretic peptide (rANP)	~74 TBq/mmol, ~2000 Ci/mmol	370 kBq, 10 μCi 1.85 MBq, 50 μCi	For use in radioimmunoassay, receptor binding, and autoradiographic studies.
IM.176	(3-[^{125}I]*Iodotyrosyl*^{4}) Angiotensin I (5-L-isoleucine) Freeze-dried solid	~74 TBq/mmol, ~2000 Ci/mmol	370 kBq, 10 μCi 1.85 MBq, 50 μCi	For use in radioimmunoassay to measure renin activity.
TRK.733	[*tyrosyl*-3,5-^{3}H]Angiotensin II (5-L-isoleucine) 0.02 M Sodium hydrogen phosphate pH 2.5 solution	1.1–2.2 TBq/mmol, 30–60 Ci/mmol	9.25 MBq, 250 μCi	Used in binding studies.
IM.177	(3-[^{125}I]*Iodotyrosyl*^{4}) Angiotensin II (5-L-isoleucine) Freeze-dried solid	~74 TBq/mmol, ~2000 Ci/mmol	370 kBq, 10 μCi 1.85 MBq, 50 μCi	Used in binding and autoradiographic studies, and in radioimmunoassay.

418

Table 13. Calmodulin/calcitonin

Code	Ligand	Specific activity	Pack sizes	Comments
IM.126	[125I]Calmodulin Freeze-dried with BSA–phosphate–azide	0.925–2.77 MBq/μg 25–75 μCi/μg	370 kBq, 10 μCi 1.85 MBq, 50 μCi	For use in radioimmunoassay and in the study of calmodulin-binding proteins.
IM.175	(3-[125I]*Iodotyrosyl*[12]) Calcitonin, human Freeze-dried solid	~74 TBq/mmol, ~2000 Ci/mmol	370 kBq, 10 μCi 1.85 MBq, 50 μCi	For use in receptor binding studies and radioimmunoassay.

Table 14. Cholecystokinin/gastrin

Code	Product	Specific activity	Pack sizes	Comments
CCK				
TRK.755	CCK-8(sulphated), [propionyl-³H]propionylated. Solution in triethylammonium phosphate pH 3.4/acetronitrile (1:1) containing 0.2% v/v mercaptoethanol	2.2–3.3 TBq/mmol, 60–90 Ci/mmol	1.85 MBq, 50 μCi	For use in the investigation of CCK receptors. Found to release pepsinogen from gastric chief cells and to stimulate acid formation in gastric parietal cells.
IM.159	CCK-8(sulphated), ¹²⁵I-labelled with Bolton and Hunter reagent. Freeze-dried solid	~74 TBq/mmol, ~2000 Ci/mmol	370 kBq, 10 μCi 1.85 MBq, 50 μCi	Used for receptor binding studies, $K_d \sim 1\,nM$ in guinea-pig cerebral cortex. For use in RIA of total CCK and gastrin using N.1591. Binds to pancreatic plasma membranes with ~5% NSB and $IC_{50} \sim 1.7\,nM$.
Gastrin				
IM.165	(3-[¹²⁵I]Iodotyrosyl¹²) Gastrin 1, (G17) human. Freeze-dried solid	~74 TBq/mmol, ~2000 Ci/mmol	370 kBq, 10 μCi 1.85 MBq, 50 μCi	Used in receptor binding studies, and is suitable for RIA (using N-1591, RPN.1651, or RPN.1653).

Table 15. Neurohypophyseal peptides

Code	Product	Specific activity	Pack sizes	Comments
IM.179	(3-[^{125}I]*Iodotyrosyl*[2])Oxytocin Freeze-dried solid	~74 TBq/mmol, ~2000 Ci/mmol	370 kBq, 10 μCi 1.85 MBq, 50 μCi	Used for the measurement of oxytocin by RIA.
Vasopressin				
TRK.776	[*tyrosyl*-3,5-^{3}H]Vasopressin [Arg^8]. 0.1 M Ammonium acetate buffer pH 5.5/acetonitrile (3:1) solution	0.37–1.1 TBq/mmol, 10–30 Ci/mmol	1.85 MBq, 50 μCi 9.25 MBq, 250 μCi	Used in binding and autoradiographic binding experiments to investigate vasopressin binding sites. $K_d \sim 0.77$ and 20.8 nM in brain membranes. Retains full biological activity.
IM.182	(3-[^{125}I]*Iodotyrosyl*[2]) Vasopressin [Arg^8] Freeze-dried solid	~74 TBq/mmol, ~2000 Ci/mmol	370 kBq, 10 μCi	Used for the measurement of vasopressin by RIA.

Table 16. Neuropeptides

Code	Product	Specific activity	Pack sizes	Comments
IM.184	(2-[^{125}I]*Iodohistidyl*[10]) Calcitonin gene related peptide, human	~74 TBq/mmol, ~2000 Ci/mmol	370 kBq, 10 μCi 1.85 MBq, 50 μCi	Used in radioimmunoassay and also in receptor binding studies.
TRK.814	Neropeptide Y,N-[*propionyl*-^{3}H] propionylated 0.15 M Triethylammonium phosphate (pH 3.2) acetonitrile (3:1) solution	2.2–3.7 TBq/mmol, 60–100 Ci/mmol	1.85 MBq, 50 μCi	For use in receptor binding studies. Shown to bind to a rat brain synaptosome preparation and to be displaced by unlabelled NPY.
IM.170	Neuropeptide Y, ^{125}I-labelled with Bolton and Hunter reagent Freeze-dried solid	~74 TBq/mmol, ~2000 Ci/mmol	370 kBq, 10 μCi 1.85 MBq, 50 μCi	Used in receptor binding studies and for radioimmunoassay with antiserum RPN.1701.
IM.163	(3-[^{125}I]*Iodotyrosyl*[3]) Neurotensin Freeze-dried solid	~74 TBq/mmol, ~2000 Ci/mmol	370 kBq, 10 μCi 1.85 MBq, 50 μCi	For use in radioimmunoassay, in receptor binding and autoradiographic studies.
IM.161	(3-[^{125}I]*Iodotyrosyl*[11]) Somatostatin-14 (tyr[11]) Freeze-dried solid	~74 TBq/mmol, ~2000 Ci/mmol	370 kBq, 10 μCi 1.85 MBq, 50 μCi	Stable analogue for use in receptor binding studies and radioimmunoassay.
IM.158	(3-[^{125}I]*Iodotyrosyl*[10]) Vasoactive intestinal polypeptide ([^{125}I]VIP) Freeze-dried solid	~74 TBq/mmol, ~2000 Ci/mmol	370 kBq, 10 μCi 1.85 MBq, 50 μCi	Specifically labelled at tyrosine residue 10. For use in receptor studies and radioimmunoassay.

Table 17. Releasing factors

Code	Ligand	Specific activity	Pack sizes	Comments
IM.189	(2-[125I]Iodohistidyl[32]) Corticotropin releasing factor (CRF), human Freeze-dried solid	~74 TBq/mmol, ~2000 Ci/mmol	370 kBq, 10 μCi	Specifically labelled at histidine-32. For use in radioimmunoassay and receptor binding studies.
IM.169	(3-[125I]Iodotyrosyl[15])Gastrin releasing peptide ([125I]GRP) Freeze-dried solid	~74 TBq/mmol, ~2000 Ci/mmol	370 kBq, 10 μCi 1.85 MBq, 50 μCi	Specifically labelled at tyrosine 15. For use in radioimmunoassay and binding studies of GRP/Bombesin receptors.
IM.180	(3-[125I]Iodotyrosyl[10]) Growth hormone-releasing factor 1–44 amide ([125I]GRF) Freeze-dried solid	~74 TBq/mmol, ~2000 Ci/mmol	370 kBq, 10 μCi	Specifically labelled at tyrosine 10. For use in radioimmunoassay and receptor binding studies. Shown to release growth hormone from cultured pituitary cells in parallel to that released by unlabelled GRF.
IM.181	(3-[125I]Iodotyrosyl[5]) Luteinizing hormone-releasing hormone (LHRH) Freeze-dried solid	~74 TBq/mmol, ~2000 Ci/mmol	370 kBq, 10 μCi	For use in radioimmunossay.

Table 18. Pancreatic hormones

Code	Ligand	Specific activity	Pack sizes	Comments
IM.166	(3-[^{125}I]Iodotyrosyl^{A14}) Insulin, human Freeze-dried solid	~74 TBq/mmol, ~2000 Ci/mmol	370 kBq, 10 µCi 1.85 MBq, 50 µCi	Specifically labelled at tyrosine A14. For use in receptor studies and analysis of insulin in immunoassays and receptor assays. HPLC purified.
IM.167	(3-[^{125}I]Iodotyrosyl^{B26}) Insulin, human Freeze-dried solid	~74 TBq/mmol, ~2000 Ci/mmol	370 kBq, 10 µCi 1.85 MBq, 50 µCi	Specifically labelled at tyrosine B26. For use in receptor studies. Shown to exhibit greater binding than native insulin in studies with extra-hepatic tissues, for example cultured human fibroblasts, IM-9 lymphocytes and human placental membranes.
IM.160	(3-[^{125}I]Iodotyrosyl^{10}) Glucagon Freeze-dried solid	~74 TBq/mmol, ~2000 Ci/mmol	370 kBq, 10 µCi 1.85 MBq, 50 µCi	Used to study receptor binding and receptor-mediated metabolism, for example by isolated canine hepatocytes. K_d ~ 4.2 nM in isolated rat hepatocytes.

Table 19. Tachykinins

Code	Product	Specific activity	Pack sizes	Comments
TRK.786	[prolyl²·⁴3,4(n)-³H] Substance P 0.15 M Triethylammonium phosphate pH 5.5/acetonitrile (3:1) solution containing 0.2% 2-mercaptoethanol	1.5–2.2 TBq/mmol, 40–60 Ci/mmol	1.85 MBq, 50 μCi	Used in binding studies to characterize Substance P receptors in brain and peripheral tissues, in autoradiography to localize receptors, as well as to study enzymatic degradation and metabolism of Substance P. $K_d \sim 0.3$ nM in rat brain membranes, $K_d \sim 1.8$ nM in guinea-pig ileum.
IM.157	Substance P, ¹²⁵I-labelled with Bolton and Hunter reagent. Freeze-dried solid	~74 TBq/mmol, ~2000 Ci/mmol	370 kBq, 10 μCi 1.85 MBq, 50 μCi	Used in binding and autoradiographic studies to investigate Substance P receptors. $IC_{50} \sim 0.46$ nM in rat whole spinal chord. Also for use in RIA in conjunction with N.1571.
IM.168	(2-[¹²⁵I]iodohistidyl¹) Neurokinin A Freeze-dried solid	~74 TBq/mmol, ~2000 Ci/mmol	370 kBq, 10 μCi 1.85 MBq, 50 μCi	For use in receptor binding and autoradiographic studies of tachykinin receptors and for use in RIA.

Table 20. Other peptides

Code	Peptide ligand	Specific activity	Pack sizes	Comments
IM.210	(3-[^{125}I]Iodotyrosyl)-tissue plasminogen activator (human)	24–30 TBq/mmol, 650–800 Ci/mmol	185 KBq, 5 μCi, 915 KBq, 25 μCi	For use in measurement of tissue plasminogen activator.
IM.215	2-[^{125}I]Iodomelatonin Water/isopropanol (75:25) solution	~74 TBq/mmol, ~2000 Ci/mmol	925 kBq, 25 μCi, 1.85 mBq, 50 μCi	For melatonin research in the study of circadian rhythms.
IM.202	(3-[^{125}I]Iodotyrosyl) human interferon-γ	25 TBq/mmol, 675 Ci/mmol	370 KBq, 10 μCi	For research into monocytic and fibroblast cell lines.
RPN.539	Azidopine	Unlabelled		For use in the characterization of the L-type Ca^{2+} channel.

Table 21. Prostanoids

Code	Ligand	Specific activity	Pack sizes	Comments
TRK.839	[³H]Iloprost	370–740 GBq/mmol, 10–20 Ci/mmol	3.7 MBq, 100 µCi	Prostanoid IP agonist
IM.254	[¹²⁵I]PTA-OH	74 TBq/mmol, 2000 Ci/mmol	370 kBq, 10 µCi 1.85 MBq, 50 µCi	Prostanoid TP antagonist

Table 22. Related ligands

Code	Ligand	Specific activity	Pack sizes	Comments
TRK.726	[G-³H]Quin-2,tetraacetoxy methyl ester Dimethylsulphoxide solution	185–555 MBq/mmol, 5–15 mCi/mmol	0.5 mg	Used to assess the quantity of Quin-2(N.239) in the cell.
N.239	Quin-2,tetraacetoxy methyl ester (non-radioactive) Solid		5 mg	Fluorescent probe used to measure intracellular free cytosolic Ca^{2+}.
RPN.508	5-FluoroBAPTA acetoxy methyl ester		5 mg	Intracellular calcium ion probe.
CES.3	Calcium-45 Calcium chloride in aqueous solution	0.37–1.48 GBq/mg, 10–40 mCi/mg	37 MBq, 1 mCi 74 MBq, 2 mCi 185 MBq, 5 mCi	Used in calcium-45 uptake experiments to study the mechanism of calcium transport.
Sodium channel ligands				
TRK.877	[11-³H]Saxitoxin Solution in 0.01 M acetic acid	0.74–1.48 TBq/mmol, 20–40 Ci/mmol	1.85 MBq, 50 µCi 9.25 MBq, 250 µCi	Used for investigating binding sites associated with the voltage-dependent sodium channels in neurones and at the neuromuscular junction. Also for radioimmunoassy required for preparation of antidotes for this toxin.
Potassium channel ligands				
IM.201	(2-[¹²⁵I]Iodohistidyl 18) Apamin	~74 TBq/mmol, ~2000 Ci/mmol	3.7 MBq, 100 µCi 1.85 MBq, 50 µCi 925 kBq, 25 µCi	For binding to Ca^{2+}-dependant K^+ channels.

Radioligands for the study of receptor–ligand interactions: Du Pont–NEN

R. L. YOUNG

Note: there have been significant additions to this list of radioligands since this appendix was compiled. Further details are available from the DuPont–NEN catalogue and from *Du Pont Biotechnology Update*.

Introduction

The comments on the following NEN™ Products ligands are intended to provide information as to relative usefulness based on affinity, specific activity, and selectivity. The experimental choices of agonist vs. antagonist and target tissues are subjectively related to the purpose of the research investigation. Where there is little or no variety for selection based on affinity, etc., no comments will be offered.

Adenosine receptor ligands

Table 1. Adenosine receptor ligands

Compound	Pharmacological action	Selectivity	Catalogue number
CGS 21680, [carboxyethyl-^{3}H(N)]-	agonist	A_2	NET-1021
Cyclohexyladenosine, N^6-[adenine-2,8-^{3}H]	agonist	A_1	NET-679
Cyclopentyladenosine, 2-chloro-N^6-[Cyclopentyl-2,3,4,5-^{3}H]— (CCPA)	agonist	A_1	NET-1026
Cyclopentyl-1,3-dipropylxanthine, 8-[dipropyl-2,3-^{3}H(N)]- (DPCPX)	antagonist	A_1	NET-974
Ethylcarboxamidoadenosine 5'-N-[adenine-2,8-^{3}H]- (NECA)	agonist	A_1, A_2	NET-811
Methyl-2-phenylethylpropyladenosine L-N^6-1-[adenine-2,8-^{3}H,ethyl-2-^{3}H]- (PIA)	agonist	A_1	NET-745
Nitrobenzylthioinosine,[benzyl-^{3}H]-	antagonist	uptake	NET-909
XAC, dipropyl, [dipropyl-2,3-^{3}H(N)]	antagonist	A_1, A_2	NET-953

Cyclohexyladenosine and ethylcarboamidoadenosine (NECA) have been the prototypical ligands for the adenosine receptor. 1,3-Diethyl-8-phenylxanthine, an antagonist to the A_1 receptor, lacks sufficient affinity for widespread use. The agonist, CCPA and the antagonist, DPCPX, provide increased selectivity and higher affinity for the A_1 receptor with very high specific activity. DPCPX has been particularly useful in myocardial membranes giving low non-specific binding (1) and dipropyl XAC has been used in autoradiographic studies (2). CCPA has a K_d of 0.4 nM for A_1 and a K_I of 3900 nM for A_2 receptors in rat brain membranes (3). Significant recent additions are [^{3}H]DPCPX (A_1 selective), [^{3}H]CCPA (A_1 selective), and [^{3}H]CGS 21680 (A_2 selective). These ligands, together with NECA, will be most useful for studies of the adenosine receptor.

References

1. Bruns, R. F., *et al.* (1987). *Naunyn-Schmiedeberg's Arch. Pharmacol.*, **335**, 59.
2. Jacobsen, K. A., Ukena, D., Kirk, K. L., and Daly, J. W. (1986). *Proc. Natl. Acad. Sci. USA*, **83**, 4089.
3. Klotz, K.-N., *et al.* (1989). *Naunyn-Schmiedeberg's Arch. Pharmacol.*, **340**, 679.

α-Adrenergic receptor ligands

Table 2. Alpha-adrenergic receptor ligands

Compound	Pharmacological action	Selectivity	Catalogue number
Desmethylimipramine, [benzene ring-10,11-^{3}H]-	inhibitor	uptake	NET-593
Aminoclonidine, *p*-[3,5-^{3}H]	agonist	alpha-2	NET 646
Clonidine HCl, [benzene ring-^{3}-H]-	agonist	alpha-2	NET-613
Dihydroergocryptine, [9,10 ^{3}H]-	mixed	mixed	NET-523
HEAT, 2-[^{125}I],	antagonist	alpha-1	NEX-182
Iodoazidoarylprazosin, [^{125}I]-	photoaffinity	alpha-1	NEX-219
p-Iodoclonidine, [^{125}I]-	agonist	alpha-2	NEX-253
Prazosin, [furoyl-5-^{3}H]-	antagonist	alpha-1	NET-661
Prazosin, [7-methoxy-^{3}H]-	antagonist	alpha-1	NET-823
Rauwolscine, [methyl-^{3}H]-	antagonist	alpha-2	NET-722
UK-14,304, [^{3}H]-	agonist	alpha-2	NET-853
WB4101, [phenoxy-3-^{3}H]-	antagonist	mixed	NET-580
Yohimbine, [methyl-^{3}H]-	antagonist	alpha-2	NET-659

Only antagonist ligands are available for the alpha-1 receptor; however, Prazosin and 2-[^{125}I]Heat are extremely high affinity with low non-specific binding and appear to be satisfactory for the current investigations of this receptor. Iodoazidoarylprazosin, a photoaffinity antagonist ligand, is an effective reagent for irreversibly labelling the alpha-1 receptor.

The imadazoline alpha-2 agonists, clonidine, aminoclonidine, and UK-14,304 have high specific activity and high affinity. UK-14,304 is the superior ligand (1), for its low non-specific binding, higher specific activity, and its full agonist properties; the clonidines are partial agonists which confuse the studies of agonist regulation of receptor function. The antagonists, yohimbine and rauwolscine are highly selective for alpha-2 over alpha-1, with rauwolscine being significantly more selective than other available ligands.

WB4101 and dihydroergocryptine are mixed alpha-1/alpha-2 ligands which find less use as research tools but are increasingly useful for screening for alpha-adrenergic activity.

Reference

1. Grant, J. A. (1980). *Brit. J. Pharmacol.*, **71**, 121.

β-Adrenergic receptor ligands

Table 3. Beta-adrenergic receptor ligands

Compound	Pharmacological action	Selectivity	Catalogue number
CGP 12177, [ring-^{3}H]-	antagonist	cell surface	NET-1006
CGP 26505, [5-phenoxy-^{3}H]-	antagonist	B-1	NET-1003
Dihydroalprenolol HCl, levo [ring, propyl-^{3}H]-	antagonist	—	NET-720
Iodocyanopindolol, (−),[^{125}I]-	antagonist	—	NEX-189
Iodohydroxybenzyl pindolol, [^{125}I]-	antagonist	—	NEX-125
Iodopindolol, (−)-[^{125}I]-	antagonist	—	NEX-211
Propranolol, L-[4-^{3}H]-	antagonist	—	NET-515

The standard ligand for the beta-receptor dihydroalprenolol (DHA), is widely used, non-selective, and an excellent ligand with low non-specific binding and high affinity. CGP 12177 is a hydrophilic ligand that labels beta-receptors only on the surface of cells, having greater beta-1 selectivity than DHA (1). (−)Iodocyanopindolol provides higher specific activity, 2200 Ci/mM, with equally good binding characteristics to DHA (2).

CGP 26505 is a new, high-affinity ligand with high beta-1 selectivity (3).

References

1. Riva, M. and Creese, I. (1989). *Mol. Pharmacol.*, **36**, 210.
2. Hoyer, D., Engel, G., and Berthold, R. (1982). *Naunyn-Schmiedeberg's Arch. Pharmacol.*, **318**, 319.
3. Dooley, D. J., Bittiger, H., and Reyman, N. C. (1986). *Eur. J. Pharmacol.*, **130**, 137.

Excitatory and inhibitory amino acid receptor ligands

Table 4. Excitatory and inhibitory acid receptor ligands

Compound	Pharmacological action	Selectivity	Catalogue number
AMPA, [5-methyl-^{3}H]-	agonist	glutamate	NET-833
CGP 39653, [propyl-1,2-^{3}H]	antagonist	NMDA	NET-1050
CGS 19755, [^{3}H]	antagonist	NMDA	NET-988
CNQX, [5-^{3}H]-,	antagonist	quisqualate	NET-1022
CPP (±)-[propyl-1,2-^{3}H]-,	antagonist	NMDA	NET-962
Dichlorokynurenic acid,-5,7[3^{3}H]	antagonist	NMDA/glycine	NET-1049
Kainic acid [vinylidine-^{3}H]	agonist	glutamate/kainate	NET-875
MK-801, (+)-[^{3}H]-	antagonist	NMDA	NET-972
NAAG, Glutamate [^{3}H]-,	endogenous	glutamate	NET-880
Phencyclidine, [piperidyl-3,4-^{3}H]-	agonist	PCP, sigma	NET-630
Strychnine, [benzene-^{3}H]	antagonist	glycine	NET-773
TCP, [piperidyl-3,4-^{3}H(N)]-	agonist	PCP, sigma	NET-886

The study of excitatory amino acid receptors is becoming very complex due to the presence of allosteric sites on the glutamate receptor that are reminiscent of the GABA receptor complex. The most effectively ligands affecting the glutamate receptor are MK-801 (1), CGP 39653 (2), TCP (3), AMPA (4)/CNQX (6), and Kainic acid (5). The MK-801 and PCP sites seem to be allosteric to the NMDA site labelled by CPP and CGS 19755. AMPA and Kainic acid label different receptor subtypes, quisqualate and kainate respectively. CNQX has an affinity of 40 nM for the quisqualate receptor (6).

References

1. Wong, E. H. F., *et al.* (1986). *Proc. Natl. Acad. Sci. USA.*, **83**, 7104.
2. Sills, M. A., *et al.* (1991). *Eur. Pharmacol.*, In press.
3. Lazdunski, M. (1983). *Brain Res.*, **280**, 196.
4. Honore, T., Lauridsen, J., and Krogsgaard-Larsen, P. (1982). *J. Neurochem.*, **38**, 173.
5. Unnerstall, J. R. and Wamsley, J. K. (1983). *Eur. J. Pharmacol.*, **86**, 361.
6. Honore, T., *et al.* (1989). *Biochem. Pharmacol.*, **38**, 3207.

Angiotensin receptor ligands

Table 5. Angiotensin receptor ligands

Compound	Pharmacological action	Selectivity	Catalogue number
Angiotensin I, [tyr-^{125}I]-	analogue	—	NEX-101
Angiotensin II, [tyr-3,5-^{3}H]-	endogenous	—	NET-446
Angiotensin II, [tyr-^{125}I]-	analogue	—	NEX-105
DUP 753, [butyryl-1,2-^{3}H]-,	antagonist	AII-1	NET-1055
Sar^1, Iodotyr^4, Ile^8-Angiotensin II, [^{125}I]- (SI AII)	antagonist	—	NEX-248

The sarcosine angiotensin II analogue has significantly higher receptor affinity, K_d 0.6 nM, than other labelled analogues and is an antagonist (1,2). DUP 753 is a non-peptide angiotensin antagonist selective for the AII-1 subtype (3).

References

1. Khosla, M. C., *et al.* (1972). *J. Med. Chem.*, **15**, 792.
2. Husain, A., *et al.* (1987). *Proc. Natl. Acad. Sci. USA*, **84**, 2489.
3. Chui, A. T., *et al.* (1990). *Biochem. Biophys. Res. Commun.*, **172**, 1195.

Benzodiazepine receptor ligands

Table 6. Benzodiazepine receptor ligands

Compound	Pharmacological action	Selectivity	Catalogue number
Diazepam, [methyl-^{3}H]-	agonist	—	NET-564
Flunitrazepam, [methyl-^{3}H]-	agonist	—	NET-567
PK11195, [*N*-methyl-^{3}H]-	antagonist	peripheral	NET-885
RO15-4513, [7,9-^{3}H]-	photoaffinity antagonist	central	NET-925
RO15-1788, [methyl-^{3}H]-	antagonist	central	NET-757
RO5-4864, [methyl-^{3}H]-	agonist	peripheral	NET-704
Zolpidem, [phenyl-2,6-^{3}H]-	agonist	central	NET-979

The standard ligands for the central benzodiazepine receptor are the agonist, flunitrazepam, and the antagonist, RO15-1788. Photoaffinity labelling of the receptor is possible using either flunitrazepam or RO15-4513 (1). RO15-4513 is of particular interest since it reverses many of the effects of alcohol in the CNS. Zolpidem is reported to label the BZ1 site exclusively with nanomolar affinity (2).

The peripheral receptor appears to be located on one of the proteins that makes up the outer mitochondrial membrane (3). RO5-4864 and PK11195 seem to label different sites on the same protein.

References

1. Mohler, H. (1982). *Eur. J. Pharmacol.*, **80**, 435.
2. Snyder, S. H., *et al.* (1985). *J. Pharmacol. Exp. Ther.*, **233**, 517.
3. Niddam, R., *et al.* (1987). *J. Neurochem.*, **49**, 890.

Bradykinin receptor ligands

Table 7. Bradykinin receptor ligands

Compound	Pharmacological action	Selectivity	Catalogue number
Bradykinin, [2,3-propyl-3,4-³H]-	endogenous	—	NET-706
Bradykinin, [tyr-¹²⁵I]-	analogue	—	NEX-097

Calcium channel receptor ligands

Table 8. Calcium channel receptor ligands

Compound	Pharmacological action	Selectivity	Catalogue number
S-(−)-BAY K 8644, [5-methyl-³H]-	agonist	dihydropyridine	NET-1002
ω-Conotoxin GVIA, [¹²⁵I]-,	antagonist	—	NEX-239
Diltiazem, *cis*-(+)-[*N*-methyl-³H]	antagonist	benzothiazepine	NET-847
Fluspirilene, [ring-³H]-	antagonist	—	NET-978
Nimodipine, [isopropyl-1,3-³H]-	antagonist	dihydropyridine	NET-891
Nitrendipine, [methyl-³H]-	antagonist	dihydropyridine	NET-741
PN 200-110, (+)-[5-methyl-³H]-	antagonist	dihydropyridine	NET-863
Ryanodine, [9,21-³H]-	—	Sarcoplasmic recticulum	NET-950
Verapamil, [methyl-³H]-	antagonist	phenylalkylamine	NET-810

Ligands are generally available for the known allosteric sites on the L subtype of voltage sensitive calcium channels. Nimodipine has the highest affinity and the highest specific activity for binding to the dihydropyridine site, while PN 200-110 is preferable when light sensitivity is a factor. The newly introduced (−)-BAY K 8644 is a pure agonist at this site and is a significant improvement over the racemic ligand. Fluspirilene seems to be a fourth site (1) for this channel. Omega-Conotoxin binds to both the N and L subtype and will be useful for studies of N receptors in the brain (2).

The binding of ryanodine is specific for the sarcoplasmic membrane and may not affect the voltage sensitive channels (3).

References

1. Galizzi, J. P., *et al.* (1986). *Proc. Natl. Acad. Sci. USA*, **83**, 7513.
2. Reynolds, I. J., *et al.* (1986). *Ibid*, **83**, 8804.
3. Fleischer, S., *et al.* (1985). *Ibid*. **82**, 7256.

Chemotactic receptor ligands

Table 9. Chemotactic receptor ligands

Compound	Pharmacological action	Selectivity	Catalogue number
Formyl-L-met-L-leu-L-phe-, N, [phe-ring-2,6-^{3}H]-	agonist	chemotactic peptide	NET-563

Cholecystokinin receptor ligands

Table 10. Cholecystokinin receptor ligands

Compound	Pharmacological action	Selectivity	Catalogue number
CCK-8, Bolton–Hunter Labelled [^{125}I]-	analogue	CNS periphery	NEX-203
CCK(26-33) (28,31-Me-Nle), [tyrosyl-3,5-^{3}H(N)]-	agonist	CNS	NET-1048
L-364,718 ($\pm$)-[*N*-methyl-^{3}H]-	antagonist	peripheral CCK	NET-971
L-365,260, [ring-^{3}H]-	antagonist	CNS	NET-1005
Pentagastrin, [B-ala-^{3}H]-	analogue	—	NET-798

The non-peptide ligands, L-364,718 for the peripheral receptor (1) and L-365,260 for the CNS receptor (2), gives investigators an opportunity to work more easily with standard ligand-binding techniques.

References

1. Chang, R. S. L., *et al.* (1986). *Mol. Pharmacol.*, **30**, 212.
2. Chang, R. S. L., *et al.* (1989). *Mol. Pharmacol.*, **35**, 803.

Cholinergic receptor ligands

Table 11. Cholinergic receptor ligands

Compound	Pharmacological action	Selectivity	Catalogue number
AH5183, L-[piperidinyl-3,4-^{3}H]-	inhibitor	vesicular acetylcholine storage	NET-964
Hemicholinium-3, [methyl-^{3}H]-	inhibitor	choline uptake	NET-884
Muscarinic			
Acetylcholine iodide, [acetyl-^{3}H]-	endogenous	—	NET-113
AF-DX 116, [piperidinyl-^{3}H]-	antagonist	M_2	NET-947
AF-DX 384, [2,3-dipropylamino-^{3}H]	antagonist	M_2	NET-1041
Atropine, [*N*-methyl-^{3}H]-	antagonist	—	NET-899
4-DAMP, [*N*-methyl-^{3}H]	antagonist	M_1, $M_3 > M_2$	NET-1040
Dioxolane, *cis*-[methyl-^{3}H]-	agonist	high affinity	NET-647
Oxotremorine-M, [methyl-^{3}H]-	agonist	high affinity	NET-671
Pirenzepine, [methyl-^{3}H]-	antagonist	M_1	NET-780
Propylbenzilylcholine mustard, [^{3}H]-	irreversible	—	NET-687
N-methyl L-QNB, [*N*-methyl-^{3}H]-	antagonist	hydrophilic QNB analogue	NET-965
QNB, L-[benzilic-^{3}H]-	antagonist	—	NET-656
Scopolamine methyl chloride, [*N*-methyl-^{3}H]-	antagonist	—	NET-636
Telenzepine, [methyl-^{3}H]-	antagonist	M_1	NET-987
Nicotinic			
α-Bungarotoxin, [^{125}I]-	antagonist	peripheral	NEX-126
Methylcarbamyl choline iodide, [*N*-methyl-^{3}H]-	agonist	CNS	NET-951
Nicotine, L-[methyl-^{3}H]-	agonist	CNS	NET-827

Methylcarbamyl choline is the superior ligand for equilibrium binding studies at the central nicotinic receptor with higher affinity than nicotine (1).

The classic muscarinic antagonist ligands, QNB and scopolamine methyl chloride, have been supplemented by pirenzepine and telenzepine (2), a more potent, analogue, which are selective for the M_1 receptor. AF-DX 116 is selective for the M_2 subtype (3). AF-DX 384 has a higher affinity than AF-DX 116 for M_2 mAChRs ($K_2 = 10 \, nM$) (4), but there are presently no completely selective ligands for the M_3, or other novel receptor subtypes, although 4-DAMP has a higher affinity for M_3 and M_1 than M_2 and M_4 (5).

The high affinity agonist binding state of the muscarinic receptor may be studied with oxotremorine-M or *cis*-dioxolane. Choline uptake sites selectively bind hemicholinium-3 and the vesicular storage site is labelled by AH5183 (6).

References

1. R. Quirion, *et al.* (1987). *Eur. J. Pharmacol.*, **139**, 323.
2. N. J. M. Birdsall, *et al.* (1988). *Eur. J. Pharmacol.*, **145**, 87.
3. R. Quirion, *et al.* (1987). *Eur. J. Pharmacol.*, **144**, 417.
4. Eberlein, W. G., *et al.* (1989). *TIPS 10-1 Suppl. Sub. Res. Muscarinic Receptors*, **IV**, 50.
5. Michel, A. D. (1989). *Eur. J. Pharmacol.*, **166**, 459.
6. S. M. Parsons, *et al.* (1986). *Proc. Natl. Acad. Sci. USA*, **83**, 2267.

Dopamine receptor ligands

Table 12. Dopamine receptor ligands

Compound	Pharmacological action	Selectivity	Catalogue number
ADTN, [5,8-^{3}H]–	agonist	—	NET-620
Cocaine, 1-[benzoyl-3,4-^{3}H(N)]-	inhibitor	uptake	NET-510
Dihydroergocryptine, [9,10-^{3}H]-	mixed	mixed	NET-523
GBR 12935, [propylene-2,3-^{3}H]-	inhibitor	dopamine uptake	NET-918
4-Iodospiperone, [^{125}I]-	antagonist	D$_2$	NEX-200
Haloperidol, [^{3}H]-	antagonist	D$_2$/σ	NET-530
Mazindol, [4'-^{3}H]-	inhibitor	dopamine uptake	NET-816
Methylspiperone, [*N*-methyl-^{3}H]-	antagonist	D$_2$	NET-856
MPP(+), [methyl-^{3}H]- Methyl-4-phenylpyridinium acetate	neurotoxin	dopamine uptake	NET-914
MPTP, [methyl-^{3}H]-	neurotoxin	—	NET-865
Propylnorapomorphine, [propyl-^{3}H]-	agonist	—	NET-619
Raclopride, [methoxy-^{3}H]-	antagonist	D$_2$	NET-975
SCH 23390, [*N*-methyl-^{3}H]-	antagonist	D$_1$	NET-930
SCH 23982, [^{125}I]-	antagonist	D$_1$	NEX-230
Spiperone, [benzene ring-^{3}H]-	antagonist	D$_2$	NET-565
Sulpiride, (−)-[methoxy-^{3}H]-	antagonist	D$_2$	NET-775
WIN 35,428, [*N*-methyl-^{3}H]-, (CFT)	inhibitor	uptake	NET-1033
YM-01951-2, (*N*-methyl-^{3}H]-	antagonist	D$_2$	NET-1004

A surfeit of ligands are available for the study of the dopamine receptor. The most useful are:

- Propylnorapomorphine—agonist, unselective
- Dihydroergocryptine—mixed, unselective with adrenergic activity

- Spiperone—antagonist, D_2, with strong serotonin binding plus other activities
- Sulpiride—antagonist, highly D_2 selective
- Raclopride—antagonist, highly D_2 selective and superior for *in vivo* experiments
- YM-01951-2—antagonist, highly D_2 selective, very low non-specific binding and very high affinity (1)
- SCH 23990/23982—antagonists, high D_1 selectivity and good binding characteristics

MPP^+, the active metabolite of MPTP, is a neurotoxin specific to dopaminergic neurons, but the mechanism of this toxicity is poorly understood.

The uptake inhibitors, GBR 12935 and mazindol, are good ligands for the study of this process. The binding of GBR 12935 is much higher affinity, but complicated by a piperazine acceptor site. Satisfactory experimental procedures have been detailed to avoid this problem in human tissue (2).

The cocaine analogue, WIN 35,428 also labels the dopamine uptake site, which is the presumed site of action of cocaine. WIN 35,428 has a much higher affinity for the receptor than cocaine, and gives excellent specific binding (3).

References

1. Niznik, H. B., *et al.* (1985). *Naunyn-Schmeideberg's Arch. Pharmacol.*, **329**, 333.
2. Marcusson, J. and Ericksson, K. (1988). *Brain Res.*, **457**, 122.
3. Madras, B. K., *et al.* (1989). *Mol. Pharmacol.*, **36**, 518.

GABA receptor ligands

Table 13. GABA receptor ligands

Compound	Pharmacological action	Selectivity	Catalogue number
Aminobutyric acid, Gamma-[2,3-^{3}H]- (GABA)	endogenous	—	NET-191X
Baclofen, (−)-[butyl-4-^{3}H]-	agonist	GABA$_B$	NET-867
Bicuculline methyl chloride, (−)-[methyl-^{3}H]-	antagonist	GABA$_A$	NET-786
Butylbicyclophosphorothionate, [^{35}S]	antagonist	Cl channel	NEG-049
Muscimol, [methylene-^{3}H]-	agonist	—	NET-574
Pregnan-3α-ol-20-one, 5α[9,11,12,-^{3}H]-	modulator	GABA$_A$	NET-1047
SR 95531, [butyryl-2,3-^{3}H]-	antagonist	GABA$_A$	NET-946

Each of the available GABA ligands is useful for studying the various sites on the receptor complex and the interactions between those sites, including the central benzodiazepine site. NET-1047, allopregnanolone, is a novel ligand interacting at a modulator site that is little understood at this time (1).

Reference

1. Morrow, A. L., Pace, J. R., Purdy, R. H., and Paul, S. M. (1990). *Mol. Pharmacol.*, **37**, 263.

Growth factors

Table 14. Growth factors

Compound	Pharmacological action	Selectivity	Catalogue number
Epidermal Growth Factor, [^{125}I]-Murine	analogue	—	NEX-160
GM-CSF, [^{125}I]-	analogue	—	NEX-249
gamma-Interferon, [^{125}I]-	analogue	—	NEX-251
Interleukin 1 alpha, [^{125}I]-Human, recombinant	analogue	—	NEX-246
Interleukin 1 beta, [^{125}I]-Human, recombinant, Bolton–Hunter labelled	analogue	—	NEX-232
Interleukin 2, [^{125}I]-, Human/recombinant	analogue	—	NEX-229
Interleukin 6, [^{125}I]- Human, recombinant	analogue	—	NEX-269
Nerve Growth Factor, [^{125}I]-	analogue	—	NEX-215
PDGF, [^{125}I]-,	analogue	—	NEX-260
Transforming Growth Factor, beta 1, [^{125}I]- Human, recombinant	analogue	—	NEX-267
Tumour Necrosis Factor, alpha- [^{125}I]- Human, recombinant	analogue	—	NEX-257

Histamine receptor ligands

Table 15. Histamine receptor ligands

Compound	Pharmacological action	Selectivity	Catalogue number
Histamine HCl, [^{3}H]-	endogenous	—	NET-732
			NET-152
Mianserin HCl, [*N*-methyl-^{3}H]-	antagonist	H$_1$, 5HT	NET-686
Pyrilamine, [pyridinyl 5-^{3}H]- (Mepyramine)	antagonist	H$_1$	NET-594
Tiotidine [methyl-^{3}H]-	antagonist	H$_2$	NET-688

Pyrilamine is high affinity and selective for the H$_1$ receptor; although tiotidine is useful as a ligand for the H$_2$ receptor, it is sticky and limited in its applications.

Inositol cycle ligands

Table 16. Inositol cycle ligands

Compound	Pharmacological action	Selectivity	Catalogue number
Cytidine, [5-^{3}H]-	DAG	—	NET-1043
Inositol 1,4,5-trisphosphate, D-[inositol-1-^{3}H]-	quantitation endogenous		NET-911
Inositol 1,4,5-trisphosphate, D-[5-^{32}p(N)]-	endogenous		NEG-066
Inositol 1,4,5-trisphosphate, D- Unlabelled, Receptor Grade	endogenous		NLP-047
Inositol 1,4,5-trisphosphorothioate, D-[^{35}S(U)]-	non-metabolizable analogue		NEG-073
Inositol 1,4,5-trisphosphorothioate, D-[inositol-1-^{3}H]-	non-metabolizable analogue		NET-1053
Inositol 1,4,5-trisphosphorothioate, D-, Unlabelled, Receptor Grade	non-metabolizable analogue		NLP-045
Inositol 1,3,4,5-trisphosphate, D-[inositol-1-^{3}H]-	endogenous		NET-941
Inositol 1,3,4,5-trisphosphate D[5-^{32}P(N)]-	endogenous		NEG-071
Inositol hexakisphosphate myo-[inositol-2-^{3}H(N)]-	endogenous		NET-1023
Phosphatidylinositol-4,5-bisphosphate, [inositol-2-^{3}H(N)]-	endogenous		NET-895

Table 16. (contd.)

Compound	Pharmacological action	Selectivity	Catalogue number
Stearoyl-2-arachidonyl-sn-glycerol 1-[arachidonyl-5,6,8,9,11,12-14,15-^{3}H(N)]-	endogenous activator	PKC	NET-896
Staurosporine,N,N-dimethyl, [dimethyl-^{3}H(N)]-	inhibitor	PKC	NET-1007

Melatonin receptor ligands

Table 17. Melatonin receptor ligands

Compound	Pharmacological action	Selectivity	Catalogue number
Melatonin, [methoxy-^{3}H]-	endogenous		NET-801
Iodomelatonin, [^{125}I]-	analogue		NEX-236

Iodomelatonin has been used to identify receptors in several animal tissues including human brain (1,2).

References

1. Laudon, M. and Zisapel, N. (1986). *Fed. Eur. Biochem. Soc.*, (*FEBS*), **197**, 9.
2. Weaver, D. R., *et al.* (1958). *ibid.*, **228**, 123.

Neurokinin receptor ligands

Table 18. Neurokinin receptor ligands

Compound	Pharmacological action	Selectivity	Catalogue number
Eledoisin, [pro-3,4-^{3}H]-	endogenous	NK-2,3	NET-901
Eledoisin, Bolton–Hunter labelled, [^{125}I]-	agonist	—	NEX-218
Neurokinin A, [^{125}I]-	agonist	NK-2	NEX-252
Senktide, [phenylalanyl-3,4-^{3}H]-	analogue	NK-3	NET-997
Substance P, [2-pro-3,4-^{3}H]-	endogenous	NK-1	NET-771
Substance P, (9-Sar, 11-Met(O$_2$))-[2-pro-3,4-^{3}H]-,	agonist	—	NET-1025
Substance P, Bolton–Hunter labelled [^{125}I]-	agonist	—	NEX-190

Neurotensin receptor ligands

Table 19. Neurotensin receptor ligands

Compound	Pharmacological action	Selectivity	Catalogue number
Neurotensin, [3,11-tyrosyl-3,5-^{3}H]-	endogenous	—	NET-605
Neurotensin, [^{125}I]-	analogue	—	NEX-198

Opiate receptor ligands

Table 20. Opiate receptor ligands

Compound	Pharmacological action	Selectivity	Catalogue number
beta-Endorphin, [^{125}I]-	analogue		NEX-143
Benzoyl-L-phe-L-ala-L-arg, N-[benzoyl-2,5-^{3}H]-	enkephalinase substrate	—	NET-797
Bremazocine, (−)-[9-^{3}H]-	agonist	kappa	NET-821
CTOP, [D-phe-3,4,5-^{3}H]- Mu opiate receptor antagonist	antagonist	mu	NET-944
DAMGO, [tyrosyl-3,5-^{3}H]-	agonist	mu	NET-902
Dihydromorphine, [N-methyl-^{3}H]-	agonist	—	NET-658
DPDPE, 4-p-Cl-Phe, [tyrosyl-3,5-^{3}H]-	agonist	delta	NET-923
DPDPE, [tyrosyl-2,6-^{3}H]-	agonist	delta	NET-922
Enkephalin-(2-D-ala-5-D-leu), [tyrosyl-3,5-^{3}H]-	agonist	delta	NET-648
Enkephalin, (2-ser-5-leu-6-thr)- [tyrosyl-3,5-^{3}H]-	agonist	delta	NET-796
Enkephalin–Leu, [tyr-3,5-^{3}H]-	endogenous	delta/sigma	NET-540
Enkephalin–Leu, [^{125}I]	analogue	—	NEX-148
Enkephalin–Met, [^{125}I]-	analogue	—	NEX-149
Ethylketocyclazocine, (−)-[9-^{3}H]-	agonist	kappa	NET-820
Morphine, [N-methyl-^{3}H]-	agonist	mu	NET-653
Naloxone, [N-allyl-^{3}H]-	antagonist	—	NET-719
PL-017, [prolyl-3,4-^{3}H, D-prolyl-3,4-^{3}H]-	agonist	—	NET-1000
U-69,593, [phenyl-3,4-^{3}H]-	agonist	kappa	NET-952

Researchers have some choices between peptides and drugs when studying the opioid receptor. The most selective and highest affinity ligands are in the chart on the next page.

Receptor	Peptide	Drug
μ	PL-017—agonist	
	CTOP—antagonist	
δ	chloro-DPDPE—agonist	
κ	— —	U-69,593—agonist

Naloxone, a none-selective antagonist, is generally used to characterize opioid receptors.

Oxytocin receptor ligands

Table 21. Oxytocin receptor ligands

Compound	Pharmacological action	Selectivity	Catalogue number
Oxytocin, [^{125}I]-	analogue	—	NEX-187
Oxytocin, [tyr-2,6-^{3}H]-	endogenous	—	NET-858
Oxytocin Analogue, [^{125}I]- (OVTA)	antagonist		NEX-254

Peptides—miscellaneous

Table 22. Peptides—miscellaneous

Compound	Pharmacological action	Selectivity	Catalogue number
Atrial Natriuretic Factor, [^{125}I]- Rat	analogue	—	NEX-228
Bombesin, [^{125}I]-	analogue		NEX-258
Complement C$_{5A}$, [^{125}I]-	analogue		NEX-250
Corticotropin Releasing Factor, [^{125}I]- (CRF)	human/rat analogue	—	NEX-216
Corticotropin Releasing Factor, [^{125}I]- (CRF)	ovine analogue	—	NEX-217
Endothelin, Big-, [^{125}I]-	human 1-38		NEX-270
Endothelin I, [^{125}I]-	analogue		NEX-259
Endothelin II, [^{125}I]-	analogue		NEX-261
Endothelin III, [^{125}I]-	analogue		NEX-262
Galanin-. [^{125}I]tyr-	analogue	CNS	NEX-243
Glucagon, [^{125}I]-	analogue	—	NEX-207
Insulin, [^{125}I]-	porcine analogue	—	NEX-196
Laminin, N-[propionate-2,3-^{3}H]- propionylated-	endogenous		NET-1016
Neuropeptide Y, Bolton–Hunter Labelled, [^{125}I]-	endogenous	—	NEX-222

Table 22. (Contd.)

Compound	Pharmacological action	Selectivity	Catalogue number
Peptide YY, [^{125}I]-	analogue	—	NEX-240
Sarafotoxin-S6b, [^{125}I]Tyr^{13}-	analogue	—	NEX-264
Transferrin, Human [^{125}I]-	analogue	—	NEX-212
Vasoactive Intestinal Contractor, [^{125}]-	analogue	—	NEX-263
Vasoactive Intestinal Polypeptide [^{125}]-	analogue	—	NEX-192

Potassium channel ligands

Table 23. Potassium channel ligands

Compound	Pharmacological action	Selectivity	Catalogue number
Apamin, [^{125}I]-	antagonist	—	NEX-242
Glibenclamide, [cyclohexyl-2,3-^{3}H(N)]-	antagonist	ATP-dependent K$^+$ channel	NET-1024

Serotinin receptor ligands

Table 24. Serotinin receptor lligands

Compound	Pharmacological action	Selectivity	Catalogue number
BRL 43694, [9-methyl-^{3}H]-	antagonist	5-HT$_3$	NET-1030
Citalopram, [N-methyl-^{3}H]-	inhibitor	uptake	NET-1039
(±)DOB, [propyl-1,2-^{3}H]-	agonist	5-HT$_2$	NET-943
(±)DOI, [^{125}]-	agonist	5-HT$_2$	NEX-255
GR 65630, [N-methyl-^{3}H]-	antagonist	5-HT$_3$	NET-1011
8-Hydroxy-DPAT, [propyl-2,3-ring-1,2,3-^{3}H]-	agonist	5-HT$_{1A}$	NET-929
Hydroxytryptamine binoxalate, 5-[1,2-^{3}H]-	endogenous	—	NET-398
Hydroxytryptamine creatine sulphate, 5-[1,2-^{3}H]-	endogenous	—	NET-498
Imipramine, [N-methyl-^{3}H]-	inhibitor	uptake	NET-710
Iodo-LSD, 2-[^{125}I]-	antagonist	5-HT$_2$	NEX-199
Ketanserin HCl, [ethylene-^{3}H]-	antagonist	5-HT$_2$	NET-791
LSD, [N-methyl-^{3}H]-	agonist	—	NET-638
Paroxetine, [phenyl-6'-^{3}H]-	inhibitor	uptake	NET-869
Quipazine, [piperizinyl-^{3}H]-	antagonist	5-HT$_3$	NET-992
Spiperone, [benzene ring-^{3}H]-	antagonist	D$_2$/5-HT$_2$	NET-565

GR 65630 and BRL 43694 are the preferred ligands for the 5-HT$_3$ receptor, having higher affinity than Quipazine (1). 8-Hydroxy-DPAT is a selective 5-HT$_{1A}$ agonist; no 5-HT$_{1A}$ antagonists have been developed. Although spiperone is used extensively for the 5-HT$_2$/D$_2$ studies, the most selective antagonists with high affinity for 5-HT$_2$ receptors are iodo-LSD and ketanserin. DOB, an agonist for this site, has sub-nanomolar affinity and gives approximately 60% specific binding (2). DOI is an iodinated ligand with similar specificity (3).

Paroxetine and Citalopram are the most potent and selective 5-HT uptake inhibitors. Imipramine binds to several different sites and is not sufficiently selective from other aminergic uptake sites (4).

References

1. Titeler, M., *et al.* (1990). *Soc. Neurosci. Abstr.*, no. 537.5.
2. Titeler, M., *et al.* (1985). *Ibid.*, **117**, 145.
3. McKenna, *et al.* (1989). *Brain Res.*, **476**, 45.
4. Marcusson, J. O., *et al.* (1988). *J. Neurochem.*, **50**, 1783.

Sigma receptor ligands

Table 25. Sigma receptor ligands

Compound	Pharmacological action	Selectivity	Catalogue number
Dextromethorphan, [*N*-methyl-^{3}H]-		sigma	NET-1032
DTG, [5-^{3}H]- (Di(2-tolyl) guanidine)	—	sigma	NET-986
Haloperidol, [^{3}H]-	agonist	sigma/D$_2$	NET-530
(+)-Pentazocine, [ring-1,3-^{3}H]-	agonist	sigma	NET-1056
3PPP, (+)-[propyl-2,3-^{3}H]-	—	sigma/PCP	NET-815
SKF-10,047, (+)-[*N*-allyl-^{3}H]-	agonist	sigma/PCP	NET-765

The binding, not displaced by dopaminergic ligands, of haloperidol has been used to define the sigma receptor. (+)-Pentazocine has both the highest affinity and the highest selectivity vs. the PCP binding site (1).

Reference

1. de Costa, B. R., *et al.* (1989). *FEBS Lett.*, **251**, 53.

Sodium channel ligands

Table 26. Sodium channel ligands

Compound	Pharmacological action	Selectivity	Catalogue number
Batrachotoxinin A 20-α-benzoate, [benzoyl-2,5-^{3}H]-	activator	—	NET-876
Brevetoxin, PbTx-3 [42-^{3}H]-	activator	site 5	NET-969

Thyrotropin-releasing hormone ligands

Table 27. Thyrotropin-releasing hormone ligands

Compound	Pharmacological action	Selectivity	Catalogue number
Thyrotropin Releasing Hormone, [L-pro-2,3,4,5-^{3}H]-	endogenous	—	NET-577
Thyrotropin Releasing Hormone, [^{125}I]-	analogue	—	NEX-153
Thyrotopin Releasing Hormone (3-methyl-his^2), [L-his-4-^{3}H, L-pro-3,4-^{3}H]-	analogue	high-affinity site	NET-705

Vasopressin receptor ligands

Table 28. Vasopressin receptor ligands

Compound	Pharmacological action	Selectivity	Catalogue number
Vasopressin V-1 antagonist, [phe-3,4,5-^{3}H]-	antagonist	V-1	NET-945
Arginine Vasopressin, [^{125}I]-	analogue	—	NEX-128
Arginine Vasopressin, [phe-3,4,5-^{3}H]-	endogenous	—	NET-800
Vasopressin V-2, antagonist, [phe-3,4,5-^{3}H]-	antagonist	V-2	NET-1010

A3

Commercially available curve-fitting programs which are suitable for the analysis of ligand-binding data

E. C. HULME

Program	Machine	Supplier	Comment
KINETIC	IBM PC	Biosoft	Calculates association and dissociation rate constants
EBDA/LIGAND	IBM PC Mac[a]	Biosoft	Analysis of equilibrium binding curves. Can fit multiple binding curves simultaneously. Provided as package with KINETIC and LOWRY. Depletion-corrected multiple sites model. Graphics not very good. £125.00.
Enzfitter (Leatherbarrow)	IBM PC	Elsevier Biosoft SIGMA	Predefined enzyme kinetic and equilibrium binding equations. New equations can be entered as explicit functions. Good graphics. Excellent manual. £163.00.
Grafit version 2.0 (Leatherbarrow)	IBM PC	Erithacus software SIGMA	Much updated, more powerful version of enzfitter. More pre-defined equations. Can be constrained to do multiple fitting. New equations entered using equation editor. Runs under Microsoft windows. Good graphics. £330.80.
LUNDON-1 LUNDON-2 KINETICS	IBM PC	Lundon software	Analysis of saturation curves, competition curves and exponential decays. £600.00 per program.
GraphPAD INPLOT	IBM PC	GraphPAD software	Built-in equations only. Excellent graphics. £300.00.
MULTIFIT	Mac.	Day computing	User-defined equations. Limited graphics. Reported to be prone to crash. £80.00.
CURVEFIT	IBM PC	IRL Press software	User-defined equations.
Fig. P V5.0	IBM PC	Biosoft	Graphics program incorporating fitting program. User equations entered using algebraic format. Fig. P. is a very popular scientific graphics package. £250.00. Fitting package can be purchased separately. £125.00. Curves can be imported into Fig. P.

Program	Machine	Supplier	Comment
Sigma Plot V4.0	IBM PC	Jandel Scientific. SIGMA	Non-linear curve-fitting and graphics program. User-defined models are readily entered. Some possibilities of multiple curve fitting. Said to be powerful and easy to use. High quality output.

[a] Mac = Apple Macintosh

This table includes information about some of the more popular commercially available programs. Further information is available from the suppliers. It should be noted that these programs do not allow the dependent variables to be calculated as the outcomes of procedures as opposed to functions.

Addresses

Biosoft, PO Box 589, Milltown, NJ 08850, USA. 22 Hills Road, Cambridge CB2 1JP, UK.

Lundon Software Inc., PO Box 1210, 14 632 Old State Road, Middlefield OH 44062, USA.

Day Computing, Cambridge, UK.

SIGMA Chemical Co., Fancy Road, Poole, Dorset, BH17 7NH, UK.

Reference

1. Leatherbarrow, R. J. (1990). *Trends Biochem. Sci.*, **15**, 455.

Receptor Biochemistry:
A Practical Approach

Edited by E. C. Hulme

Receptor–Effector Coupling: A Practical Approach

Edited by E. C. Hulme

Index

453

Index